T0181868

Texts and
Monographs
in Physics

W. Beiglböck
series editor

Wolfgang Rindler

Essential Relativity

Special, General, and Cosmological

Second Edition

Springer-Verlag
New York   Heidelberg   Berlin

Wolfgang Rindler
The University of Texas at Dallas
Box 688
Richardson, Texas 75080/USA

Wolf Beiglböck
Institut für Angewandte Mathematik der Universität
Im Neuenheimer Feld 294
6900 Heidelberg
Federal Republic of Germany

With 44 Figures

ISBN 978-3-540-10090-4   ISBN 978-3-642-86650-0 (eBook)
DOI 10.1007/978-3-642-86650-0

Library of Congress Cataloging in Publication Data

Rindler, Wolfgang, 1924–
    Essential relativity.
    (Texts and monographs in physics)
    Includes bibliographical references and index.
    1. Relativity (Physics)   I. Title.

QC173.55.R56   1977   530.1'1   76-28816

Preface to the Second Edition

In retrospect, the first edition of this book now seems like a mere sketch for a book. The present version is, if not the final product, at least a closer approximation to it. The table of contents may show little change. But that is simply because the original organization of the material has been found satisfactory. Also the basic purpose of the book remains the same, and that is to make relativity come alive *conceptually*. I have always felt much sympathy with Richard Courant's maxim (as reported and exemplified by Pascual Jordan) that, ideally, proofs should be reached by comprehension rather than computation. Where computations are necessary, I have tried to make them as transparent as possible, so as not to hinder the progress of comprehension.

Among the more obvious changes, this edition contains a new section on Kruskal space, another on the plane gravitational wave, and a third on linearized general relativity; it also contains many new exercises, and two appendices: one listing the curvature components for the diagonal metric (in a little more generality than the old "Dingle formulas"), and one synthesizing Maxwell's theory in tensor form. But the most significant changes and additions have occurred throughout the text. Many sections have been completely rewritten, many arguments tightened, many "asides" added, and, of course, recent developments taken into account. Yet I am keenly aware that, by being overloaded with information, a distinctly non-encyclopedic book like this might easily lose whatever immediacy and transparency it possesses. Thus, ultimately, the new material was determined by the enthusiasm with which I felt I could present it.

My thanks go out to all those who have taken the trouble to comment on the first edition, particularly to Banesh Hoffmann, Kenneth Jacobs, and Robert Gowdy. The new book owes much to them. I again owe a special debt of gratitude to Jürgen Ehlers, for his continued support of this project, and for reading portions of the new manuscript in critical detail. I am also grateful to Ivor Robinson, Roman Sexl, and Beatrice Tinsley for extensive conversations which contributed new ideas to specific parts of the book.

Dallas, February 1977   Wolfgang Rindler

Preface to the First Edition

This book is an attempt to bring the full range of relativity theory within reach of advanced undergraduates, while containing enough new material and simplifications of old arguments so as not to bore the expert teacher. Roughly equal coverage is given to special relativity, general relativity, and cosmology. With many judicious omissions it can be taught in one semester, but it would better serve as the basis of a year's work. It is my hope, anyway, that its level and style of presentation may appeal also to wider classes of readers unrestricted by credit considerations. General relativity, the modern theory of gravitation in which free particles move along "straightest possible" lines in curved spacetime, and cosmology, with its dynamics for the whole possibly curved universe, not only seem necessary for a scientist's balanced view of the world, but offer some of the greatest intellectual thrills of modern physics. Nevertheless, considered luxuries, they are usually squeezed out of the graduate curriculum by the pressure of specialization. Special relativity escapes this tag with a vengeance, and tends to be taught as a pure service discipline, with too little emphasis on its startling ideas. What better time, therefore, to enjoy these subjects for their own sake than as an undergraduate? In spite of its forbidding mathematical reputation, even general relativity is accessible at that stage. Anyone who knows the calculus up to partial differentiation, ordinary vectors to the point of differentiating them, and that most useful method of approximation, the binomial theorem, should be able to read this book. Its mathematical level rises very gradually. Four-vectors are introduced half way through special relativity, and are then used sufficiently to leave no doubt that they are tools and not just ornaments. And, of course, no serious approach to general relativity is possible without tensors. Accordingly, these are introduced at the appropriate point (in Chapter 8), but as unobtrusively as possible, so that those who habitually skip a few paragraphs will perhaps hardly notice them. Yet the more ambitious reader, who is prepared to pause and think, will find enough tensor theory outlined here to give him a basic understanding of all relativistic tensor arguments. The experience of seeing tensors and Riemannian geometry in action may motivate him to learn more about these subjects eventually.

In its order of presentation the book breaks with tradition. With few exceptions, the usual procedure has been to accept Newton's definition of inertial frames for special relativity, and not to introduce the equivalence principle until just before it is needed in general relativity. It is then that the student suddenly learns that he has been working against the wrong background: that Newton's frames are not, after all, the frames relevant to special relativity. This is excusable only as long as general relativity, and, in particular, the equivalence principle, are regarded as doubtful. Once we accept them, as do most experts today, we should paint a consistent picture from the beginning. And so, in this book, not only the equivalence principle, but even Mach's principle (with due reservations) and some cosmology precede special relativity in order to set it in its proper perspective. Special relativity is then developed rigorously, with an eye on the physics at all times, but emphasizing ideas, not experimental detail. There is some unavoidable overlap of subject matter with my previous book *Special Relativity** (referred to hereafter as RSR), but the actual duplication is minimal, and many common topics are treated quite differently. I trust that friends of RSR can read even this part of the present book with profit. There follows an "easy" chapter on general relativity, which represents about the limit to which one can go without the use of tensors. The next chapter (Chapter 8) is probably the hardest in the book, but after Section 8.4 it can at first be omitted, except for a "light" reading of Section 8.10. The last chapter, on cosmology, though rigorous and detailed, should be easier again. Those whose primary interest lies in this field can read it almost independently of the rest of the book. At the end of the text there is a collection of over 130 exercises, put together rather carefully. I would urge even the most casual reader not quite to ignore these. Though their full solution often requires ingenuity, even a mere perusal should provide some useful insights. In addition, students should give their curiosity full rein and develop the habit of continually inventing and answering their own problems. Finally a warning: in several sections the units are chosen so as to make the speed of light unity, which should be borne in mind when comparing formulas.

These remarks would not be complete without an expression of gratitude to my friend Professor Jürgen Ehlers for reading the manuscript in detail and making innumerable suggestions for improvement, which I was only

* Oliver and Boyd, Edinburgh and London; New York: Interscience Publishers, Inc., a Division of John Wiley & Sons, Inc., Second Edition, 1966.

too glad to accept. Furthermore, it is a pleasure to have this opportunity of thanking Professor Carlo Cattaneo and members of the Istituto Matematico "Guido Castelnuovo" of the University of Rome, where part of this book was written, for their warm hospitality and most stimulating discussions, and the authorities of the Southwest Center for Advanced Studies for granting me leave.

Rome, April 1969   Wolfgang Rindler

Contents

3 Einsteinian Optics 54

4 Spacetime and Four-Vectors 61

5 Relativistic Particle Mechanics 75

Appendices   245

Exercises   253

Subject Index   277

Abbreviations

RSR Rindler, *Special Relativity*, 2nd ed., 1966
 SR Special relativity
 GR General relativity
 AS Absolute space
 GT Galilean transformation
 LT Lorentz transformation
 RP Relativity principle
 EP Equivalence principle
LIF Local inertial frame
 CM Center of momentum
 CP Cosmological principle
PCP Perfect cosmological principle
 RW Robertson–Walker
 EH Event horizon
 PH Particle horizon

CHAPTER 1

The Rise and Fall of Absolute Space

1.1 Definition of Relativity

Originally, relativity in physics meant the abolition of absolute space. More particularly, it has come to mean either of Einstein's famous two theories: his *special relativity* (SR) of 1905, and his *general relativity* (GR) of 1915. SR abolished absolute space in its Maxwellian role as the "ether"—the carrier of light waves and of electromagnetic fields in general—whereas GR abolished absolute space also in its Newtonian role as the standard of nonacceleration. (Even more importantly, though not by design, Einstein's theories abolished the concept of absolute time; *that* we shall discuss in the next chapter.) Since these ideas are fundamental, we devote the first chapter to a brief discussion centered on the three questions: What is absolute space? Why should it be abolished? How can it be abolished?

A more modern and positive definition of relativity has evolved *ex post facto* from the actual relativity theories; according to this view, the relativity of any physical theory expresses itself in the group of transformations which leave the laws of the theory invariant and which therefore describe symmetries, for example of the space and time arenas of these theories. Thus, as we shall see, Newton's mechanics possesses the relativity of the so-called Galilean group, SR possesses the relativity of the Poincaré (or "general" Lorentz) group, GR possesses the relativity of the full group of smooth one-to-one transformations, and the various cosmologies possess the relativity of the various symmetries with which the large-scale universe is credited. Even a theory valid only in one absolute Euclidean space, provided *that* is physically homogeneous and isotropic, would possess a relativity, namely the group of rotations and translations.

1.2   Newton's Laws

We recall Newton's three laws of mechanics, of which the first (Galileo's law) is really a special case of the second:

(i) Free particles move with constant vector-velocity (i.e., with zero acceleration, or, in other words, with constant speed along straight lines).

(ii) The vector-force on a particle equals the product of its mass into its vector-acceleration: $\mathbf{f} = m\mathbf{a}$. (This is more than a mere *definition* of force, since there are independent laws governing the origin of forces, e.g., Hooke's law, Coulomb's law, and indeed Newton's third law.)

(iii) The forces of action and reaction are equal and opposite; e.g., if a particle A exerts a force $\mathbf{f}$ on a particle B, then B exerts a force $-\mathbf{f}$ on A. (The assumption of absolute time, i.e., of absolute simultaneity, is implicit here if the action is at a distance.)

Physical laws are usually stated relative to some *reference frame*, which allows physical quantities like velocity, acceleration, etc., to be defined. Preferred among reference frames are *rigid* frames, and preferred among these are the *inertial* frames. Newton's laws apply in the latter.

A classical rigid reference frame is an imagined extension of a rigid body. For example, the earth determines a rigid frame throughout all space, consisting of all those points which remain "rigidly" at rest relative to the earth and to each other. We can associate an orthogonal Cartesian coordinate system S with such a frame in many ways, by choosing three mutually orthogonal planes within it and measuring x, y, z as distances from these planes. Also a time coordinate t must be defined in order that the system can be used to catalog events. A rigid frame, endowed with such coordinates, we shall call a *Cartesian* frame. The description given presupposes that the geometry in such a frame is Euclidean, which was taken for granted until 1915.

Newton's first law serves as a test to single out inertial frames among rigid frames: a rigid frame is called inertial if free particles move without acceleration relative to it. However, since all particles are affected by gravity, no particle near a heavy mass is (force-)"free." So the criterion of uniform motion can be applied only in regions of the frame which are "sufficiently removed from all attracting matter."

Empirically, the frame of the "fixed stars" was long recognized as inertial to considerable accuracy, and it was taken as the basic reference frame for Newton's laws. Today, when our galaxy is known to rotate, and the universe is known to expand, we might substitute for the "fixed stars" that rigid frame in which all other galaxies appear to us to recede radially, i.e., the frame which is symmetrical relative to the distant universe. When we talk of the frame of the "fixed stars," we shall really mean the latter.

1.3   The Galilean Transformation

Now consider two Cartesian frames $S(x, y, z, t)$ and $S'(x', y', z', t')$ in "standard configuration," as shown in Figure 1.1: S' moves in the x-direction of S with uniform velocity v and the corresponding axes of S and S' remain parallel throughout the motion, having coincided at time $t = t' = 0$. It is assumed that the same units of distance and time are adopted in both frames. For example, clocks can be synchronized to the vibrations of caesium atoms in both frames, and rulers can be calibrated to the wavelength of an agreed spectral line, as determined by interference methods. Suppose an *event* (like the flashing of a light bulb, or the collision of two point-particles) has coordinates (x, y, z, t) relative to S and (x', y', z', t') relative to S'. Then the classical (and "common sense") relations between these two sets of coordinates are given by the standard *Galilean transformation* (GT):

$$x' = x - vt, \quad y' = y, \quad z' = z, \quad t' = t, \tag{1.1}$$

since vt is the distance between the spatial origins. The last of these relations is usually taken for granted; it expresses the universality, or absoluteness, of time.

Differentiating (1.1) with respect to $t = t'$ immediately leads to the classical velocity transformation, which relates the velocity components of a moving particle in S with those in S':

$$u'_1 = u_1 - v, \quad u'_2 = u_2, \quad u'_3 = u_3, \tag{1.2}$$

where $(u_1, u_2, u_3) = (dx/dt, dy/dt, dz/dt)$ and $(u'_1, u'_2, u'_3) = (dx'/dt', dy'/dt', dz'/dt')$. Thus if I walk forward at 2 mph (u'_1) in a bus traveling at 30 mph (v), my speed relative to the road (u_1) will be 32 mph. In relativity this will no longer be true.

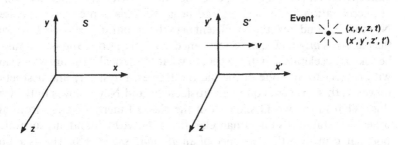

Figure 1.1

1.4   The Set of All Inertial Frames

If the frame S in Figure 1.1 is inertial, so is S'—since the linear equations of motion of free particles in S are transformed by (1.1) into similar linear equations in S', and a Cartesian reference frame relative to which free particles have linear equations of motion is inertial. Conversely, any inertial frame must move "uniformly" (i.e., with constant vector-velocity and without rotation) relative to any other such frame, since a nonuniform relative motion would make it impossible for Newton's first law to be satisfied in both frames. Hence the class of inertial frames consists precisely of the basic frame of the fixed stars plus all other frames moving uniformly relative to that one. According to Newton, all these inertial frames have infinite spatial and temporal extent, i.e., x, y, z, t all range from minus to plus infinity.

1.5   Newtonian Relativity

Not only Newton's first law but also his second and third are valid in *all* inertial frames. For in Newton's theory, both f and m are invariant between inertial frames by axiom (and experience confirms this—at least approximately) and a can be seen to be invariant under the Galilean transformation. Hence the mechanics based on these three laws is identical in all inertial frames. Newton, as Galileo before him, illustrated this fact with the familiar example of a ship, in which all motions and all mechanics happen in the same way whether the ship is at rest or is moving uniformly. This property of Newtonian mechanics is often referred to as *Newtonian* (or *Galilean*) *relativity*.

1.6   Newton's Absolute Space

Because of this "relativity," the uniform motion of one inertial frame relative to another cannot be detected by internal mechanical experiments subject to Newton's theory. But *acceleration* is absolute: although I may be in doubt whether my ship has sailed or is still tied to the pier, I shall be in no doubt if it suddenly hits an iceberg. According to Newton, a particle does not resist uniform motion, of whatever speed, but it does resist any change in its velocity, i.e., acceleration, both positive and negative. This is precisely expressed by Newton's second law: the coefficient m in the equation $f = ma$ is a measure of the particle's *inertia*, i.e., of its resistance to acceleration. And here it may well be asked: acceleration with respect to what? Practically the answer is simple: with respect to any one of the inertial frames. Physically and aesthetically, however, this answer is quite unsatisfactory, and Newton was fully aware of this. What in the world singles out the class of inertial frames from all the others as standards of nonacceleration? Newton found no answer, and postulated instead the existence of an *absolute space* (AS). This is supposed to interact with every particle so as to resist its acceleration. Newton chose to identify AS with the center-of-mass frame of the solar system, and his successors identified it with the frame of the fixed stars. But logically one could identify it with any other inertial frame just as well.

1.7 Objections to Newton's Absolute Space

Newton's concept of an absolute space has never lacked critics. From Huyghens and Leibniz, from Bishop Berkeley, a near-contemporary of Newton, to Ernst Mach in the nineteenth century, to Einstein in the twentieth, cogent arguments have been brought against AS:

(a) It is purely *ad hoc* and *explains* nothing. It can be likened to Kepler's early suggestion that angels were responsible for pushing the planets along their paths or to Sciama's modern analogy of "demons" appropriately active in metals to explain what was until recently the mystery of superconductivity.

(b) There is no unique way of locating Newton's AS within the infinite class of inertial frames.

(c) "It conflicts with one's scientific understanding to conceive of a thing [AS] which acts but cannot be acted upon." The words are Einstein's, but he attributes the thought to Mach. In any case, this is perhaps the most powerful objection of all. It not only questions AS (which Newton's theory can do without) but also the set of all inertial frames.

The fact remains that inertial frames are very real: they play a central role in Newton's theory, and, indeed, in our experience. But Newton's theory offers no satisfactory explanation for their existence. If we fix our attention on two elastic spheres suspended on a common axis, one of which rotates and bulges, while the other is at rest and undeformed, it is hard to understand how the spheres "know" which of them rotates and so must bulge.

1.8 Maxwell's Ether

For a time it appeared as though Maxwell's "ether" could be identified with Newton's AS and so provide an answer to all the above objections. As is well known, in Maxwell's theory there occurs a constant c with the dimensions of a speed, which was originally defined as a ratio between electrostatic and electrodynamic units of charge, and which can be determined by simple laboratory experiments involving charges and currents. Moreover, Maxwell's theory predicted the propagation of disturbances of the electromagnetic field in vacuum with this speed c—in other words, the existence of electromagnetic waves. The amazing thing was that c coincided precisely with the known vacuum speed of light, which immediately led Maxwell to conjecture that light must be an electromagnetic wave phenomenon. (At that time "c" had not yet invaded the rest of physics; Maxwell would have been unlucky had light turned out to be *gravitational* waves!) To serve as a carrier for such waves, and for electromagnetic "strains" in general, Maxwell resurrected the old idea of a "luminiferous ether." It seemed reasonable to assume that the frame of "still ether" coincided with the frame of the "fixed stars."

1.9   Where Is Maxwell's Ether?

The great success of Maxwell's theory since about 1860 put considerable pressure on experimenters to find direct evidence for the existence of the ether. In particular, they tried to determine the terrestrial "ether drift," i.e., the speed of the earth through the ether, as the earth circles the sun. The best known of all these experiments is that of Michelson and Morley (1887). They sent a light signal from a source to a mirror and back again, first in the supposed direction of the ether drift and then at right angles to it, attempting to measure a time difference for the two directions by delicate interference methods. The well-known analogy of the swimmer in the river makes the principle of the experiment clear: it takes less time to swim across the river and back than an equal distance downstream and back. (See Exercises 1.1, 1.2.) Michelson and Morley could detect no time difference at all. Since the earth's orbital speed is 18 miles per second, one could expect the ether drift at *some* time during the year to be at least that much, no matter how the ether streamed past the solar system. Moreover, a drift of this magnitude was well within the capability of the apparatus to detect. Many later and equally ingenious experiments also all failed to find any ether drift whatsoever. The facile explanation that the earth completely "dragged" the ether along with it in its neighborhood could be ruled out because of the observed aberration of starlight. Thus it slowly became apparent that Maxwell's ether was as "useless" in explaining observed phenomena, or predicting new ones, as Kepler's angels.

Additionally, electromagnetic theory was left with a serious puzzle: no matter how one "chased" a light-wave, one could apparently not alter its speed relative to oneself. Such behavior is totally absurd from the point of view of classical kinematics.

1.10   Lorentz's Ether Theory

One way out of these difficulties was provided by the Lorentz ether theory of about 1909.[1] Logically it should have preceded Einstein's SR (1905), but in a sense, only half of it did.[2] It was based on the assumptions of *length contraction* and *time dilation*. Fitzgerald in 1889[3] and, independently, Lorentz in 1892, hypothesized that bodies moving through the ether with

[1] H. Erlichson, *Am. J. Phys. 41*, 1068 (1973).
[2] W. Rindler, *Am. J. Phys. 38*, 1111 (1970).
[3] S. G. Brush, *Isis, 58*, 230 (1967).

velocity v suffer a contraction in the direction of motion by a factor $\gamma = (1 - v^2/c^2)^{-1/2}$. This "explained" the Michelson–Morley experiment. (See Exercises 1.1 and 1.2.) Then Larmor in 1898 and, again independently, Lorentz sometime before 1904—in their search for the transformation that leaves Maxwell's equations invariant—discovered a time transformation equation which *could* have been interpreted as saying that clocks moving through the ether at velocity v go slow (i.e., their rates dilate) by the same factor γ. This would explain the null results of experiments like the Kennedy–Thorndike experiment—similar to that of Michelson and Morley but with unequal arms and kept going for several weeks to test for *changes* in the ether drift. (See Exercise 1.3.) Such experiments, however, were not performed until much later (1932). And Larmor and Lorentz missed this physical interpretation of their equation, apparently regarding it as a mere mathematical artifice. Only after Einstein discovered length contraction and time dilation in a quite different context did Lorentz adopt these as the twin hypotheses upon which his ether theory was based. This does not belittle the work of Einstein's great precursor. But it illustrates the barriers to really revolutionary ideas.

For a while the Lorentz theory provided an alternative to Einstein's theory, equivalent to it observationally and less jolting to classical prejudices. But it was also infinitely less elegant and, above all, less suggestive of new results. Though apparently based on a preferred ether frame, the Lorentz theory yields symmetry, as far as observable predictions go, between all inertial frames. For from the assumed length contraction and time dilation relative to the ether frame it *follows* that rods and clocks moving at speed v through *any* inertial frame appear, respectively, to be shortened and to go slow *relative to that frame* by the same factor $(1 - v^2/c^2)^{-1/2}$. So the theory is open to an objection analogous to (b) in Section 1.7.

1.11 The Relativity Principle

Einstein's solution of the ether puzzle was more drastic: it was like cutting the Gordian knot. In his famous *relativity principle* (RP) he asserted that "all inertial frames are totally equivalent for the performance of all physical experiments." Note that this is a generalization to the whole of physics of Newton's purely mechanical relativity principle.[4] At first sight, Einstein's principle appears to be no more than a whole-hearted acceptance of the null-results of all the ether-drift experiments. But by ceasing to look for special "explanations" of these results and using them rather as the empirical evidence for a new fundamental principle of nature, Einstein had turned the tables: predictions could be made. Soon a whole new theory based on Einstein's principle (and on the experimental fact that light has the same

---

[4] In fact, it seems to have been Huyghens who recognized this principle as something deeper in mechanics than a mere property of Newton's laws. See H. Stein, *Texas Quarterly 10*, 174 (1967), especially p. 183.

velocity in all inertial frames) was in existence, and this theory is called special relativity. Its aim is to modify all the laws of physics, where necessary, so as to make them equally valid in all inertial frames. Today, over seventy years later, the enormous success of this theory has made it impossible to doubt the wide validity of its basic premises. It has led, among other things, to the relativity of simultaneity, to a new mechanics in which mass increases with speed, to the formula $E = mc^2$, and to de Broglie's association of waves with particles.

The RP is an overall principle which "explains" the failure of all the ether drift experiments in much the same way as the principle of the conservation of energy "explains" the failure of all attempts to construct a perpetual motion machine. By declaring it necessarily unobservable, the RP abolished Maxwell's ether forever. The phenomena of contracting bodies and slow-going clocks (on which the Lorentz ether theory is based) also occur in SR, though from quite a different viewpoint. Like all good theories, however, SR did more than illuminate facts already known. It opened up entirely new fields. The situation can be compared to that in astronomy at the time when the intricate geocentric system (corresponding to the Lorentz theory) gave way to the liberating ideas of Copernicus, Galileo, and Newton. In both cases progress was achieved by a change of reference frames. But whereas, at the time, it had taken civil courage to advocate the heliocentric view, it now took intellectual courage to uphold the RP. For, when applied to light propagation, it immediately conflicts with our classical ideas of space and time (cf. end of Section 1.9). It is a mark of Einstein's genius that he saw these ideas as dispensable.

1.12 Arguments for the RP

Although the weightiest argument today for the principle of relativity is the success of special relativity theory, at least three different types of argument could and can be made for it *ab initio*:

(a) The null results of all the ether drift experiments. Einstein did not even bother to make much of this, as can be seen from the quotation under (b) below. But we do know that in his youth he was influenced by the positivists' view that "unobservables" (like the ether) have no place in physics.

(b) The evident "relativity" of Maxwell's theory, if not in spirit, yet in fact. This, to Einstein's mind, carried a great deal of weight. He realized the economy of thought that could be introduced into Maxwell's theory by the RP. Here, in a standard translation, is the beginning of his famous paper [*Annalen der Physik 17*, 891 (1905)]:

> It is known that Maxwell's electrodynamics—as usually understood at the present time—when applied to moving bodies, leads to asymmetries which do not appear to be inherent in the phenomena. Take, for example, the reciprocal electrodynamic action of a magnet and a conductor. The observable phenomenon here depends only

on the relative motion of the conductor and the magnet, whereas the customary view draws a sharp distinction between the two cases in which either the one or the other of these bodies is in motion. For if the magnet is in motion and the conductor at rest, there arises in the neighbourhood of the magnet an electric field with a certain definite energy, producing a current at the places where parts of the conductor are situated. But if the magnet is stationary and the conductor in motion, no electric field arises in the neighbourhood of the magnet. In the conductor, however, we find an electromotive force, to which in itself there is no corresponding energy, but which gives rise—assuming equality of relative motion in the two cases discussed—to electric currents of the same path and intensity as those produced by the electric forces in the former case.

Examples of this sort, together with the unsuccessful attempts to discover any motion of the earth relative to the "light medium," suggest that the phenomena of electrodynamics as well as of mechanics possess no properties corresponding to the idea of absolute rest. They suggest rather that ... the same laws of electrodynamics and optics will be valid for all frames of reference for which the equations of mechanics hold good. We will raise this conjecture (hereafter called the Principle of Relativity) to the status of a postulate....

(c) The unity of physics. This is an argument of more recent origin. It has become increasingly obvious that physics cannot be separated into strictly independent branches; for example, no electromagnetic experiment can be performed without the use of mechanical parts, and no mechanical experiment is independent of the electromagnetic constitution of matter, etc. If there exists a *strict* relativity principle for mechanics, then a large part of electromagnetism must be relativistic also, namely that part which has to do with the constitution of matter. But if part, why not all? In short, if physics is indivisible, either all of it or none of it must satisfy the relativity principle. And since the RP is so strongly evident in mechanics, it is at least a good guess that it is generally valid.

1.13   Maxwellian Relativity

There is an element of irony in that, according to SR, the so-called Lorentz transformations rather than the Galilean transformations relate *actual measurements* made in inertial frames (as we shall see in Section 2.6). Lorentz transformations were already known to be those transformations which *formally* leave Maxwell's equations invariant; but their *physical* significance had remained unrecognized. Since Galilean transformations were long thought to relate inertial frames, Maxwell's theory had been regarded as "unrelativistic." Now it turned out to be fully relativistic after all. Alone among physical theories it needed no modification in the light of SR. Newton's theory, on the other hand, which had always contained a relativity principle, conflicts with the Lorentz transformations and had to be modified.

1.14   Origins of General Relativity

Although SR completely abolished the ether concept, namely AS in its Maxwellian role, it still provided neither an explanation nor a substitute for AS in its Newtonian role as the cause of the absolute inertial structure of the world. SR strongly depends on the concept of the inertial frames. But *why* these frames constitute a privileged class in nature, serving as a standard of nonacceleration (and also as the arena for the simplest formulation of all physical laws) remained as much a mystery as before. It was reserved for GR to solve, or at least to shed much light upon, this problem.

It is often said that GR grew out of the failure of the various attempts to modify Newton's gravitational (inverse square) theory so as to fit it satisfactorily into the framework of SR. Certainly, GR is the modern theory of gravitation which has supplanted Newton's—in principle, if not always in computational practice. Nevertheless, it is clear that Einstein was led to GR primarily by his philosophic desire to abolish totally the role of absolute space from physics. He would probably not have stopped at a special-relativistic theory of gravitation, however satisfactory, since SR begins by taking the inertial frames for granted. Einstein thus had to go beyond SR. In this task he loyally and repeatedly acknowledged his debt to the physicist–philosopher Mach (1836–1916). It is probably fair to say that Mach, in turn, is indebted to Einstein for honing, elaborating, and perpetuating his relativistic ideas on inertia.

1.15   Mach's Principle

Mach's ideas on inertia, whose germ was already contained in the writings of Bishop Berkeley, are roughly these: (a) space is not a "thing" in its own right; it is merely an abstraction from the totality of distance-relations between matter; (b) a particle's inertia is due to some (unfortunately unspecified) interaction of that particle with all the other masses in the universe; (c) the local standards of nonacceleration are determined by some average of the motions of all the masses in the universe; (d) all that matters in mechanics is the *relative* motion of *all* the masses. Thus Mach wrote: "... it does not matter if we think of the earth as turning round on its axis, or at rest while the fixed stars revolve around it. ... The law of inertia must be so conceived that exactly the same thing results from the second supposition as from the first." It is perhaps significant that even before Einstein, Mach referred to himself and his followers as "relativists."

A spinning elastic sphere bulges at its equator. To the question of how the sphere "knows" that it is spinning and hence must bulge, Newton might have answered that it "felt" the action of absolute space. Mach would have answered that the bulging sphere "felt" the action of the cosmic masses rotating around it. To Newton, rotation with respect to AS produces centrifugal (inertial) forces, which are quite distinct from gravitational forces; to

Mach, centrifugal forces *are* gravitational, i.e., caused by the action of mass upon mass.

Einstein coined the term *Mach's principle* for this whole complex of ideas. Of course, with Mach these ideas were still embryonic in that a quantitative theory of the proposed "mass induction" effect was totally lacking. Einstein, at one stage in his progress towards GR, conjectured that Newton's inverse square theory probably differed as much from a complete gravitational theory as a simple electric theory based only on Coulomb's law differed from Maxwell's ultimate theory. Indeed, D. W. Sciama[5] in 1953 revived and extended an 1872 Maxwell-type gravitational theory of F. Tisserand and found that it largely includes Mach's principle: inertial forces correspond to the gravitational "radiation field" of the universe and are proportional to the inverse *first* power of the distance. But, unfortunately, this theory violates relativity on other grounds. For example, whereas charge is necessarily invariant in Maxwell's theory, mass varies with speed in SR. Also, because of the relation $E = mc^2$, the gravitational binding energy of a body has (negative) mass; thus the total mass of a system cannot equal the sum of the masses of the parts, whereas in Maxwell's theory charge (the analog of mass) is strictly additive, as a direct consequence of the linearity of the theory.

Einstein's solution to the problem of inertia, GR, turned out to be much more complicated than Maxwell's theory. However, in "first approximation" it reduces to Newton's theory, and in "second approximation" it actually has Maxwellian features. But in what sense GR is truly "Machian" is still a matter of debate. (See Section 9.12.) In any case it must be noted that (i) Mach's principle is rooted in classical kinematics and (ii) it ignores "fields" as a possible content of space. Thus its very formulation within modern physics is problematic.[6]

1.16   Consequences of Mach's Principle

It is sometimes held that Mach's principle is physically empty: since we cannot "experiment" with the universe, we cannot test the principle; we can never decide whether it is absolute space or the cosmic masses that determine inertia, and thus the choice is philosophical rather than physical.

Yet this is not so. Even without the help of a detailed theory, Mach's principle leads to certain *testable* non-Newtonian predictions:

(a) The following remark is due to Sciama. It was verified in 1926 that our galaxy rotates, somewhat like a huge planetary system, with a period of about 250 million years in the neighborhood of the sun. Such a rotation was already postulated by Kant to account for the flattened shape of the galaxy

---

[5] *Mon. Not. R. Astron. Soc. 113*, 34 (1953); see also D. W. Sciama, *The Unity of the Universe*, New York, Doubleday and Co., Inc., 1959, especially Chapters 7–9.

[6] But see C. W. Misner, K. S. Thorne, and J. A. Wheeler, *Gravitation*, W. H. Freeman and Co., 1973, §21.12.

(as evidenced by the Milky Way in the sky). Also, without orbiting, the sun would fall into the center of the galaxy in about 20 million years. Had Mach been aware of this rotation, he could have been led by his principle to postulate the existence of a whole vast extragalactic universe—which was not discovered observationally until much later—simply in order to make the standard of rest in the galaxy come out right. A strict Newtonian, on the other hand, would see nothing remarkable in the galactic rotation.

(b) Consider a Foucault pendulum suspended, for the sake of simplicity, from a tripod at the earth's North Pole. It swings in a plane that remains fixed relative to the universe while the earth turns. Now suppose we could remove all the cosmic masses except the earth. According to Newton, the pendulum experiment would not be affected. According to Mach, however, the pendulum would now swing in a plane fixed relative to the earth—which has become the obvious and only standard of nonacceleration. Next let us reintroduce the cosmic masses gradually, until *their* inertia-producing effect predominates again. But, by continuity, it will never predominate totally. The earth will always make *some* contribution to the local "compass of inertia" and thus, however minutely, drag the plane of the Foucault pendulum around in the sense of its rotation. Given sufficiently accurate apparatus, this would be a measurable, non-Newtonian effect. It has recently been suggested that in place of the Foucault pendulum, one could perhaps observe the orbital plane of an artificial satellite in polar orbit.

(c) Consider the following pair of diagrams, representing the system earth–moon–universe, the universe being schematically represented by a massive shell. Figure 1.2a represents the "conventional" view. Figure 1.2b represents a *relatively* equivalent view, with the moon at rest and the universe revolving. According to Mach, as we have seen, "the law of inertia must be so conceived that exactly the same thing results from the second supposition as from the first." Hence the rotating universe must provide a centrifugal force f to counteract the earth's attraction h. (It is this force, too, which causes the earth's equatorial bulge.) The universe must also provide a Coriolis force which has the cumulative effect of turning the plane of the Foucault pendulum. By a straightforward analogy, we would therefore expect minute centrifugal and

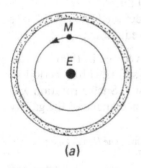

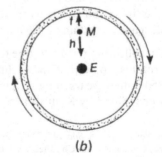

(a)   (b)

Figure 1.2

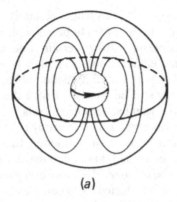   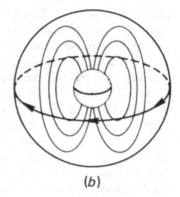

(a)   (b)

Figure 1.3

Coriolis forces to arise inside *every* rotating massive shell—another quite non-Newtonian effect.[7] Furthermore, with reference to Figure 1.2a, some force (e.g., a rocket) would be needed to accelerate the earth from rest in a given direction. But this could be reinterpreted as the earth needing a force to stand still inside a universe that accelerates past it. By extension, therefore, we would expect a particle to experience a force whenever masses in its neighborhood are accelerated, the force being in the direction of the acceleration.

(d) By its denial of AS, Mach's principle actually implies that not only gravity but *all* physics should be formulated without reference to preferred inertial frames. It advocates nothing less than the total relativity of physics. As a result, it even implies interactions between inertia and electromagnetism. Consider, for example, a positively charged, nonconducting sphere which rotates. Each charge on it gives rise to a circular current and thus to a magnetic field. Figures 1.3a and 1.3b again represent the conventional view and a Mach-equivalent view, respectively, of the sphere, the universe, and the corresponding magnetic field. By analogy again, a minute magnetic field should arise within any massive rotating shell with stationary charges inside it.[8]

(e) A single test particle in an otherwise empty universe would, by Mach's principle, not be expected to have any appreciable inertia (merely "self inertia"). The introduction of other masses into the universe would gradually bestow inertia upon the particle. Thus again, by extension, one would expect any particle's inertia to increase in the presence of large masses. Moreover, one might well expect this inertia to have directional properties, related to

---

[7] See end of Section 8.12. This problem was investigated, on the basis of GR, by H. Thirring in *Phys. Zeits. 19*, 33 (1918); *22*, 29 (1921); and that of (b), above, by H. Thirring and J. Lense in *Phys. Zeits. 19*, 156 (1918). Effects somewhat like those conjectured were indeed found. On the basis of an earlier version of GR, Einstein had already found similar effects, and communicated them to Mach in a letter dated 1913. (See C. W. Misner *et al.*, *loc. cit.*)

[8] This conjecture was examined on the basis of GR by J. Ehlers and W. Rindler in *Phys. Rev. D, 4*, 3543 (1971). A magnetic field was found, though somewhat at variance with the Machian expectation.

the mass distribution. (In a Maxwell-type theory, this would certainly be so.) Now to a very high degree of accuracy, inertia is known to be isotropic in our part of the universe. From this, some people have drawn certain conclusions: First, since the matter in our *immediate* neighborhood (sun, planets, etc.) is patently nonisotropic, the overwhelming part of the induction effect must come from *distant* matter: a "$1/r$" law would be consistent with this. And secondly, the distant matter, i.e., the universe, must be isotropic relative to us. This would add weight to the optically observed rough isotropy of the universe and speak against the various nonisotropic world-models that have been proposed. On the other hand, it can be argued that directional variation, or indeed any variation, of inertia would be *unobservable*, since every apparatus designed to measure it would itself be so affected as to completely mask the effect. Such would certainly be the case if Einstein's strong equivalence principle is true. (See Section 1.19.)

Even allowing this last disclaimer, examples (a)–(d) should make it amply clear that Mach's principle has physical content. On the other hand, it is so far still without experimental verification. Also, as we have mentioned, its modern formulation is problematic. In spite of its aesthetic appeal, one should therefore keep an open mind about the principle. In particular, Mach's way to resolve the objections against absolute space, namely abolishing space as a "thing" altogether, may be too radical. The main objection against absolute space, that it acts but cannot be acted upon, can also be met while retaining the separate existence of space, but allowing it to interact with matter: this is actually what happens in GR. Strangely enough, though, and as we mentioned before, the logical status of GR vis-a-vis Mach's principle has still not been totally clarified. One thing is certain: Mach's principle has proved a fruitful source of conjectures—often vindicated—for GR to investigate quantitatively. Most of these have to do with the "dragging of inertia" by accelerating or rotating masses. Mach's principle provides a way to understand such phenomena intuitively whereas the GR approach is highly mathematical.

1.17   Cosmology

Mach's principle suggests that *the whole universe matters locally*. It will therefore be useful even at this early stage to review briefly the main features of the universe as they are known today. Our galaxy contains about 10^{11} stars—which account for most of the objects in the night sky that are visible to the naked eye. Beyond our galaxy there are other more or less similar galaxies, shaped and spaced roughly like dimes three feet apart. The "known" part of the universe, which stretches to a radius of about 10^9 light years, contains about 10^{11} such galaxies. They recede from each other in such a way that, had the presently observed recession been uniform in time, the observable universe would have been a dense ball some 10^{10} years ago. However, there are good reasons for believing that the recession has *not* been uniform in time. For example, gravity would slow the expansion after a "big-bang" origin of the universe. These facts necessitate some revision of our original definition of inertial frames.

As a very simplified possible model of the universe, consider an infinite array of galaxies, scattered more or less uniformly throughout space, and mutually receding, much like an array of knots in an infinite rubber sponge that is being stretched at a variable rate, but equally in all directions. Our model satisfies the so-called *cosmological principle*, according to which all galaxies stand in the same relation to the whole universe. This principle is adopted, partly for empirical but mainly for simplistic reasons, by practically all modern cosmologies. It excludes, for example, a finite "island" universe immersed in infinite space, since that contains atypical "outermost" galaxies.

Now, how can we determine infinitely extended Newtonian inertial frames in this universe? If the center of *our* galaxy were at rest in one such inertial frame, would not the center of each other galaxy also be at rest in one such frame, by the cosmological principle or simply by symmetry? Yet those other inertial frames *do not move uniformly* relative to ours! Moreover, where in *our* inertial frame do free particles obey Newton's first law? At most, in the neighborhood of our own galaxy. Far out among the distant galaxies a test particle is affected by the same gravitational acceleration that pulls all galaxies towards each other and, in particular, towards us. There simply are *no* regions "sufficiently removed" from all attracting matter in which free particles move uniformly relative to us, except in our immediate vicinity. And so, extended inertial frames would not exist in such a universe.

Mach's principle obviously suggests that under these conditions the center of each galaxy provides a basic *local* standard of nonacceleration, and the lines of sight from this center to the other galaxies (rather than to the stars of the galaxy itself, which may rotate) provide a local standard of nonrotation: together, a *local inertial frame*. Inertial frames would no longer be of infinite extent, and they would not all be in uniform relative motion. A frame which is locally inertial would cease to be so at a distance, if the universe expands nonuniformly. Nevertheless, *at each point* there would still be an infinite set of local inertial frames, all in uniform relative motion.

1.18 Inertial and Gravitational Mass

Much the same conclusion was reached by Einstein in his *equivalence principle* (EP) of 1907, though rather more generally and from a different point of view. (The expansion of the universe was unknown then.) Einstein began with a closer look at the concept of "mass." It is not always stressed that at least two quite distinct types of mass enter into Newton's theory of mechanics and gravitation. These are (i) the *inertial mass*, which occurs as the ratio between force and acceleration in Newton's second law and thus measures a particle's resistance to acceleration, and (ii) the *gravitational mass*, which may be regarded as the gravitational analog of electric charge, and which occurs in the equation

$$f = \frac{Gmm'}{r^2} \tag{1.3}$$

for the attractive force between two masses (G being the gravitational constant.)

One can further distinguish between *active* and *passive* gravitational mass, namely between that which causes and that which yields to a gravitational field, respectively. Because of the symmetry of Equation (1.3) (due to Newton's third law), no essential difference between active and passive gravitational mass exists in Newton's theory. In GR, on the other hand, the concept of passive mass does not arise, only that of active mass—the creator of the field.

It so happens in nature that for *all* particles the inertial and gravitational masses are in the same proportion, and in fact they are usually made equal by a suitable choice of units, e.g., by designating the same particle as unit for both. The proportionality was carefully verified by Eötvös, first in 1889, and finally in 1922 to an accuracy of five parts in 10^9. Eötvös suspended two equal weights of different material from the arms of a delicate torsion balance pointing west–east. Everywhere but at the poles and the equator the earth's rotation would produce a couple if the inertial masses of the weights were unequal—since centrifugal force acts on inertial mass. By an ingenious variation of Eötvös's experiment, using the earth's orbital centrifugal force which changes direction every 12 hours and so lends itself to amplification by resonance, Roll, Krotkov, and Dicke (Princeton, 1964) improved the accuracy to one part in 10^{11}, and Braginski and Panov (Moscow, 1971) even to one part in 10^{12}.

The proportionality of gravitational and inertial mass is sometimes called the "weak" equivalence principle. A fully equivalent property is that *all* particles experience the same acceleration in a given gravitational field: for the field times the passive mass gives the force, and the force divided by the inertial mass gives the acceleration. Hence the path followed by a particle in space and time is entirely independent of the kind of particle chosen. This path unicity in a gravitational field is usually referred to as *Galileo's principle*, by a slight extension of Galileo's actual findings. (Recall his alleged experiments on the Leaning Tower of Pisa!)

It was once questioned whether the mass-equivalent of the atomic binding energy (via $E = mc^2$) satisfies the weak EP. Southerns in 1910 was able to answer this affirmatively by repeating Eötvös's experiment with a uranium oxide weight, whose binding energy is particularly high. Today the same conclusion can be inferred from the work of Dicke et al. with "ordinary" weights, since the accuracy has increased so much. More recently the question arose whether antimatter obeys Galileo's principle, or falls *up*. Direct experiments with positrons etc. are extremely difficult. But quantum-mechanical calculations by Schiff[9] have shown that there are enough virtual positrons in

[9] L. I. Schiff, *Phys. Rev. Lett.*, *1* (1958) 254.

ordinary matter to be detected in the Eötvös–Dicke experiments *if* positrons fall up. Thus it seems established that even antimatter satisfies the weak EP.

The proportionality of inertial and gravitational mass for different materials is really a very remarkable fact. It is totally unexplained in Newton's theory, taken as an axiom, and apparently fortuitous. Newton's theory would work perfectly well without it: it would then resemble a theory of motion of electrically charged particles under an attractive Coulomb law, where particles of the same (inertial) *mass* can carry different (gravitational) *charges*. GR, on the other hand, contains Galileo's principle as a primary ingredient, and could not survive without it.

It may be noted that Mach's principle goes some way towards explaining the *identity* of inertial and passive gravitational mass. Consider, for example, the situation illustrated in Figure 1.2b, where at the moon's location the rotating universe provides an inductive field f to counteract the attraction of the earth. Now it is conceivable that induction acts on one kind of "charge," namely inertial mass, and attraction on another, namely gravitational mass. But it is far more natural, once both kinds of mass are recognized as "charge," to think of them as identical—as they are in Maxwell's theory, where the same charge submits to Coulomb *and* induction fields. In Figure 1.2b the universe and the earth then simply combine to produce a zero field at the moon's location, in which *every* particle would remain at rest.

1.19   The Equivalence Principle

In conjunction with Newton's three laws, the "weak" equivalence principle discussed in the last section both implies, and is implied by, another "relativity" embodied in Newton's theory, different from the one discussed in Section 1.5, *but one which also invites generalization to the whole of physics.* It is this: Consider an elevator cabin which is severed from its supporting cable and allowed to fall freely down a long shaft under the earth's gravitational pull; Newton's theory predicts that mechanics will take *precisely* the same course in this freely falling "laboratory" as in a laboratory that is unaccelerated and far away from all attracting masses, i.e., as in a strict inertial frame.

It is easy to see why: For a particle being pushed around arbitrarily, let $\mathbf{f}$ and $\mathbf{f}_G$ be the total and the gravitational force, respectively, relative to the earth (here treated as a Newtonian inertial frame), and m_I and m_G the inertial and gravitation mass. Then $\mathbf{f} = m_I \mathbf{a}$ and $\mathbf{f}_G = m_G \mathbf{g}$, where $\mathbf{a}$ is the acceleration of the particle and $\mathbf{g}$ the gravitational field, and thus the acceleration of the cabin. The acceleration of the particle relative to the cabin is $\mathbf{a} - \mathbf{g}$ and so the force relative to the cabin is $(\mathbf{a} - \mathbf{g})m_I$. This equals the nongravitational force $\mathbf{f} - \mathbf{f}_G$ *if* $m_I = m_G$; hence Newton's second law (incuding the first) holds in the cabin. And the same is true of the third law. Gravity has been "transformed away" in the cabin.

Einstein, in his EP, assumed once again that the rest of physics goes along with mechanics. He postulated that "all local, freely falling, nonrotating laboratories are fully equivalent for the performance of all physical experiments." (The test for nonrotation can be any one of the obvious mechanical ones, e.g., the walls of the laboratory must be unstressed during the free fall, or, if made of loosely laid bricks, they must not come apart.) Justification for the EP is discussed in Section 1.20. According to Einstein, then, a freely falling laboratory, even near a strongly gravitating mass, is *fully* equivalent to a laboratory floating motionlessly relative to the fixed stars out in space. Each such freely falling, nonrotating laboratory constitutes a *local inertial frame* (LIF). All LIF's at the same event are necessarily in uniform relative motion, but LIF's far apart may mutually accelerate. Thus Einstein, by a different line of argument, arrived at conclusions similar to those we reached in Section 1.17.

The class of uniformly moving LIF's at any event is clearly the one discussed by SR, which now appears as a *locally* applicable theory. (In SR one uses different inertial frames to look at the *same* situation, and thus one never needs widely separated reference systems.) Before the recognition of the EP, the RP could only be understood to refer to the infinitely extended Newtonian inertial frames, i.e., to the frame of the fixed stars and all others in uniform motion relative to that one. To be safe, one considered only the parts of these frames in which gravity was essentially absent, i.e., strictly inertial regions in which free particles really obey Newton's first law. For it is in the inertial frames that each physical law is supposed to assume maximal simplicity, including, for example, isotropy; and in regions of the frames in which there is an "overlay of gravity," as on earth, isotropy could hardly be assumed *a priori* even for "nongravitational" physics. (In Section 1.21 we shall see the justification for this suspicion.) Thus, strictly speaking, special-relativity physics seemed to apply in hypothetical interstellar laboratories only. The EP changed all that. It both widened and narrowed the RP: it brought many more frames within its scope, but it limited their extent; it made all regions equally accessible to SR, but it made SR a local theory. In the future when we speak of the RP, we shall implicitly identify it with the EP.

As a consequence of the equivalence principle, not only can we *eliminate* gravity by free fall, we can also *create* it by acceleration. Consider, for example, a rocket sitting on its pad on earth. This should be indistinguishable, as a physics laboratory, from a rocket that just hovers over its pad. The latter, in turn, should be indistinguishable from any other rocket that moves with acceleration g through its LIF, e.g., one in outer space, since all LIF's are equivalent. And this proves our assertion. As Sciama has pointed out, this version of the EP is directly supported by Mach's principle. The rocket in outer space sees the universe accelerate past it. The accelerating universe creates a gravitational field inside the rocket. So it is not really surprising if all physical processes in the rocket go on in the same way as they do in the earth's gravity: one gravitational field is just like another.

On the other hand, situations can be imagined where the EP apparently conflicts with Mach's principle. For example, if two identical spheres spin at identical rates in their respective LIF's, one, say A, in tenuous outer space and the other, B, near a heavy mass, should not B bulge more, since by Mach's principle its inertia might be expected to be greater? And would not this contradict the EP? There is a way out: time actually runs more slowly near heavy masses (itself a consequence of the EP, as we shall see in Section 1.21), and thus A will *see* B rotate more slowly than itself. But since it sees B bulge as much as itself, it concludes that B has more inertia, or a lesser coefficient of elasticity, or both. Similar arguments can be made in all other cases of apparent conflict.

Of course, gravity can only be fully eliminated from a rigid laboratory by free fall if the field is parallel. In all other cases the tolerable extent in space *and* time of a LIF depends on the accuracy desired. For example, if an elevator cabin were allowed to fall to the center of the earth, two free particles initially at rest in opposite corners of the floor would gradually approach each other and finally meet at the center. Similarly, two free particles initially at rest in the cabin, one above the other, would gradually come together. Nevertheless, such particle pairs can be treated as being mutually at rest to any desired accuracy *provided* we suitably limit the dimensions of the cabin and the duration of the experiment.

The RP *denies* the existence of a preferred standard of rest; yet such a standard exists *de facto* everywhere locally in the universe, namely, that which is essentially determined by the local set of galaxies. And although, in accordance with the RP, no known law of physics assumes a special form in that particular frame, logically we could not be surprised if one did (as Bondi has pointed out). And similarly, in a cabin freely hurtling towards the earth, there is obviously a *de facto* preferred direction. Nevertheless, in accordance with the EP, all the known laws of physics appear to be isotropic within the cabin.

1.20   The Semistrong Equivalence Principle

Einstein's EP is sometimes called the "strong" EP to distinguish it from its restriction to Newtonian mechanics, known as the "weak" EP. There is also a "semistrong" (or "local") EP: it asserts the full equivalence of all nonrotating free laboratories *locally*, and this implies a *local* SR everywhere, with its own numerical content. But it envisages the possibility of different numerical contents (speed of light, gravitational "constant," fine structure "constant," etc.) at different regions and times in the universe. The "real" laws whose local approximations are recognized in the various local SR's would presumably involve also the derivatives of these variable "constants." But in this connection it must be said that *every* law of physics could always be generalized and made more complicated while still remaining within all theoretical and observational bounds, and no progress in physics can be made without simplicity assumptions. Complications should therefore not be countenanced except for very good reasons. It is true, however, that all the arguments usually advanced for the EP, *except simplicity*, are in fact arguments only for the semistrong version:

(a) As for the RP before, one can appeal to the unity of physics, according to which all physics should share the transformation properties of mechanics.

(b) Often cited are the successful experimental checks on two direct predictions made on the basis of the EP (see Section 1.21).

(c) GR is still the most satisfactory modern theory of gravitation, and GR incorporates the EP. GR is based on the *strong* EP. As we shall see later, the EP allows GR to be built up from all the local SR's in a certain way—much as a curved surface can be built up, approximately, from a lot of plane elements. But GR can be generalized fairly easily to accommodate the semistrong EP, if necessary. Such a generalization has been made, for example, by Pascual Jordan, who believes that certain geological facts indicate a secular variation in the "constant" of gravitation. Jordan's theory was later extended by Brans and Dicke.[10]

It must be said in fairness that the *empirical* evidence for the (strong or semistrong) EP is very poor. The checks mentioned under (b) above concern two aspects of light propagation, and it can be maintained that what they really test is only the *weak* EP, once we grant the corpuscular nature of light (photons). Nor can the success of GR in reproducing the well-established classical gravitational results be considered specific evidence for the EP, since Newton's theory, of course, does the same without it. Best evidence is provided by the few tested "post-Newtonian" results of GR, but even these can be duplicated by theories denying the EP. The appeal of the EP is thus mainly theoretical. But that is nevertheless so strong that most experts accept it.

1.21   Consequences of the Equivalence Principle

The EP leads directly to two interesting predictions. First, it implies that light bends in a gravitational field, much as though it were made up of particles traveling at speed c. For consider a freely falling cabin, say in an elevator shaft on earth. Consider a flash of light emitted in the cabin at right angles to its motion. By the EP this flash travels along a straight line inside the cabin, but since the latter accelerates while the flash within it travels uniformly, the light path must in fact be curved parabolically relative to the earth, just like the path of a projectile. This is really a very remarkable argument, since from the mere fact that light travels at finite speed one deduces that "light has weight." We have made absolutely no other assumptions about it. *All* phenomena (gravitational waves, ESP?) that propagate with finite velocity in an inertial frame would thus be forced by gravity into a locally curved path. This suggests

[10] Various recent observations (e.g. six years of lunar laser ranging) seem to speak directly against a significant variation of the constant of gravitation. Other data (e.g. very accurate light bending observations—see Section 1.21) increasingly favor GR as compared to the theory of Brans and Dicke.

rather strongly that what we have discovered here is not so much a new property of light, but, instead, a new property of space in the presence of mass, namely *curvature*: if space itself were curved, *all* naturally straight phenomena would thereby be forced onto curved "rails." (Such curvature of space—actually of space and time—forms the basis of GR; it also "explains" Galileo's principle.)

Secondly, the EP implies that light traveling down a gravitational field suffers a blueshift. For consider now a vertical ray of light entering the cabin's ceiling at the moment when the cabin is dropped. By the EP, an observer A on the cabin's floor, observing the ray just as he passes a stationary observer B outside, observes no Doppler shift between ceiling and floor. But by the time A observes the ray, the cabin is already moving, and, relative to A, B moves *into* the light waves. Since A sees no Doppler shift, B, observing the same light, must see a blueshift. This proves our assertion. Conversely, light traveling *against* a gravitational field suffers a redshift. (The same result can be obtained from the photon theory by appeal to Planck's relation—cf. Exercise 7.10.) As a corollary, if we regard vibrating (light-emitting) atoms as "clocks," it follows that clocks fixed at a lower potential go slower than clocks fixed at a higher potential. This is called "gravitational time dilation."

Indeed, owing to this effect, the U.S. atomic standard clock kept since 1969 at the National Bureau of Standards at Boulder, Colorado, at an altitude of 5400 ft, gains about five microseconds each year relative to a similar clock kept at the Royal Greenwich Observatory, England, at an altitude of only 80 ft, both clocks being intrinsically accurate to one microsecond per year.

The predicted gravitational Doppler shift as such has been observed by Pound and Rebka in 1960, and with improved accuracy (to 1 %) by Pound and Snider in 1964, in highly sensitive experiments which involved sending light down a 70-foot (!) tower at Harvard. (See Exercise 1.11.) It had previously been observed with lesser accuracy in the light from very dense stars, following a suggestion of Einstein.

The gravitational deflection of light has not yet been observed in purely terrestrial experiments. These would test the EP directly. Over astronomical distances the theory of light deflection suffers from a complication: it is necessary to specify how the local 3-spaces along the light path are "patched" together. GR does this in a "curved" way which leads to a deflection around a central mass twice as large as that which would result from a "flat" patching of local 3-spaces.[11] On the other hand, an earlier relativistic theory (Nordström's), which also contains the EP, patched the local 3-spaces around a central mass with the opposite curvature from GR, so that *no* overall deflection of light resulted.[12]

The traditional method of testing the GR prediction consisted in observing the apparent positions of a set of stars near the sun during a solar eclipse, and

[11] See Section 8.3, paragraph containing Equation (**8.45**) *et seq.*

[12] This, incidentally, provides a useful counterexample to those attempts that have been made, from time to time, to obtain the full GR bending of light from the EP alone. For another, and for references on this topic, see W. Rindler, *Am. J. Phys. 36*, 540 (1968).

comparing these with their "real" positions in the night sky six months later. The accuracy of this method (about 20 %) has hardly improved since Eddington's memorable first in 1919. However, recent measurements by Fomalont and Sramek (1975) of the deflection in the sun's field of radio waves from three quasars which "pass" close to the sun once a year (and which can be observed without eclipse) have confirmed the GR bending to an accuracy of 1 %. Further quasar work is in progress and is expected to yield even greater accuracy soon.

The two consequences of the EP discussed in this section illustrate the danger in regarding gravitational regions of extended Newtonian inertial frames as strictly inertial (e.g., isotropic, homogeneous) for "nongravitational" phenomena. They indicate that gravity cannot be divorced from the rest of physics—or, in a colorful phrase, gravity cannot be "painted onto" the rest of physics. We have already found reason to suspect this in (d) of Section 1.16.

CHAPTER 2

Einsteinian Kinematics

2.1 Basic Features of Special Relativity

In this chapter we begin to develop the consequences of Einstein's "first postulate," the relativity principle (RP)—a principle of venerable standing in mechanics, now newly extended to all of physics. Einstein chose to ignite it with a spark from electromagnetic theory: his "second postulate," according to which light travels rectilinearly with constant speed c in vacuum in *every* inertial frame. After the blaze, the old relativity principle showed its new mathematical core: the Lorentz transformation (LT). Previously "common sense" had shown that the core "must" be the Galilean transformation (GT).

For if we assume perfect symmetry between inertial frames in accordance with the RP, and accept *Newton's* "second postulate" of an absolute time covering all inertial frames, then inevitably we find the GT as that transformation which relates *actually measured* Cartesian coordinates in two different inertial frames. But if, instead, we insist against all "common sense" on accommodating Einstein's law of light propagation in all inertial frames, then inevitably the LT must relate the coordinates. It is of course evident that this law of light propagation cannot be reconciled with the GT: for that implies the classical velocity addition law according to which a light signal traveling with velocity c in a frame S has anything *but* the same velocity in a second frame moving uniformly relative to S [cf. Equation (**1.2**)].

The second postulate is reducible to the following three statements: in each inertial frame light travels rectilinearly, at a finite speed, which is independent of the motion of the source. The frame-invariance of that speed then follows from the RP. That it is finite was established by Roemer as early as 1675. (He found that the times between successive observations of the eclipsing of its moon Io by Jupiter grew shorter as the earth in its orbit around the sun approached Jupiter, and longer as the earth moved away.) Rectilinear propagation is also an empirical fact. It cannot be simply predicted from symmetry: light could consist of spinning and swerving particles. Lastly, the source-independence of the light velocity seemed to have been established by de Sitter's analysis (1913) of double stars in orbit around a common center, where the maximum redshift in the light of one star is seen at the same instant as the maximum blueshift in the light of the other. But that particular argument has been challenged by Fox[1] who believes that double stars may be enveloped in a gas cloud which would re-emit their light anyway with equal velocity (by Ewald and Oseen's "extinction theorem"). Direct laboratory evidence for the source-independence of the light velocity had to wait until 1963.[2] However, the strongest support for the second postulate, i.e., for the three statements at the beginning of the present paragraph, comes from the overall success of Maxwell's theoretical treatment of light as an electromagnetic wave phenomenon.

The physics based on Einstein's two postulates is called special relativity (SR). The second postulate, however, has served its purpose as soon as the LT's are derived. The first postulate can then be re-expressed mathematically: for a physical law to be valid in *all* inertial frames, its formal mathematical statement in *one* inertial frame must transform into iself under a LT. Hence, according to the first postulate, the new laws must be "Lorentz-invariant." SR, then, is Lorentz-invariant physics. Newton's mechanics, for example, is Galileo-invariant but not Lorentz-invariant, and thus it is inconsistent with SR. It is the task of SR to review *all* existing laws of physics and to subject them to the test of the RP with the help of the LT's. Any law found to be lacking must be modified accordingly. These modifications, though highly significant in many modern applications, are negligible in most classical circumstances, which explains why they were not discovered earlier. It is rather remarkable that, in the mathematical formalism of SR, most of the new laws were neither very difficult to find nor in any way less elegant than their classical counterparts.[3]

We shall show at the end of this chapter that the RP by itself (together with certain "reasonable" assumptions) is consistent with only two possible transformations and no others: the GT and the LT. Hence logically one can replace Einstein's second postulate by *any* phenomenon—and today we know many—which is peculiar to SR, i.e., consistent with the RP but not with the GT. The resulting theory, in every case, must be SR. Thus the detailed behavior of light (e.g., the source-independence of its velocity) is not *the* pillar

---

[1] J. G. Fox, *Am. J. Phys. 30*, 297 (1962).

[2] See T. A. Filippas and J. G. Fox, *Phys. Rev. 135*, B1071 (1964), where further references can be found.

[3] However, in some modern areas such as the quantum theory of interacting systems, there still remain fundamental difficulties with the relativistic formulation.

on which SR stands or falls. It is no more crucial to the validity of the theory than any other of its many predictions.

Apart from leading to new laws, SR leads to a useful technique of problem solving, namely the possibility of switching inertial reference frames. This often simplifies a problem. For although the totality of laws is the same, the configuration of the problem may be simpler, its symmetry enhanced, its unknowns fewer, and the relevant subset of laws more convenient, in a judiciously chosen inertial frame.

If we accept the equivalence principle, then SR is necessarily a local and approximate theory, since it makes statements about strict inertial frames, and since these are realized in nature only locally and approximately. We must insist on *strict* inertial frames, since we wish to assume perfect isotropy and homogeneity for such frames. For example, we wish to make such assumptions as that light travels rectilinearly in all directions, or that a source moving at given velocity shows the same Doppler shift no matter in what direction it recedes from us. As we have seen in Section 1.21, these would not be safe assumptions in an extended Newtonian inertial frame. It is in the strict inertial frames—and the EP tells us how to find them locally in practice—that we assume the laws of physics to hold in their simplest form. These laws are idealizations of those observed to hold in our terrestrial laboratories, which of course are *not* strictly inertial.

The reader may now ask: if Newton's theory is incompatible with SR, and SR is concerned with inertial frames, and inertial frames are defined by Newton's first law, is there not something wrong somewhere? The answer is no: Newton's *first* law *is* compatible with SR, and is in fact accepted by SR. Nevertheless, as an economy, Newton's first law can be replaced by the law of light propagation in singling out the inertial frames among all other rigid frames.

2.2   On the Nature of Physical Laws

This may be a convenient point at which to make a comment on the "truth" of physical laws. According to modern thought—largely influenced by Einstein—even the best of physical laws do not assert an absolute truth, but rather an approximation to the truth. Moreover, they are not mere summaries of experimental facts, available to any diligent seeker "cleansed of prejudices," as was thought by Bacon, and still by Mach. Human invention necessarily enters into the systematization of these facts.[4] A physical theory, then, is a subjective amalgam of concepts, definitions, and laws, regarded as a *model* for a certain part of nature, and asserting not so much what nature *is*, but rather what it *is like*. Agreement with experiment is the most obvious

---

[4] As Einstein wrote in 1952: "There is, of course, no logical way to the establishment of a theory . . ." (cf. p. 35 of R. S. Shankland, *Am. J. Phys. 32*, 16 (1964)). See also pp. 11, 12 of Einstein's Autobiographical Notes in *Albert Einstein: Philosopher-Scientist* (ed. P. A. Schilpp), 1949.

requirement for the usefulness of such a theory. However, no amount of experimental agreement can ever "prove" a theory, partly because no experiment (unless it involves counting only) can ever be infinitely accurate, and partly because we can evidently not test all relevant instances. Experimental disagreement, on the other hand, does not necessarily lead to the rejection of a physical theory, unless an equally simple one can be found to replace it. Such disagreement may simply lead to a narrowing of the known "domain of sufficient validity" of the model. We need only think of Newton's laws of particle mechanics, which today are known to fail in the case of very fast-moving particles, or Newton's gravitational theory, which today is known to fail in the finer details of planetary orbit theory. The "truer" relativistic laws are also mathematically more complicated, and so Newton's laws continue to be used whenever their known accuracy suffices.

Although SR is today one of the most firmly established theories in physics, spectacularly verified in literally millions of experiments, it is well to keep an open mind even here. To orders of accuracy so far unattained or unimagined, some law of SR may one day be found to fail—for example, along the lines mentioned at the end of Section 1.19. We shall not go wrong if we remember that every theory is only a model, with perhaps four prime desiderata: simplicity (or what Einstein called "beauty"), internal consistency, compatibility with other scientific concepts of the day, and good experimental "fit." There is also a secondary desideratum, stressed by Popper, and that is the possibility of experimental disproof. Theories should not stagnate in complacency: the better a theory, the more forecasts it will make by which it can be disproved. And, as a corollary, a theory should not allow indefinite *ad hoc* readjustments to take care of each new counter-result that may be found.

2.3   An Archetypal Relativistic Argument

With the following brief example we shall try to illustrate the flavor and the power of many of the relativistic arguments to follow. To make it simple and transparent, we consider a Newtonian example; an Einsteinian version of the same problem will be given later. (See Section 5.8.) We wish to prove the result, familiar to billiard players, that if a stationary and perfectly elastic sphere is struck by another similar one, then the diverging paths of the two spheres after collision will subtend a right angle. If the incident sphere travels at velocity $2\mathbf{v}$, say, relative to the table, let us transfer ourselves to an inertial frame traveling in the same direction with velocity $\mathbf{v}$. In this frame the two spheres approach each other symmetrically with velocities $\pm\mathbf{v}$ (see Figure 2.1a), and the result of the collision is clear: by momentum conservation, or simply by symmetry, the rebound velocities must be equal and opposite ($\pm\mathbf{u}$, say), and by energy conservation they must be numerically equal to $\mathbf{v}$ (i.e., $u = v$). With this information, we can revert to the frame of the table, by adding $\mathbf{v}$ to all the velocities in Figure 2.1a, thus arriving at Figure 2.1b.

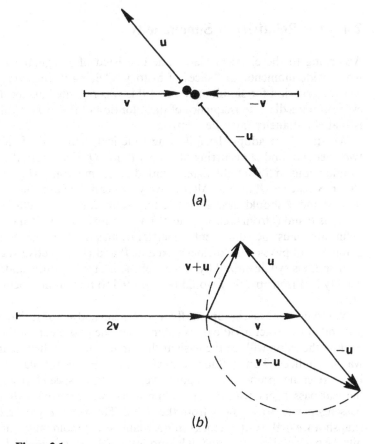

(a)

(b)

Figure 2.1

Here the rebound velocities are evidently $\mathbf{v} \pm \mathbf{u}$, and the simple expedient of drawing a semicircle centered at the tip of the arrow representing $\mathbf{v}$ makes the desired result self-evident, by elementary geometry. Alternatively, we have $(\mathbf{v} + \mathbf{u}) \cdot (\mathbf{v} - \mathbf{u}) \equiv v^2 - u^2 = 0$, which also shows that the vectors $\mathbf{v} \pm \mathbf{u}$ are orthogonal. (For further examples of Newtonian relativity, see Exercises 2.1–2.3.)

It was Einstein's recognition of the fact that arguments of a similar nature were apparently possible also in electromagnetism that significantly influenced his progress towards SR. We have already quoted (in Section 1.12) the beginning of his 1905 paper, where he discusses the apparent relativity of electromagnetic induction. As late as 1952 (in a letter to a scientific congress), we find him writing: "What led me more or less directly to the special theory of relativity was the conviction that the electromagnetic force acting on a [charged] body in motion in a magnetic field was nothing else but an electric field [in the body's rest frame]."[5]

[5] See p. 35 of R. S. Shankland, *Am. J. Phys. 32*, 16 (1964).

2.4   The Relativity of Simultaneity

According to the classical view, time consisted of a regular succession of world-wide moments, or "slices of history." "Being in the same moment" was the criterion for simultaneity, and this was regarded as absolute. However, an immediate consequence of the adoption of the *two* postulates of SR is that simultaneity must be relative.

We shall here adopt the following practical definition of simultaneity: two events $\mathcal{P}$ and $\mathcal{Q}$ occurring at points P and Q of an inertial frame S are simultaneous in S if a light signal emitted at the midpoint M of the segment PQ in S reaches $\mathcal{P}$ and $\mathcal{Q}$. Alternatively, we can demand that light signals *from* $\mathcal{P}$ and $\mathcal{Q}$ should meet at M; or that cannon balls simultaneously fired towards P and Q from identical guns at M should arrive at $\mathcal{P}$ and $\mathcal{Q}$. All these definitions must be equivalent if inertial frames are homogeneous and isotropic for all physics. Two clocks fixed at P and Q, respectively, could have their settings synchronized by any experiment of this nature, and, by homogeneity and isotropy, they would be expected to remain in synchrony thereafter.

Now consider a fast aeroplane flying overhead and suppose that a flashbulb goes off at the exact middle of its cabin. Then the passengers at the front and back of the cabin will see the flash at the same time, say when their clocks or watches indicate 3 units. Suppose next that we observe this flash from earth. In our reference frame too the light travels with equal speed fore and aft. Thus the rear passengers, who travel *into* the signal, will receive it *before* the front passengers who travel away from the signal. The top of Figure 2.2a shows a snapshot *we* make of the clocks in the plane at the moment when the signal hits the back of the cabin. Since it has not yet reached the front, the front clocks will read less than 3, say 1 (in units very much smaller than seconds!). These two *different* clock readings are simultaneous events in *our* frame. Thus simultaneity is relative!

Now add a second plane to the argument, traveling at the same speed, but in the opposite direction. If, when we take the snapshot, it is just level with the first plane—as in Figure 2.2a—then by symmetry *its* clocks will also read two units apart, say again 3 and 1 if the zero settings are suitably adjusted. Now observe: the front end of the second plane passes beyond the first plane at time 3 (by the first plane's reckoning) but its rear end reached the first plane earlier, at time 1. Thus the first plane must consider the second shorter than itself. Here we have the phenomenon of relativistic *length contraction*. Note its perfect symmetry: the second plane too considers the other to be shorter than itself.

Lastly, consider another snapshot, taken at the moment when the rear ends of the cabins meet. Clearly the clocks in each plane will still read exactly two units apart, but the actual readings will depend on the speed and the length of the planes. Suppose they are now 4 and 2, as in Figure 2.2b. Again observe: in snapshot (a) the rear clock of the first plane is two units ahead of the clock in the other plane which it just passes, while in snapshot

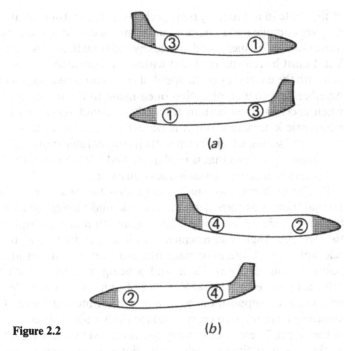

Figure 2.2   (b)

(b) that same clock is not ahead at all. So, compared with the synchronized clocks of the second plane, the rear clock of the first plane has lost time. This is the phenomenon of relativistic *time dilation*. Note again its perfect symmetry relative to the two planes.

We may note that while cannon balls, or sound signals in still air, etc., can all be used instead of light to define simultaneity in any inertial frame, the above arguments would be inconclusive had we used such methods. Light is special: it has the same velocity in all frames, and so the *same* flash can be used to synchronize clocks in *different* frames.

We should also note that the synchronization of clocks (or, equivalently, the definition of simultaneity) is, in a sense, a matter of convention. We could, for example, adopt in *all* inertial frames the time of a specific *one* (cf. Exercise 1.4). But that has its practical disadvantages: the *coordinate* description of physics in the general inertial frame would then not be iso-tropic. For example, the coordinate speed of light, or of cannon balls, would vary with direction. We therefore prefer the unique synchronization adapted to each frame which makes the physics isotropic not only in fact but also in its coordinate description.

2.5   The Coordinate Lattice

Before we actually consider *transformations* of coordinates, it will be well to clarify how coordinates are defined in a single inertial frame. The most obvious way to measure distances is to use rigid scales. But rigid scales are of ill repute in relativity: justly, perhaps, because from the atomic viewpoint they are complicated structures and it may seem more fundamental and even more practical to measure distances by radar methods, using clocks and light. Yet it must be remembered that without the existence of *some* rigid standard of length the constancy of the speed of light would become a mere convention. Another, less cogent, objection often made to the use of rigid scales is that, when accelerated in certain ways, they cannot *remain* rigid, according to relativistic kinematics. But it is perfectly logical to assume the existence of "resilient" scales, which return to their original shape soon after the acceleration ceases, provided that is moderate. And the use of such scales as distance indicators in inertial frames is above suspicion.

Be this as it may, we shall assume that the observer at the origin of an inertial frame possesses a standard clock, and that he measures the distance to any particle by bouncing a light signal off it and multiplying the elapsed time by $\frac{1}{2}c$. Angle measurements with a theodolite will then furnish the relevant (x, y, z). We may place free test particles at rest at all the "lattice points" $(\pm m, \pm n, \pm p)$—m, n, and p being integers—which may be made arbitrarily close by a suitable choice of units. These particles will *remain* at rest, since, by supposition, we are in a *strict* inertial frame. They may conveniently be thought to carry standard clocks, all replicas of the master clock at the origin. These clocks can by synchronized by a single light signal emitted at the origin at time t_0: when this signal is received at any lattice clock, that clock is set to read $t_0 + r/c$, r being its predetermined distance from the origin. On the *classical* theory it is clear that this process would satisfactorily synchronize all the clocks in the "ether frame" so that any two clocks indicate the same time at simultaneous events, according to the definition of simultaneity of the last section. But none of the relevant classical laws is affected by relativity, except that *each* inertial frame is now as good as the ether frame. Hence the process is a valid one for clock synchronization in any inertial frame. Once this is achieved, i.e., once the "lattice" is set up and is spatially and temporally calibrated, we can discard theodolites and light signals: the coordinates of all events can now be read off locally—i.e., directly where the events occur—by suitable auxiliary observers.[6] Of course, any two lattice clocks could also be permanently connected with a rigid scale, if such scales are contemplated.

It is well to re-emphasize that other "signals," e.g., cannon balls, can be used *just* as well as light signals to calibrate and synchronize the coordinate lattice. One need merely require that the observer at the origin shoot standard projectiles from standard guns in all directions at time t_0, say. The speed u of the projectiles may be taken to be an arbitrary number, whereby one merely fixes the unit of distance. (Alternatively, the muzzle velocity can be measured at the origin, or even calculated theoretically.) Angle measurements are to be made of the direction of the gun barrels. When a projectile from the origin passes an auxiliary observer, he is to shoot back a similar one from a

---

[6] Cf. RSR (1960).

similar gun. If that gets to the origin at time t, the auxiliary observer was at distance $\frac{1}{2}u(t - t_0)$ and his clock should have read $\frac{1}{2}(t + t_0)$ when he received the "control signal." All this, of course, presupposes Newton's first law.

2.6 The Lorentz Transformation

Now consider two *arbitrary* inertial frames S and S' in which Cartesian coordinate lattices have been set up. (We shall usually identify an inertial frame with the collection of free particles at rest in it and regard the coordinates as an additional structure.) We wish to find the relation between the S-coordinates (x, y, z, t) and the S'-coordinates (x', y', z', t') corresponding to an arbitrary event. First, this relation must be linear—as is, for example, the Galilean transformation (**1.1**). This follows directly (though not trivially) from the definition of inertial frames: only under a linear transformation can the linear equations of motion of free particles in S go over into linear equations of motion also in S'. Actually, this requirement by itself only implies that the transformations are necessarily projective, i.e., that the S'-coordinates are ratios of linear functions of the S-coordinates, all with the same denominator.[7] However, if we reject, for physical reasons, the existence of finite events in S which have infinite coordinates in S', then the denominator must be constant, and thus the transformation linear.[8]

Because of this linearity, in particular, any fixed values of x', y', z' imply unique constant values of dx/dt, dy/dt, dz/dt. Thus each inertial frame is in uniform translatory motion relative to every other. Linearity also implies that the axes of S' remain parallel to themselves relative to S, and that, as in the classical theory, one can choose the coordinates in S and S' in the "standard configuration" described in Section 1.3. (This is fairly obvious; see, for example, RSR, page 17.)

Another basic property is that S and S' assign equal and opposite velocities to each other (say $\pm v$). For, by continuity, it is clear that there exists a frame S'' "between" S and S' such that S and S' have equal and opposite velocities *relative to* S''. Now the manipulation performed in S to determine the velocity of S' can be regarded as an experiment in S''. The corresponding manipulation in S' will be the mirror image of that experiment in S''. And thus, by the assumed isotropy of S'', the two experiments must yield the same numerical result, which establishes our assertion.

Now choose coordinates in S and S' in the standard configuration of Figure 1.1. Then the coordinate planes $y = 0$ and $y' = 0$ coincide permanently.

---

[7] See, for example, V. Fock, *The Theory of Space Time and Gravitation*, Pergamon Press, 1959, Appendix A.

[8] The same limitation results from the demand that uniformly moving *bodies* transform into uniformly moving bodies (conservation of parallelism). Another argument for linearity can be made from the constancy of the speed of light—see, for example, RSR, page 17.

But y' is of the form $Ax + By + Cz + Dt + E$, by linearity, whence $A = C = D = E = 0$, and

$$y' = By, \tag{2.1}$$

where B is a constant possibly depending on v. Let us reverse the directions of the x and z axes in S and S'. This cannot affect (2.1), but it interchanges the roles of S and S': Figure 1.1 now holds with the unprimed and primed symbols interchanged. Thus we must have

$$y = By',$$

and so $B = \pm 1$. The negative value can at once be dismissed, since $v \to 0$ must lead to $y \sim y'$ continuously. The argument for z is similar, and so we arrive at

$$y' = y, \quad z' = z, \tag{2.2}$$

the two "trivial" members of the transformation.

Next, since x' is linear in the unprimed coordinates, and since $x = vt$ must imply $x' = 0$, we find that x' must be of the form

$$x' = \gamma(x - vt), \tag{2.3}$$

where γ is a constant, again possibly depending on v. Similarly, since $x' = -vt'$ must imply $x = 0$, we have

$$x = \gamma'(x' + vt'), \tag{2.4}$$

γ' being another such constant. Let us again reverse the directions of the x and z axes in S and S'. Then, replacing x and x' by their negatives in (2.3), we have

$$x' = \gamma(x + vt).$$

But also, by the reversal of roles, we find from (2.4) that

$$x' = \gamma'(x + vt),$$

whence

$$\gamma = \gamma'. \tag{2.5}$$

Evidently γ must be positive, since x and x' increase together at $t = 0$.

Now, by the "second postulate" (constancy of the speed of light), we know that $x = ct$ must imply $x' = ct'$, and vice versa. Substituting these expressions in (2.3) and (2.4) and using (2.5), we get

$$ct' = \gamma t(c - v), \quad ct = \gamma t'(c + v).$$

Multiplying these equations together, and canceling tt', yields

$$\gamma = \gamma(v) = \frac{1}{(1 - v^2/c^2)^{1/2}}. \tag{2.6}$$

(As we have seen above, we need the *positive* root here.) This particular function of v is the famous "Lorentz factor" which plays an important role in the theory. Eliminating x' between (2.3) and (2.4), we now finally obtain

$$t' = \gamma(t - vx/c^2).$$

Thus, collecting our results, we have found the *standard* Lorentz transformation equations

$$x' = \gamma(x - vt), \quad y' = y, \quad z' = z, \quad t' = \gamma(t - vx/c^2). \tag{2.7}$$

If a law of physics is invariant under these transformations, *and* under spatial rotations, spatial translations, and time translations, then it is invariant between *any* two inertial coordinate systems. For it is easily seen that the general transformation between two inertial frames, whose coordinates are standard but not mutually in standard configuration, consists of the following: (1) a space rotation and translation (to make the x axis of S coincide with the line of motion of the S' origin); (2) a time translation (to make the origins coincide at $t = 0$); (3) a standard LT; and, finally, another rotation and time translation to arrive at the coordinates of S'. The resultant transformation is called a *general* LT, or a *Poincaré transformation*. Since each link in this chain of transformations is linear, so is the resultant.

2.7   Properties of the Lorentz Transformation

What we have proved in the preceding section is in fact only this: *if* there is a transformation satisfying the requirements of SR, then it must be (2.7). It remains to be verified that the LT (2.7) really *does* satisfy these requirements. First, in our derivation we have made use only of light signals in the x direction; thus we must still verify that *arbitrary* light signals in S correspond to light signals in S'. Secondly, the assumption of Euclidean space geometry in S and S' must be respected by the transformation. And thirdly, in order that no one frame be preferred, we require that if any two frames are related to a third by LT's, they must also be so related to each other. These properties of the LT's will be examined in (i)–(iii) below. Other important properties are discussed in (iv)–(vii).

(i) If Δx, Δy, etc., denote the finite coordinate differences $x_2 - x_1$, $y_2 - y_1$, etc., corresponding to two events $\mathcal{P}_1$ and $\mathcal{P}_2$, then by substituting the coordinates of $\mathcal{P}_1$ and $\mathcal{P}_2$ successively into (2.7) and subtracting, we get the following transformation:

$$\Delta x' = \gamma(\Delta x - v\Delta t), \quad \Delta y' = \Delta y, \quad \Delta z' = \Delta z, \quad \Delta t' = \gamma(\Delta t - v\Delta x/c^2). \tag{2.8}$$

If, instead of forming differences, we take differentials in (2.7), we obtain equations identical with the above but in the differentials:

$$dx' = \gamma(dx - vdt), \quad dy' = dy, \quad dz' = dz, \quad dt' = \gamma(dt - vdx/c^2). \tag{2.9}$$

Thus the finite coordinate differences, as well as the differentials, satisfy the same transformation equations as the coordinates themselves. This, of course, is always the case with linear homogeneous transformations.

(ii) A purely algebraic consequence of the transformation (2.8) is the identity

$$c^2\Delta t'^2 - \Delta x'^2 - \Delta y'^2 - \Delta z'^2 = c^2\Delta t^2 - \Delta x^2 - \Delta y^2 - \Delta z^2. \quad (2.10)$$

[The same identity in the differentials can be deduced from (2.9).] The common value of these quadratic forms is defined as the *squared interval* Δs^2 between the events $\mathscr{P}_1$ and $\mathscr{P}_2$:

$$\Delta s^2 = c^2\Delta t^2 - \Delta x^2 - \Delta y^2 - \Delta z^2. \quad (2.11)$$

It may be positive, zero, or negative. For two events on a light signal it is zero, since the distance traveled by the signal, $(\Delta x^2 + \Delta y^2 + \Delta z^2)^{1/2}$, equals $c\Delta t$. The identity (2.10) now shows that arbitrary light signals in S indeed correspond to light signals in S', and vice versa. We note that the Euclidean "metric" $\Delta x^2 + \Delta y^2 + \Delta z^2$ is *consistent* with the property that light travels at speed c in all directions.

The squared interval Δs^2 is invariant even under the *general* LT's defined at the end of Section 2.6, which relate arbitrary inertial frames with standard coordinates, and which are compounded of standard LT's, spatial rotations, and space and time translations. For it is invariant under standard LT's, also under rotations (which preserve $\Delta x^2 + \Delta y^2 + \Delta z^2$ and Δt^2 separately), and finally under translations (which leave *each* Δ-term unchanged). Conversely, it can be shown that the general LT's are the most general transformations—excluding space and time reflections ($t \to -t$, etc.)—which preserve Δs^2.

(iii) Direct algebraic solution of (2.7) for x, y, z, t gives

$$x = \gamma(x' + vt'), \quad y = y', \quad z = z', \quad t = \gamma(t' + vx'/c^2), \quad (2.12)$$

and thus the inverse of (2.7) is a LT with parameter $-v$ instead of v, as indeed one would expect from symmetry considerations. [Of course, (2.12) holds also in differential and Δ-form.] The resultant of two LT's with parameters v_1 and v_2, respectively, is also found to be of type (2.7), with parameter $v = (v_1 + v_2)/(1 + v_1 v_2/c^2)$. (The direct verification of this is a little tedious, but an easy way of seeing it will be given in Section 2.8.) These two properties—symmetry and transitivity—imply that the LT's constitute, technically speaking, a "group." As a consequence of them, if S, S', and S'' are three coordinate systems such that S is related to both S' and S'' by LT's, then—by symmetry—S' is so related to S and therefore—by transitivity—also to S''. Not surprisingly, it can be shown that the general LT's also form a group, and thus put *all* inertial standard coordinate systems on the same footing.

In the Lorentz ether theory (cf. Section 1.10) it follows from the hypotheses of length contraction and time dilation *relative to the ether* that any inertial frame is related to the ether frame by a LT. The group property then shows that in spite of the conceptual pre-eminence of the ether frame it is kinematically equivalent to all other inertial frames.

(iv) For $v \neq 0$ the Lorentz factor γ is always greater than unity, though not much so when v is small. For example, as long as $v/c < 1/7$ (at which speed the earth is circled in one second), γ is less than 1.01; when $v/c = \sqrt{3/2} = 0.866$, $\gamma = 2$; and when $v/c = 0.99 \ldots 995$ ($2n$ nines), γ is approximately 10^n. Thus, as comparison of (**2.7**) with (**1.1**) shows, the GT approximates well to the LT when v is small. Alternatively, if we formally let $c \to \infty$, the LT goes over into the GT.

(v) The appearance of the space coordinate x in the transformation (**2.7**)(iv) of the time is the mathematical expression of the relativity of simultaneity. It implies that two events with the same t do not necessarily have the same t'.

It has been pointed out by M. v. Laue that a cylinder rotating uniformly about the x' axis in the frame S' will seem *twisted* when observed instantaneously in the usual second frame S in which it not only rotates but also travels forward. This can be seen in many ways, but in particular as a simple illustration of the relativity of simultaneity. Think of the cylinder as made up of a lot of circular slices, each slice by its rotation serving as a clock. All these clocks are synchronized in S' by having arbitrary but parallel radii designated as clock "hands." In S the clocks are *not* synchronous, hence the hands are not parallel, hence the cylinder is twisted! [If the angular speed of the cylinder in S' is ω, we find from Equation (**2.7**)(iv), on putting $t = 0$ and $x = 1$, that in S the twist per unit length is $\gamma\omega v/c^2$.]

(vi) *Any* effect whose speed in vacuum is always the same could have been used to derive the LT's, as light was used above. Since only *one* transformation can be valid, it follows that *all* such effects (gravitational waves, ESP?) must propagate at the speed of light. In particular, in contrast to their various speeds in translucent media, electromagnetic waves of all frequencies must travel at exactly speed c in vacuum.

(vii) When $v = c$, γ becomes infinite, and $v > c$ leads to imaginary values of γ. This shows that the relative velocity between two inertial frames must be less than the speed of light, since finite real coordinates in one frame must correspond to finite real coordinates in any other frame.

Indeed, we can show that the speed of particles, and more generally, of all physical "signals," is limited by c, *if* we insist on the invariance of causality. For consider any process whatever, whereby an event $\mathscr{P}$ causes an event $\mathscr{Q}$ (or, whereby information is sent from $\mathscr{P}$ to $\mathscr{Q}$) at a "superlight" speed $U > c$ relative to some frame S. Choose coordinates in S so that these events both occur on the x axis, and let their time and distance separations be $\Delta t > 0$ and $\Delta x > 0$. Then in the usual second frame S' we have, from (2.8),

$$\Delta t' = \gamma\left(\Delta t - \frac{v\Delta x}{c^2}\right) = \gamma\Delta t\left(1 - \frac{vU}{c^2}\right). \tag{2.13}$$

For $c^2/U < v < c$, we would then have $\Delta t' < 0$. Hence there would exist inertial frames in which $\mathscr{Q}$ precedes $\mathscr{P}$, i.e., in which cause and effect are reversed, or in which information flows from receiver to transmitter.

Among other embarrassments, this would allow us to know our future. For if a signal can be sent from a point P to a point Q (in a frame S') so as to reach Q *before* it left P, then, by symmetry, the signal can be returned at once from Q to P, reaching P before it left Q, and consequently before the original signal left P. A man at P can therefore have foreknowledge of events at himself, and could deliberately foil them, which leads to grave contradictions.

On the other hand, the speed limit c *does* guarantee invariance of causality. For if two events happen on a line making an angle θ with the x axis in S (thus not restricting their generality relative to S and S'), and are connectible in S by signal with speed $u \leq c$, we see on replacing U by $u \cos \theta$ in (2.13), that for all v between $\pm c$, Δt and $\Delta t'$ have indeed the same sign.

We may note that, relative to any frame S, two particles or photons may have a *mutual* velocity up to $2c$. This velocity is the time rate of change of the connecting vector $\mathbf{r}_2 - \mathbf{r}_1$ between the particles, which we assume to have position vectors and velocities $\mathbf{r}_1, \mathbf{u}_1$ and $\mathbf{r}_2, \mathbf{u}_2$, respectively: $(d/dt)(\mathbf{r}_2 - \mathbf{r}_1)$ $= \mathbf{u}_2 - \mathbf{u}_1$, as in classical kinematics. For example, the mutual velocity of two photons traveling in opposite directions along a common line is precisely $2c$. *Arbitrarily* large velocities are possible for signals that carry no information—e.g., the sweep of a searchlight spot on high clouds. (See also Section 2.9.)

One consequence of the relativistic speed limit is that "rigid bodies" and "incompressible fluids" have become impossible objects, even as idealizations or limits. For, by definition, they would transmit signals instantaneously. In relativity, a body being acted on by various forces at various points simultaneously will yield to *each* force initially as though all the others were absent; for at each point it takes a finite time for the effects of the other forces to arrive. Hence, in relativity, a body has infinitely many degrees of freedom. Again, a body which retains its shape in one frame may appear *deformed* in another, if it accelerates. We have already seen an example of this in Laue's cylinder. As an even simpler example consider a rod, in a frame S', which remains parallel to the x' axis while moving with constant acceleration a in the y' direction. Its equation of motion, $y' = \frac{1}{2}at'^2$, translates by the LT into $y = \frac{1}{2}a\gamma^2(t - vx/c^2)^2$, and so the rod has the shape of part of a parabola at each instant $t = $ constant in the usual second frame S. However, even accelerating bodies may undergo "rigid motion"—a concept which will be defined in Section 2.16.

2.8   Hyperbolic Forms of the Lorentz Transformation

For certain purposes, it is convenient to replace the "velocity parameter" v by an alternative "hyperbolic parameter" ϕ in the LT, where ϕ is defined by any one of the following four equations:

$$\cosh \phi = \gamma, \quad \sinh \phi = \frac{v}{c}\gamma, \quad \tanh \phi = \frac{v}{c}, \tag{2.14}$$

$$e^{\phi} = \gamma\left(1 + \frac{v}{c}\right) = \left[\frac{1 + (v/c)}{1 - (v/c)}\right]^{1/2} \tag{2.15}$$

(for the uniqueness of ϕ, the first of these must be augmented by the requirement that ϕ and v have the same sign). The reader is reminded of the relations $\cosh \phi = \cos i\phi$, $i \sinh \phi = \sin i\phi$, whereby any trigonometric identity can be conveniently converted into an identity in the hyperbolic functions. In particular, $\cosh^2 \phi - \sinh^2 \phi = 1$, whence the equivalence of (2.14)(i) and (ii). Recall also that $\cosh \phi \pm \sinh \phi = \exp(\pm \phi)$, whence (2.15)(i). Substituting (2.14)(i) and (ii) into the standard LT (2.7), we obtain

$$x' = x \cosh \phi - ct \sinh \phi, \quad y' = y, \quad z' = z, \tag{2.16}$$
$$ct' = -x \sinh \phi + ct \cosh \phi,$$

which is reminiscent of a "rotation" in x and ct. [In fact, it is formally a rotation in x and ict, by an angle $i\phi$; as such, it preserves $x^2 + (ict)^2$.]

By adding and subtracting the two nontrivial equations in (2.16) we obtain an even more useful form of the LT's:

$$ct' + x' = e^{-\phi}(ct + x), \quad y' = y, \quad z' = z, \tag{2.17}$$
$$ct' - x' = e^{\phi}(ct - x).$$

Now multiplying together the two nontrivial members of (2.17) and squaring the other two, we find

$$c^2 t'^2 - x'^2 - y'^2 - z'^2 = c^2 t^2 - x^2 - y^2 - z^2. \tag{2.18}$$

Since (2.17) holds equally in differential or Δ-form, we can similarly deduce from it Equation (2.10) or its differential version.

The form of Equations (2.17) shows—without labor—that the inverse of a LT with hyperbolic parameter ϕ is a LT with parameter $-\phi$ [and thus with velocity parameter $-v$ instead of v, as we already found in (2.12)]; also that the resultant of two LT's with hyperbolic parameters ϕ_1 and ϕ_2 is a LT with parameter $\phi = \phi_1 + \phi_2$. The corresponding relation between the velocity parameters of these transformations then follows at once from (2.14)(iii) by substitution into the analog of the well-known identity for $\tan(\phi_1 + \phi_2)$:

$$\tanh(\phi_1 + \phi_2) = (\tanh \phi_1 + \tanh \phi_2)/(1 + \tanh \phi_1 \tanh \phi_2). \tag{2.19}$$

It is

$$v = (v_1 + v_2)/(1 + v_1 v_2/c^2). \tag{2.20}$$

This, of course, must be the velocity at which the third frame moves relative to the first, and thus it represents the "relativistic sum" of the (collinear) velocities v_1 and v_2.

This may be as good a moment as any to make a remark of very general usefulness. In working a SR problem or calculation, especially for the first time, one can often omit the c's, i.e., one can work in units in which $c = 1$. The c's can easily be inserted later, either throughout the work, or directly in the answer by simple dimensional arguments. For example, had we worked toward (2.20) without the c's, it would be quite obvious where to put a c^2, since by use of c alone the dimensions must be made to balance.

2.9   Graphical Representation of the Lorentz Transformation

In this section we concern ourselves solely with the transformation behavior of x and t under standard LT's, ignoring y and z which in any case are unchanged. What chiefly distinguishes the LT from the classical GT is the fact that space and time coordinates *both* transform, and, moreover, transform partly into each other: they get "mixed," rather as do x and y under a rotation of axes in the Cartesian x, y plane. We have already remarked on the formal similarity of (2.16) to a rotation. Physically, however, the character of a LT differs significantly from that of a rotation. This is brought out well by the graphical representation.

Recall first that there are two ways of regarding any transformation of coordinates (x, t) into (x', t'). *Either* we think of the point (x, t) as moving to a new position (x', t') on the same set of axes, i.e., we regard the transformation as a motion in x, t space; this is the "active" view. *Or* we regard (x', t') as merely a new label of the old point (x, t); this is the "passive" view. In SR the passive view is the more relevant, and the standard graphical representation is chosen accordingly. (See Figure 2.3.) The events, once marked relative to a set of x, t axes, remain fixed; only the coordinate axes change. (For an active representation see Exercise 2.13.) *For convenience we choose units in which $c = 1$.* We usually draw the x and t axes (corresponding to the frame S) orthogonal, but this is a convention without physical significance. "Moments" in S have equation $t = $ constant and correspond to horizontal lines, while the history (or "worldline") of each fixed point on the spatial x axis of S corresponds to a vertical line, $x = $ constant. (We must distinguish between the x axis in Figure 2.3 and the "spatial" x axis of x, y, z space; the latter can be thought of as one of three mutually orthogonal straight wires.) Moments in S' have equation $t' = $ constant and thus, by (2.7), $t - vx = $ constant; so in our diagram they correspond to lines with slope v. In particular, the x' axis ($t' = 0$) corresponds to $t = vx$. Again, worldlines of fixed points on the spatial x' axis have equation $x' = $ constant, and thus, by (2.7), $x - vt = $ constant. In our diagram they are lines with slope v relative to the t axis. In particular, the t' axis ($x' = 0$) corresponds to $x = vt$. Thus the axes of

S' subtend equal angles with their counterparts in S; but whereas in rotations these angles have the same sense, in LT's they have opposite sense.

For calibrating the primed axes, we draw the hyperbolae $x^2 - t^2 = \pm 1$. Since they coincide with $x'^2 - t'^2 = \pm 1$, they cut *all* the axes at the relevant unit time or unit distance from the origin. The units can then be repeated along the axes, by linearity. The diagram shows how to read off the coordinates (a', b') of a given event relative to S': we must go along lines of constant x' or t' from the event to the axes.

As an alternative view of the LT, we note from (2.17) that it simply effects a stretching and shrinking, respectively, of the $\pm 45°$ axes of $\xi = t + x$ and $\eta = t - x$, by a factor e^{ϕ}. The axes of ξ, η (ξ', η') can be calibrated by dropping onto them perpendiculars from the calibration marks of the x, t (x', t') axes, since $t - x$ or $t + x$ ($t' - x'$ or $t' + x'$) is constant along these perpendiculars.

Diagrams like Figure 2.3 are called *Minkowski diagrams*. They can be extremely helpful and illuminating in certain types of relativistic problems. For example, one can use them to get a rough preliminary idea of the answer. But one should beware of trying to use them for *everything*, for their utility is limited. Analytic or algebraic arguments are generally much more powerful.

As a simple example on the use of the diagram, we shall look at a non-informative superlight signal such as may occur in reality. For instance, if a slanting guillotine blade falls past a level block, the intersection point of blade and block travels at an arbitrarily large speed if the slant is small enough; or, a searchlight spot can be made to travel over high clouds at arbitrarily large speeds by simply rotating the searchlight fast enough. In Figure 2.3 such a signal along the spatial x axis of S is shown as a dotted line, to suggest that it be considered as a sequence of events rather than as a moving object. Since the x' and t' axes may subtend any angle between $0°$

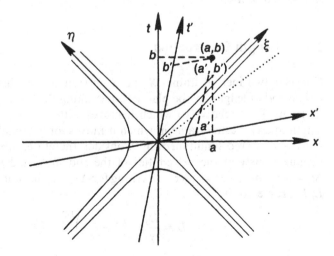

Figure 2.3

and 180°, there is a frame S' in which the signal has infinite velocity (i.e., coincides with the x' axis in the diagram) and others in which it moves in the opposite spatial direction. If the signal were informative, it would have to be regarded in this latter case as traveling with unchanged spatial sense into the past.

2.10 World-picture and World-map

In relativity, it is important to distinguish between what an observer *sees* and what he knows *ex post facto*. What he actually or potentially sees or photographs at any one instant is called his *world-picture* at that instant. This is a practically not very important and theoretically quite complicated concept, since what is seen at any one instant is a composite of events that occurred progressively earlier as they occurred further and further away. (Only in cosmology does the world-picture assume importance; there it is our prime datum.) A much more useful concept is the *world-map*. This, as the name implies, may be thought of as a mapping of events into an observer's instantaneous space $t = t_0$: a kind of three-dimensional life-sized snapshot exposed everywhere simultaneously, or a frozen instant in the observer's spatial reference frame. It could be produced as a joint effort of the auxiliary observers at the "lattice points" of a given inertial frame—each mapping *his* neighborhood at a predetermined instant $t = t_0$.

When we loosely say "the length of an object in S," or "a snapshot taken in S," or "a moving cylinder seems twisted," etc., we invariably think of the world-map, unless the contrary is stated explicitly. The world-map is generally what matters. These remarks are relevant in the next section, where we show that moving bodies shrink. The shrinkage refers to the world-map. How the eye actually *sees* a moving body is rather different, and not very significant, except that in relativity some of the facts of vision are rather amusing, as we shall see in Section 3.3.

2.11 Length Contraction

Consider two inertial frames S and S' in standard configuration. In S' let a rigid rod of length $\Delta x'$ be placed at rest along the x' axis. We wish to find its length in S, relative to which it moves with velocity v. To measure its length in any inertial frame in which it moves longitudinally, its end-points must be observed simultaneously. Consider, therefore, two events occurring simultaneously at the extremities of the rod in S, and use (2.8)(i). Since $\Delta t = 0$, we have $\Delta x' = \gamma \Delta x$, or, writing for Δx, $\Delta x'$ the more specific symbols L, L_0, respectively,

$$L = \frac{L_0}{\gamma} = \left(1 - \frac{v^2}{c^2}\right)^{1/2} L_0. \tag{2.21}$$

This shows, quite generally, that *the length of a body in the direction of its motion with uniform velocity v is reduced by a factor* $(1 - v^2/c^2)^{1/2}$.

Evidently the greatest length is ascribed to a uniformly moving body in its *rest frame*, i.e., the frame in which its velocity is *zero*. This length, L_0, is called the *rest length* or *proper length* of the body. On the other hand, measured in a frame in which the body moves with a velocity approaching that of light, its length approaches zero.

The statement following (**2.21**) happens to be identical with that proposed by Fitzgerald and Lorentz to explain the null-results of the ether drift experiments—except that they qualified "uniform velocity v" with the phrase "*relative to the ether*." According to Lorentz, the mechanism responsible for the contraction was a certain increment in electrical cohesive forces which tightened the atomic structure. Relativity, on the other hand, bypasses all explanations in terms of forces or the like, yet it predicts the phenomenon as inevitable. (This is comparable to some of the predictions made on the basis of energy conservation.) In relativity the effect is essentially a geometric "projection" effect,[9] quite analogous to looking at a *stationary* rod which is not parallel to the plane of the retina. Imparting a uniform velocity in relativity corresponds to making a pseudorotation in "spacetime" (see Section 2.9). Our eye, though essentially designed to receive two-dimensional impressions, has long been trained to sense, and to make allowance for, a third spatial dimension. We are thus not surprised at the foreshortening of a spatially rotated rod. If the speed of light were small, and significant length contractions could thus arise at "ordinary" speeds, our eyes would also have learned to sense the fourth dimension, namely time. Just as by bringing the eye into a suitable spatial position we can restore the full length of a rotated stationary rod, so by moving the eye along with the moving rod we could undo the relativistic length contraction. The eye would soon learn to look at this effect "geometrically."

By the relativity principle, it is *a priori* evident that if two observers A and B compare yardsticks along their common line of motion, and if A considers B's stick to be shorter than his own, then B considers A's stick to be the shorter. The projection analogy makes this unsurprising: a corresponding situation occurs when two relatively stationary observers hold their yardsticks at the same angle relative to their line of sight. The symmetry of length contraction is particularly well illustrated by the example of Section 2.4, Figure 2.2. We have already remarked on how fortuitous this symmetry must seem in the Lorentz theory where it can only be discovered by calculation, since the *cause* of length contraction resides in the preferred ether frame.

2.12   Length Contraction Paradoxes

The relativistic length contraction is no "illusion": it is real in every sense. Though no direct experimental verification has yet been attempted, there is no

---

[9] The following remarks to the end of the present paragraph are intended to be suggestive only, and no very accurate understanding is called for at this stage.

question that in principle it could be done. Consider the admittedly un-realistic situation of a man carrying horizontally a 20-foot pole and wanting to get it into a 10-foot garage. He will run at speed $v = 0.866c$ to make $\gamma = 2$, so that his pole contracts to 10 feet. It will be well to insist on having a sufficiently massive block of concrete at the back of the garage, so that there is no question of whether the pole finally stops in the inertial frame of the garage, or vice versa. Thus the man runs with his (now contracted) pole into the garage and a friend quickly closes the door. In principle we do not doubt the feasibility of this experiment, i.e., the reality of length contraction. When the pole stops in the rest frame of the garage, it is, in fact, being "rotated in spacetime" and will tend to assume, if it can, its original length relative to the garage. Thus, if it survived the impact, it must now either bend, or burst the door.

At this point a "paradox" might occur to the reader: what about the symmetry of the phenomenon? Relative to the runner, won't the garage be only 5 feet long? Yes, indeed. Then how can the 20-foot pole get into the 5-foot garage? Very well, let us consider what happens in the rest frame of the pole. The open garage now comes towards the stationary pole. Because of the concrete wall, it keeps on going even after the impact, taking the front end of the pole with it. But the back end of the pole is still at rest: it cannot yet "know" that the front end has been struck, because of the finite speed of propagation of *all* signals. Even if the "signal" (in this case the elastic shock wave) travels along the pole with the speed of light, that signal has 20 feet to travel against the garage front's 15 feet, before reaching the back end of the pole. This race would be a dead heat if v were $0.75c$. But v is $0.866c$! So the pole *more* than just gets in. (It could even get into a garage whose length was as little as 5.4 feet at rest and thus 2.7 feet in motion: the garage front would then have to travel 17.3 feet against the shock wave's 20 feet, requiring speeds in the ratio 17.3 to 20, i.e., 0.865 to 1 for a dead heat.)

There is one important moral to this story: whatever result we get by correct reasoning in any one frame, must be true; in particular, it must be true when viewed from any other frame. As long as the set of physical laws we are using is Lorentz-invariant, there *must* be an explanation of the result in every other frame, although it may quite a different explanation from that in the first frame. Recall Einstein's "hunch" that the force experienced by an electric charge when moving through a magnetic field is equivalent to a simple electric force in the rest frame of the charge.

Consider, as another example, a "rigid" rod of rest length L sliding over a hole of diameter $\frac{1}{2}L$ on a smooth table. When its Lorentz factor is 10, the length of the rod is $\frac{1}{5}$ of the diameter of the hole, and in passing over the hole, it will fall into it under the action of gravity[10] (at least slightly: enough to be stopped). This *must* be true also in the rest frame of the rod—in which, how-ever, the diameter of the hole is only $\frac{1}{20}L$! The only way in which this can

---

[10] We are here violating our resolve to work in *strict* inertial frames only! The conscientious reader may replace the force of gravity acting down the hole by a sandblast from the top—the result will be the same. For a fuller discussion of this paradox, see W. Rindler, *Am. J. Phys.* **29**, 365 (1961).

happen is that the front of the "rigid" rod *bends* into the hole (cf. the end of Section 2.7). Moreover, even after the front end strikes the far edge of the hole, the back end keeps coming in (not yet "knowing" that the front end has been stopped), as it must, since it does so in the first description.

2.13 Time Dilation

Let us again consider two inertial frames S and S' in standard configuration. Let a standard clock be fixed in S' and consider two events at that clock at which it indicates times t'_1 and t'_2 differing by $\Delta t'$. We inquire what time interval Δt is ascribed to these events in S. From the Δ-form of **(2.12)**(iv) we see at once, since $\Delta x' = 0$, that $\Delta t = \gamma \Delta t'$, or, replacing Δt and $\Delta t'$ by the more specific symbols T and T_0, respectively,

$$T = \gamma T_0 = \frac{T_0}{(1 - v^2/c^2)^{1/2}}. \tag{2.22}$$

We can deduce from this quite generally that *a clock moving uniformly with velocity v through an inertial frame S goes slow by a factor $(1 - v^2/c^2)^{1/2}$ relative to the stationary clocks in S.* Clearly, then, the fastest rate is ascribed to a clock in its rest frame, and this is called its *proper rate*. On the other hand, at speeds close to c, the clock rate would be close to zero.

If an *ideal* clock moves *nonuniformly* through an inertial frame, we shall *assume* that acceleration as such has no effect on the rate of the clock, i.e., that its instantaneous rate depends only on its instantaneous speed v according to the above rule. This we call the *clock hypothesis*. It can also be regarded as the definition of an "ideal" clock. By no means all clocks meet this criterion. For example, a spring-driven pendulum clock whose bob is connected by two coiled springs to the sides of the case (so that it works without gravity) will clearly increase its rate as it is accelerated "upward." On the other hand, as stressed by Sexl, the absoluteness of acceleration ensures that ideal clocks *can* be built, in principle. We need only take an arbitrary clock, observe whatever effect acceleration has on it, then attach to it an accelerometer and a servomechanism that exactly cancels the acceleration effect! By contrast, the velocity effect **(2.22)** can *not* be eliminated. As we shall see, certain natural clocks (vibrating atoms, decaying muons) conform very accurately to the clock hypothesis. Generally this will happen if the clock's internal driving forces greatly exceed the accelerating force.

For the nonuniform motion of an *infinitesimal* body a similar assumption is usually made, namely that the relation between its length in the direction of motion and its proper length depends only on its instantaneous velocity v according to Formula **(2.21)**. This we call the *length hypothesis*.

Time dilation, like length contraction, must *a priori* be symmetric: if one inertial observer considers the clocks of a second inertial observer to run slow, the second must also consider the clocks of the first to run slow. Figure

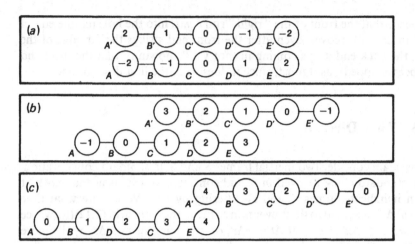

Figure 2.4

2.4—which is really an extension of Figure 2.2—shows in detail how this happens. Synchronized standard clocks $A, B, C, \ldots$ and $A', B', C', \ldots$ are fixed at certain equal intervals along the x axes of two frames S and S' in standard configuration. The figure shows three snapshots taken, at convenient intervals, by an auxiliary observer in a frame S'', relative to which S and S' have equal and opposite velocities. In each snapshot, the clocks of S and S' will, of course, be all seen to indicate different times, since simultaneity is relative. Suppose the clocks in the diagram indicate seconds. As can be seen, A' reads 4 seconds ahead of A in Figure 2.4a, only 2 seconds ahead of C in Figure 2.4b, and equal with E in Figure 2.4c. Thus A' loses steadily relative to the clocks in S. Similarly, E loses relative to the clocks in S', and indeed all clocks in the diagram lose at the same rate relative to the clocks of the other frame. The interested reader can easily work out the relative velocity between S and S' and the spacing of the clocks in the situation illustrated by Figure 2.4. Only one set of values fits. (See Exercise 2.11.)

Time dilation, like length contraction, is *real*. And this *has* been confirmed experimentally. For example, certain mesons (muons) reaching us from the top of the atmosphere in cosmic rays are so short-lived that, even had they traveled with the speed of light, their travel time in the absence of time dilation would exceed their lifetime by factors of the order of 10. Rossi and Hall (1941) timed such muons between the summit and foot of Mt. Washington, and found their lifetimes were indeed dilated in accordance with (**2.22**). Equivalent experiments with muons (at $\gamma \approx 12$) in "storage rings" at the CERN laboratory in 1968[11] have refined these results to an impressive accuracy of 1% and have additionally shown that, to such accuracy, accelerations up to 10^{19} g (!) do not contribute to the muon's

[11] D. H. Perkins, *Introduction to High Energy Physics*, Addison-Wesley Pub. Co. (1972), p. 192.

time dilation. It may be objected that muons are not clocks—but the time dilation argument applies to *any* temporal change or process and therefore also to muon decay and even human life times. (To see this, we need only imagine a standard clock to travel *with* the muon or the space traveler.)

Another striking instance of time dilation is provided by "relativistic focusing" of electrically charged particles, which plays a role in the operation of high-energy particle accelerators. Any stationary cluster of electrons (or protons) tends to expand at a characteristic rate because of mutual electrostatic repulsion. But electrons in a fast-moving beam are observed to spread at a much slower rate. If we regard the stationary cluster as a kind of clock, we have here an almost visible manifestation of the slowing down of a moving clock. Yet another manifestation, the so-called transverse Doppler effect, is discussed in Section 3.2. (It, too, has led to a confirmation of the clock hypothesis for certain natural clocks.) But perhaps the most direct proof of time dilation—though only to an accuracy of about 10%—was given in 1971 by Hafele and Keating, who simply took some very accurate caesium clocks on two trips around the world in commercial airliners![12]

2.14   The Twin Paradox

Like length contraction, so also time dilation can lead to an apparent paradox when viewed by two different observers. In fact this paradox, the so-called twin or clock paradox (or paradox of Langevin), is the oldest of all the relativistic paradoxes. It is quite easily resolved, but its extraordinary emotional appeal keeps debate alive as generation after generation goes through the cycle of first being perplexed, then elated at understanding (sometimes mistakenly), and then immediately rushing into print as though no one had understood before. The articles that have been published on this one topic are practically uncountable, while their useful common denominator would fill a few pages at most. But while no one can get very excited about pushing long poles into short garages and the like, the prospect of going on a fast trip through space and coming back a few years later to find the earth aged by a few thousand years—this modern *elixir vitae*—keeps stirring the imagination.

As we have seen, a standard clock A moving in any way through the synchronized standard lattice clocks of an inertial frame loses steadily relative to those clocks. Consequently, if A is taken on a complete roundtrip through an inertial frame, let us say from the origin O back to O, it will read slow compared with the clock B that has remained at O. If, of two twins, one travels with A while the other stays with B, the B-twin will be older than the A-twin when they meet again, for each ages at the same rate relative to *his* clock.

[12] J. C. Hafele and R. Keating, *Science 177*, 166 (1972). The flights were made in easterly and westerly directions, respectively, which allowed a separation of the velocity effect from the "gravitational time dilation" effect mentioned in Section 1.21.

Now the paradox is this: cannot A (we shall identify clocks with persons) say with equal right that it was *he* who remained where he was, while B went on a round-trip, and that, consequently, B should be the younger when they meet? The answer, of course, is no, and this resolves the paradox: B has remained at rest in a single inertial frame, while A, in the simplest case of a uniform to-and-fro motion—say from earth to a nearby star and back— must at least be accelerated briefly out of B's frame into another, decelerated again briefly to turn around, and finally decelerated to stop at B. These accelerations (positive and negative) are *felt* by A, who can therefore be under no illusion that it was he who remained at rest. Of course, the first and last of these accelerations are nonessential, since age comparisons can be made "in passing," but the middle one is unavoidable.

Still, it may be argued in the above to-and-fro case that there *is* symmetry between A and B for "most of the time," namely during the times of A's free fall. The three asymmetric accelerations can be confined to arbitrarily short periods (as measured by B—they are even shorter as measured by A). How is it, then, that a large asymmetric effect can build up, and, moreover, one that is proportional to the symmetric parts of the motion? But the situation is no more strange than that of two drivers α and β going from O to P to Q (three points in a straight line), β going directly, while α deviates from P to a point R off the line, and thence to Q. They behave quite similarly except that α turns his steering wheel and readjusts his speed briefly at P and again at R. Yet when they meet at Q, their odometers may indicate a large difference!

In one way, the twins' eventual age difference can be seen to arise during A's initial acceleration away from B. During this period, however brief, if his γ factor gets to be 2, say, A finds he has accomplished more than half his outward journey! For he has transferred himself to a frame in which the distance between the earth and his celestial target is halved (length contraction), and this halving is real to him in every possible way. Thus he accomplishes his outward trip in about half the time which B ascribes to it, and the same is true of his return trip.

In another way, the eventual age difference can be seen to arise during a certain period after A's turning-round point, namely between then and the time that B *sees* A turn around. For suppose A makes a flying start and sets his clock to read the same as B when he passes B. Then A and B are totally symmetric until A turns around. At that point, A immediately sees B's clock tick faster, since he is now traveling *into* B's signals; B, on the other hand, cannot see A's clock tick faster until he sees A turn around. From then on the situation is again symmetric to the end. Thus A has seen slow ticks for half his time and fast ticks for the other half, while B has seen slow ticks for more than half his time. If the *same* total time had elapsed at A and B, this would imply that B has seen fewer ticks of A's clock than A has seen of B's—a contradiction. The same contradiction arises *a forteriori* if less time elapses at B than at A. Hence more time must elapse at B than at A. [It is interesting to note why this reasoning fails in classical "ether" optics: there, the long and short ticks are not of equal duration for A and B.]

Arguments of this nature are perhaps instructive or amusing, but not essential. They simply illustrate the self-consistency of the theory and embroider the lack of symmetry between twin A and twin B. But the paradox is disposed of as soon as the asymmetry is pointed out. Sciama has made perhaps the most significant remark about this paradox: it has, he said, the same status as Newton's experiment with the two buckets of water—one, rotating, suspended below the other, at rest. If these were the whole content of the universe, it would indeed be paradoxical that the water surface in the one should be curved and that in the other flat. But inertial frames have a real existence too, and relative to the inertial frames there is no symmetry between the buckets, and no symmetry between the twins, either.

It should be noted carefully that the clock paradox is entirely independent of the clock hypothesis. Whatever the effects of the accelerations as such may be on the moving clock or organism, these effects can be dwarfed by simply extending the periods of free fall.

Finally, a word of caution against a superstition, a semantic confusion, and an oversight. The superstition is that SR does not apply to phenomena involving accelerations: this has caused some authors to assert that GR is needed to resolve the clock paradox. In fact, SR applies to *all* physics in inertial frames, and GR simply reduces to SR in such frames. The semantic confusion is that some authors call the mere application of GR-("covariant") *mathematics*, even in inertial frames, "GR." In that sense, of course, the use of "GR" in the clock paradox is unexceptionable. The exceptionable case occurs when authors introduce GR to avoid the clock hypothesis and overlook the fact that even in GR the clock hypothesis is but a hypothesis.

2.15   Velocity Transformation

Consider, once again, two frames S and S' in standard configuration. Let $\mathbf{u}$ be the vector velocity of a particle in S. We wish to find its velocity $\mathbf{u}'$ in S'. If the particle moves uniformly, we have

$$\mathbf{u} = (u_1, u_2, u_3) = (\Delta x/\Delta t, \Delta y/\Delta t, \Delta z/\Delta t) \qquad (2.23)$$

$$\mathbf{u}' = (u_1', u_2', u_3') = (\Delta x'/\Delta t', \Delta y'/\Delta t', \Delta z'/\Delta t'), \qquad (2.24)$$

where the increments refer to any two events at the particle. Substitution from (2.8) into (2.24), division of each numerator and denominator by Δt, and comparison with (2.23), immediately yields the velocity transformation formulae:

$$u_1' = \frac{u_1 - v}{1 - u_1 v/c^2}, \quad u_2' = \frac{u_2}{\gamma(1 - u_1 v/c^2)}, \quad u_3' = \frac{u_3}{\gamma(1 - u_1 v/c^2)}. \qquad (2.25)$$

These formulae, of course, apply equally well to sub- and superlight velocities, and it would be quite proper to use them to transform, for example, the speed of the searchlight spot of Section 2.9. They also apply to the instantaneous velocity in a nonuniform motion, as can be seen by going to the limit

$\Delta x/\Delta t \to dx/dt$, etc., in the final step of the argument, or by using the differential version (2.9) of (2.8).

Note how (2.25) reduces to the classical formula (1.2) when either $v \ll c$ or $c \to \infty$ formally.

We can pass to the inverse relations without further effort simply by interchanging primed and unprimed symbols and replacing v by $-v$. (For if we replace unprimed by primed, and primed by doubly primed symbols, we evidently get a transformation from S' to a frame S'' which moves with velocity v relative to S'; replacing v by $-v$ then makes S'' into S.) Thus,

$$u_1 = \frac{u_1' + v}{1 + u_1' v/c^2}, \quad u_2 = \frac{u_2'}{\gamma(1 + u_1' v/c^2)}, \quad u_3 = \frac{u_3'}{\gamma(1 + u_1' v/c^2)}. \quad (2.26)$$

Equations (2.26) can be interpreted, alternatively, as giving the "resultant" $\mathbf{u} = (u_1, u_2, u_3)$ of the two velocities $\mathbf{v} = (v, 0, 0)$ and $\mathbf{u}' = (u_1', u_2', u_3')$, and they take the place of the classical formula $\mathbf{u} = \mathbf{v} + \mathbf{u}'$. In this role they are referred to as the *relativistic velocity addition formulae* [cf. (2.20)]. For any such addition to be meaningful, of course, the axes of S and S' must be equally oriented, though not necessarily in standard configuration. It is perhaps worth remarking that the relativistic additions of $\mathbf{u}'$ to $\mathbf{v}$ and of $\mathbf{v}$ to $\mathbf{u}'$ are, in general, not equivalent—though, in fact, the magnitudes are the same in both cases. For example, if $\mathbf{v} = (v, 0, 0)$ and $\mathbf{u}' = (0, u', 0)$, it is easily seen that $\mathbf{v} + \mathbf{u}' = (v, u'/\gamma(v), 0)$ while $\mathbf{u}' + \mathbf{v} = (v/\gamma(u'), u', 0)$, where "$+$" denotes "relativistic sum." That the magnitudes of $\mathbf{u}' + \mathbf{v}$ and $\mathbf{v} + \mathbf{u}'$ are always equal follows from the symmetry of (2.27) below (since $u_1' v = \mathbf{u}' \cdot \mathbf{v}$).

A straightforward computation yields from (2.26) the following relation between the magnitudes $u = (u_1^2 + u_2^2 + u_3^2)^{1/2}$ and $u' = (u_1'^2 + u_2'^2 + u_3'^2)^{1/2}$ of corresponding velocities in S and S':

$$c^2 - u^2 = \frac{c^2(c^2 - u'^2)(c^2 - v^2)}{(c^2 + u_1' v)^2}. \quad (2.27)$$

(We shall indicate an easier way to obtain this identity in Exercise 4.2.) If $u' < c$ and $v < c$, the right member is positive, whence $u < c$. Hence the resultant of two velocities less than c is always a velocity less than c. This shows that, however many velocity increments (less than c) a particle receives in its instantaneous rest frame, it can never attain the velocity of light relative to a given inertial frame. Thus the velocity of light plays the role in relativity of an infinite velocity, inasmuch as no "sum" of lesser velocities can ever equal it. More generally, (2.27) shows that if $v < c$ (as is the case for any two inertial frames), $u \gtreqless c$ implies $u' \gtreqless c$, respectively, and vice versa.

Rewriting (2.27), we have

$$\frac{1 - u'^2/c^2}{1 - u^2/c^2} = \frac{(1 + u_1' v/c^2)^2}{1 - v^2/c^2},$$

which, on taking square roots, yields the first of the following two useful relations,

$$\frac{\gamma(u)}{\gamma(u')} = \gamma(v)\left(1 + \frac{u'_1 v}{c^2}\right), \quad \frac{\gamma(u')}{\gamma(u)} = \gamma(v)\left(1 - \frac{u_1 v}{c^2}\right), \tag{2.28}$$

while the second results again from the generally valid process of interchanging primed and unprimed symbols and replacing v by $-v$. [Cf. the remark before (2.26).] These relations show how the γ factor of a moving particle transforms.

2.16   Proper Acceleration

A useful concept in the study of nonuniform motion is that of *proper acceleration*. This is defined as the acceleration of a particle relative to its instantaneous rest frame. We here restrict ourselves to a discussion of one-dimensional motion, say along the x axis of S. Let S' (in standard configuration with S) be the instantaneous rest frame of a particle moving with velocity u, so that $u' = 0$ and $u = v$ momentarily, but v is constant whereas u and u' are not. Then, from (2.26)(i), we find after some calculation,

$$du = \gamma^{-2}(u)du'. \tag{2.29}$$

If α denotes the proper acceleration, we have $du' = \alpha dt'$, and, by time dilation, $dt' = dt/\gamma(u)$. Thus $du = \alpha dt/\gamma^3(u)$, i.e.,

$$\alpha = \left(1 - \frac{u^2}{c^2}\right)^{-3/2}\frac{du}{dt} = \frac{d}{dt}[\gamma(u)u]. \tag{2.30}$$

This gives the transformation of the acceleration from the rest frame to an arbitrary frame. Consequently we have, for *two* arbitrary frames S and S' (whose relative direction of motion coincides with that of the particle),

$$\left(1 - \frac{u^2}{c^2}\right)^{-3/2}\frac{du}{dt} = \left(1 - \frac{u'^2}{c^2}\right)^{-3/2}\frac{du'}{dt'}. \tag{2.31}$$

We note that acceleration is *not* invariant under a LT.

If α is constant, we can integrate (2.30) at once with respect to t (choosing $t = 0$ when $u = 0$), solve for u, integrate once more and, by suitably adjusting the constant of integration, obtain the following equation of motion:

$$x^2 - c^2 t^2 = c^4/\alpha^2. \tag{2.32}$$

Thus, for obvious reasons, rectilinear motion with constant proper acceleration is called "hyperbolic" motion (see Figure 2.5). (The corresponding classical calculation gives $x = \frac{1}{2}\alpha t^2 + $ constant, i.e., "parabolic motion.") Note that $\alpha = \infty$ implies $x = \pm ct$, hence the proper acceleration of a photon can be taken to be infinite.

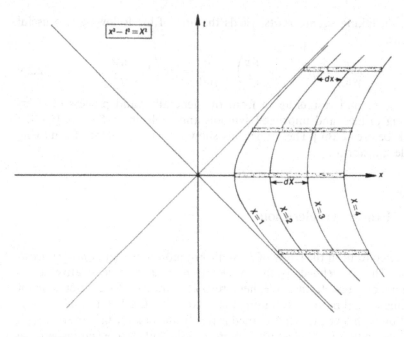

Figure 2.5

Consider next the equation

$$x^2 - c^2t^2 = X^2, \tag{2.33}$$

for various values of the parameter X. For each fixed X it represents a particle moving with constant proper acceleration c^2/X. Altogether it represents, as we shall show, the equations of motion of all points of a rod "moving rigidly" in the x direction.

By the rigid motion of a body one understands a motion during which every small volume element of the body shrinks always in the direction of its motion in proportion to its instantaneous Lorentz factor relative to a given inertial frame. Thus every small volume element preserves its dimensions in its instantaneous rest frame, which shows that the definition is intrinsic, i.e., frame-independent. It also shows that during rigid motion no elastic stresses arise. A body moving rigidly cannot start to rotate, since circumferences of circles described by points of the body would have to shrink, while their radii would have to remain constant, which is impossible. In general, therefore, the motion of one point of a rigidly moving body determines that of all others.

Going back to Equation (2.33), we can find (by implicit differentiation) the velocity u and the corresponding γ factor of a particle moving so that $X =$ constant:

$$u = \frac{dx}{dt} = \frac{c^2t}{x}, \quad \gamma(u) = \frac{x}{X}. \tag{2.34}$$

Now consider the motion of two such particles, whose parameters X differ by dX; at any fixed time t we have, again from (2.33) and then from (2.34),

$$dx = \frac{X dX}{x} = \frac{dX}{\gamma(u)}. \tag{2.35}$$

Hence at every instant $t = $ constant the two particles are separated by a coordinate distance dx inversely proportional to their γ factor, and consequently they "move rigidly"; moreover, dX is recognized as their *proper* separation. Since this applies to any two neighboring particles in the aggregate represented by (2.33), that whole aggregate "moves rigidly," like an unstressed rigid rod. Figure 2.5 shows the position of such a rod at various instants $t = $ constant. Units are chosen to make $c = 1$. The rod cannot be extended to negative values of X, since the asymptotes in the diagram represent photon paths; if continued right up to $X = 0$, the rod "ends in a photon."

It is interesting to observe that, according to the equivalence principle, an observer traveling with this rod at some fixed $X = X_0$ can be under the illusion that he is in a static gravitational field of strength c^2/X_0. A whole bundle of such rods, say of the length of a skyscraper, can give comoving observers the illusion of living in a rigid skyscraper immersed in a parallel and static gravitational field whose strength, however, falls off inversely as distance. A *uniform* gravitational field can *not* be constructed by this method.

2.17   Special Relativity without the Second Postulate

We shall now investigate the full implications of the relativity principle by itself, without the "second postulate," i.e., without assuming the invariance of the speed of light. We shall, however, demand Euclidicity and isotropy. We recall that in Section 2.6 we derived Equations (2.2)–(2.5) (with $\gamma > 0$) *without* the second postulate, appealing merely to the Euclidicity and isotropy of inertial frames. We were careful to point out that for calibrating the coordinate lattice, light signals were dispensable. Let us now return to Equations (2.2) through (2.5). Elimination of x' between (2.3) and (2.4) gives

$$t' = \gamma\left(t - \frac{\gamma^2 - 1}{\gamma^2 v} x\right) = \gamma\left(t - \frac{v}{V^2} x\right), \tag{2.36}$$

where we have simply written V^2 for $v^2 \gamma^2/(\gamma^2 - 1)$. (Note that V^2 could be negative.) Thus

$$0 < \gamma = (1 - v^2/V^2)^{-1/2}. \tag{2.37}$$

Equation (2.36), together with our previous equations

$$x' = \gamma(x - vt), \quad y' = y, \quad z' = z, \tag{2.38}$$

constitute the complete transformation. This is a Lorentz-type transformation with V^2 in place of the usual c^2; we shall call it a V^2-LT.

Next we show that if *any* two inertial frames are related by a V^2-LT, all inertial frames are related by a V^2-LT; i.e., V^2, which could *a priori* depend on v, in fact does not. For suppose frame S' is in standard configuration with frame S at relative velocity v and is related to S by a V^2-LT. Consider *any* other frame S''' in standard configuration with S' at relative velocity u. Suppose S' and S'' are related by a U^2-LT. Then

$$x' = \gamma(x - vt), \quad t' = \gamma(t - vx/V^2), \tag{2.39}$$

$$x'' = \gamma'(x' - ut'), \quad t'' = \gamma'(t' - ux'/U^2), \tag{2.40}$$

where γ and γ' are the Lorentz factors corresponding to V and U, respectively. If we substitute from (2.39) into (2.40), we find the transformation between S and S'':

$$x'' = \gamma\gamma'\left(1 + \frac{uv}{V^2}\right)\left(x - \frac{u + v}{1 + uv/V^2}t\right)$$

$$t'' = \gamma\gamma'\left(1 + \frac{uv}{U^2}\right)\left(t - \frac{u/U^2 + v/V^2}{1 + uv/U^2}x\right). \tag{2.41}$$

Now this is not *any* kind of LT unless $U^2 = V^2$. But it must be a LT! Hence $U^2 = V^2$, and thus the transformation between S' and S'' is a V^2-LT, and V^2 is indeed independent of the relative velocity since u was arbitrary.

If $V^2 = \infty$ (i.e., $V^{-2} = 0$—we could have worked with V^{-2} throughout to avoid infinities), we have the Galilean transformation group. The case $V^2 < 0$ corresponds to real rotations in x and Kt (where $K^2 = -V^2$):

$$x' = x\cos\theta - Kt\sin\theta,$$

$$Kt' = x\sin\theta + Kt\cos\theta, \tag{2.42}$$

with $\cos\theta = \gamma$, $\sin\theta = \gamma v/K$. The corresponding group has many unphysical properties if x and t have the significance of the present context. For example, repeated application of a low-velocity transformation (small θ) would lead to a resultant θ between $\pi/2$ and π, and hence to $\gamma < 0$, contrary to our requirement. The group also has an infinite velocity discontinuity at $\theta = \pi/2$ ($v = K\tan\theta$), and permits a reversal of time ($t' = -t$) and thus of causality. Clearly, the case $V^2 < 0$ must be ruled out.

Thus the RP by itself (together with Euclidicity and isotropy) necessarily implies that either all inertial frames are related by GT's, or all are related by LT's with the same positive V^2. The role of a "second postulate" in relativity is now clear: it merely has to isolate one or the other of these transformation groups. *Any* second postulate consistent with the RP but not with the GT isolates the LT group. However, in order to determine the universal constant V^2 the postulate must be quantitative. For example, a statement like "simultaneity is not absolute," while implying *a* Lorentz group, fails to determine V^2. On the other hand, the statement "at speed $3c/5$ there is a time dilation by a factor $5/4$," not only implies *a* Lorentz group, but *the* Lorentz group (with $V^2 = c^2$). We shall see later that the relativistic mass increase, or the famous formula $E = mc^2$, and others, could all equally well serve as second postulates.

We may also approach the topic of the present section from a different, more physical, point of view. This is based on the following dichotomy: Either particles can be accelerated to arbitrarily large speeds, or they can not. Suppose they can not. Then there must exist, mathematically speaking, a least upper bound c to particle speeds in any one inertial frame. By the RP, this bound must be the same in all inertial frames. Moreover, the speed c— whether attained or not by any physical effect—must transform into itself. It will be sufficient to prove this for motion along the common x axes of the frames S and S'. If c in S transformed into $c' > c$ in S', then, by continuity, particle speeds just short of c in S would correspond to speeds in excess of c in S'. If, on the other hand, c transformed into $c' < c$ in S', then particle speeds greater than c' in S' (but less than c) would correspond to speeds greater than c in S; for the velocity transformation must be monotonic in order to be one-valued both ways, at least on the x axes. On the x axes, therefore, c is an invariant velocity, and we can now derive the LT as in Section 2.6. If *no* bound exists on the particle velocities, then in S *any* event with $t > 0$ is causally connectible to the origin-event $x = y = z = t = 0 = x' = y' = z' = t'$. Thus any event with $t > 0$ must also have $t' > 0$. Similarly any event with $t < 0$ must have $t' < 0$. Consequently $t = 0$ must correspond to $t' = 0$, from which it follows, as in the argument for (2.2), that $t' = t$; and this leads directly to the GT.

[Early derivations of the LT without the postulate of light propagation were given by W. v. Ignatowsky in *Phys. Zeits. 11*, 972 (1910) and by L. A. Pars in *Philos. Mag. 42*, 249 (1921). However, like numerous others that followed, these have gone largely unnoticed.]

CHAPTER 3

Einsteinian Optics

3.1 The Drag Effect

Relativity provided an ideally simple solution to a problem that had considerably exercised the ingenuity of theoreticians before. The question is, to what extent a flowing liquid will "drag" light along with it. Flowing air, of course, drags sound along totally, but the optical situation is different: on the basis of an ether theory, it would be conceivable that there is no drag at all, since light is a disturbance of the ether and not of the liquid. Yet experiments indicated that there *was* a drag: the liquid seemed to force the ether along with it, but only partially. If the speed of light in the liquid *at rest* is u', and the liquid is set to move with velocity v, then the speed of light relative to the outside was found to be of the form

$$u = u' + kv, \quad k = 1 - 1/n^2, \tag{3.1}$$

where k is the "drag coefficient," a number between zero and one indicating what fraction of its own velocity the liquid imparts to the ether within, and n is the refractive index c/u' of the liquid. Fifty years before Einstein, Fresnel succeeded in giving a plausible ether-based explanation of this. From the point of view of special relativity, however, the result (3.1) is nothing but the relativistic velocity addition formula! The light travels relative to the liquid with velocity u', and the liquid travels relative to the observer with velocity v, and therefore [cf. (2.26)(i)]

$$u = u' \dotplus v = \frac{u' + v}{1 + u'v/c^2} = \frac{(c/n) + v}{1 + v/cn} \approx \frac{c}{n} + v\left(1 - \frac{1}{n^2}\right), \tag{3.2}$$

neglecting terms of order v/c in the last parenthesis. Einstein already gave the velocity addition formula in his 1905 paper, but, strangely, it took two more years before Laue made this beautiful application of it.

3.2 The Doppler Effect

Even on prerelativistic theory, if I watch a clock moving away from me, it will *appear* to go more slowly than a clock at rest. For when it indicates "two," say, it is further away than when it indicated "one," and light from it then takes longer to reach me in addition to having been emitted one unit of time later. If instead of a clock I watch a receding vibrating atom, then for the same reason the frequency that I observe will be smaller than that of the atom at rest, or, in other words, the spectrum of the atom will be redshifted. The reverse happens when these "clocks" approach, rather than recede. Similar effects exist for sound waves in air; all are named after the Austrian physicist Doppler.

Relativity has added a correction to the optical Doppler effect: the receding clock will appear to go even more slowly, and the receding atom will be even more redshifted, because of the time dilation of the moving clock or atom. (In principle, the same correction applies to the acoustical Doppler effect; but whereas vibrating atoms often move at time-dilating speeds, sound emitters do not, and so an acoustical correction is not needed in practice.)

Let a light-source travel uniformly through an inertial frame S with speed u, and let it have an instantaneous radial velocity component u_r relative to the origin-observer O. Then in the time Δt_0 which a comoving observer C measures between the emission of successive "wavecrests" (or ticks of a clock), the source has increased its distance from O by $\Delta t_0 \gamma(u)u_r$ (time dilation!). Consequently, the times between wavecrests as observed by O and C are $\Delta t_0 \gamma(u) + \Delta t_0 \gamma(u)u_r/c$ and Δt_0, respectively. The ratio of these times gives the Doppler shift

$$D = \frac{\lambda}{\lambda_0} = \frac{1 + u_r/c}{(1 - u^2/c^2)^{1/2}} = 1 + \frac{u_r}{c} + \frac{1}{2}\frac{u^2}{c^2} + O\left(\frac{u^3}{c^3}\right), \qquad (3.3)$$

λ_0 being the proper wavelength (as observed by C) and λ that observed by O. (The prerelativistic formula simply had the Lorentz factor missing, but it was considered valid only in the ether frame. In other frames the formula, and its derivation, were more complicated.) Our series expansion separates the "pure" Doppler effect $1 + u_r/c$ from the contribution $\frac{1}{2}u^2/c^2$ of time dilation, to the order shown.

Note that, when the motion of the source is purely radial, $u_r = u$ and Equation (3.3) reduces to

$$D = \frac{\lambda}{\lambda_0} = \left(\frac{1 + u/c}{1 - u/c}\right)^{1/2}. \qquad (3.4)$$

It is also useful to have a formula relating the wavelengths λ and λ' ascribed by *two* observers O and O' at the same event to an incoming ray of unspecified origin. Let O and O' have the usual frames S and S', respectively. Let α be the angle which the negative direction of the ray makes with the x axis of S. Also assume, without loss of generality, that the ray originated from a source at rest in S'. Then (3.3) applies with the specializations $\lambda_0 = \lambda'$, $u = v$, $u_r = v \cos \alpha$:

$$D = \frac{\lambda}{\lambda'} = \frac{1 + (v/c)\cos \alpha}{(1 - v^2/c^2)^{1/2}}. \tag{3.5}$$

In relativity there is a Doppler shift even for a purely transversely moving source, due, of course, *entirely* to time dilation. This gives us yet another way of verifying the existence in nature of time dilation, e.g., with rotor experiments. But there, either source or "observer" (receiver) do not move uniformly. If the source moves nonuniformly, the argument leading to Formula (3.3) remains valid, if we adopt the length and clock hypotheses. The proper wavelength λ_0 will then coincide with that measured in the instantaneous inertial rest frame of the source, and u, u_r will be its *instantaneous* velocity and radial velocity in the observer's frame. If the observer, too, accelerates, the formula still applies, for he, too, will make the same local length and time measurements as an inertial observer momentarily comoving with him. Then u and u_r must be measured at the moment of emission in the observer's rest frame at reception. (The reader should bear in mind the possibility of similar generalizations of relativistic formulae in the future.) Alternatively, we may evaluate the Doppler shift in *any* convenient frame at the reception event and transform to the rest frame of the observer in question by use of Equation (3.5).

With these methods it is now trivial to calculate the Doppler shift from the rim to the center of a rotor, or vice versa. It is more tedious so to calculate the Doppler shift between *arbitrary* points on the rotor—a problem of possible practical interest. We shall circumvent the tedium with a trick. Consider a large disc which rotates at uniform angular velocity ω in an inertial frame S, and which has affixed to it a light source and receiver, at points P_0 and P_1, at distances r_0 and r_1 from the center, respectively. Since each signal from P_0 to P_1 clearly takes the same time in S, two successive signals are emitted and received with the same time difference in S, say Δt. These differences correspond to *proper* time differences $\Delta t/\gamma(\omega r_0)$ and $\Delta t/\gamma(\omega r_1)$ at P_0 and P_1, respectively. But those are proportional to the locally observed wavelengths λ_0 and λ_1, respectively, and so

$$D = \frac{\lambda_1}{\lambda_0} = \frac{\gamma(\omega r_0)}{\gamma(\omega r_1)}. \tag{3.6}$$

The most obvious experiment for measuring the transverse Doppler effect (i.e., time dilation) would be to place the source at the rim and the receiver at the center of a rotor; in practice, however, the exact opposite was done (thus essentially measuring the time dilation of the receiver rather than of the source) by Hay, Schiffer, Cranshaw, and Engelstaff in 1960, using Mössbauer resonance. Agreement with the theoretical predictions was obtained to within an expected experimental error of a few percent. This, incidentally, furnished some validation of the clock hypothesis, for the "clock" at the receiver was clearly accelerated (up to about 6×10^4 g, in fact), and no measurable effect of this acceleration could be detected.

Prior to the rotor experiments, it was difficult to ensure exact transverseness in the motion of the sources (e.g., fast-moving hydrogen ions). The slightest radial component would swamp the transverse effect. Ives and Stilwell (in 1938) cleverly used a to-and-fro motion whereby the first-order Doppler effect cancelled out, and thus they were able to verify the contribution of time dilation to considerable accuracy.

A similar canceling of the first-order contribution occurs in the so-called "thermal" Doppler effect. Radioactive nuclei bound in a hot crystal move thermally in a rapid and random way. Because of this randomness, their first-order (classical) Doppler effects average out, but not the second-order effect due to their mean square velocities. These velocities were calculated and the Doppler shift was observed, once again by use of Mössbauer resonance (in 1960, by Rebka and Pound). Again the agreement with theory was excellent, to within an expected experimental error of about ten percent. One by-product of this experiment was an impressive validation of the clock hypothesis: in spite of accelerations up to 10^{16} g (!), these nuclear "clocks" were slowed simply by the velocity factor $(1 - v^2/c^2)^{1/2}$.

3.3 Aberration and the Visual Appearance of Moving Objects

If I drive into the rain, it seems to come at me obliquely. For similar reasons, if two observers measure the angle which an incoming ray of light makes with their relative line of motion, their measurements will generally not agree. This phenomenon is called aberration, and of course it was well known before relativity. Nevertheless, as in the case of the Doppler effect, the relativistic formula contains a correction, but it applies to all pairs of observers, whereas the prerelativistic formula was simple only if one of the observers was at rest in the ether frame.

To obtain the basic aberration formulae, consider an incoming light signal whose negative direction makes angles α and α' with the x axes of the usual two frames S and S', respectively. The velocity transformation formula (2.25) (i) can evidently be applied to this signal, with $u_1 = -c \cos \alpha$ and $u'_1 = -c \cos \alpha'$, yielding

$$\cos \alpha' = \frac{\cos \alpha + v/c}{1 + (v/c)\cos \alpha}. \tag{3.7}$$

Similarly, from (2.25)(ii) we obtain the alternative formula (assuming temporarily, without loss of generality, that the signal lies in the x, y plane)

$$\sin \alpha' = \frac{\sin \alpha}{\gamma[1 + (v/c)\cos \alpha]}. \tag{3.8}$$

But the most interesting version of the aberration formula is obtained by substituting from Equations (3.7) and (3.8) into the trigonometric identity

$$\tan \tfrac{1}{2}\alpha' = \sin \alpha'/(1 + \cos \alpha'),$$

which gives

$$\tan \tfrac{1}{2}\alpha' = \left(\frac{c - v}{c + v}\right)^{1/2} \tan \tfrac{1}{2}\alpha. \tag{3.9}$$

For rays going *out* at angles α and α', we merely replace c by $-c$ in all the above formulae.

Aberration implies, for example, that as the earth travels along its orbit, the apparent directions of the fixed stars trace out small ellipses in the course of each year (with major axes of about 41 seconds of arc). Aberration also causes certain distortions in the visual appearance of extended uniformly moving objects. For, from the viewpoint of the rest frame of the object, as the observer moves past the conical pattern of rays converging from the object to the observer's eye, rays from its different points are unequally aberrated. Alternatively, from the viewpoint of the observer's rest frame, the light from different parts of the moving object takes different times to reach his eye, and thus it was emitted at different past times; the more distant points of the object consequently appear displaced relative to the nearer points in the direction opposite to the motion.

It is in connection with the visual appearance of uniformly moving objects that the relativistic results are a little unexpected. Following an ingenious argument of R. Penrose, let us draw a sphere of unit diameter around each observer's space-origin (see Figure 3.1), cutting the negative and positive x axis at points P and Q, respectively. All that an observer sees at any instant can be mapped onto this sphere (his "sky"). Let it further be mapped from this sphere onto the tangent plane at Q (his "screen") by stereographic projection from P. We recall that the angle subtended by an arc of a circle at the circumference is half of the angle subtended at the center, and we have accordingly labeled the diagram (for a single incident ray). Thus the significance of (3.9) is seen to be precisely this: whatever the two momentarily coincident observers see, the views on their "screens" are identical except for scale.

Now consider, for example, a solid sphere Z at rest somewhere in the frame of an inertial observer O'. He sees a circular outline of Z in his sky, and projects a circular outline on his screen (for under stereographic projection, circles on the sphere correspond to circles or straight lines on the plane). Relative to the usual second observer O, of course, Z moves. Nevertheless, according to our theory, his screen image will differ from that of O' only in size, and thus it will also be circular; consequently, his "sky" image of Z must be circular too.

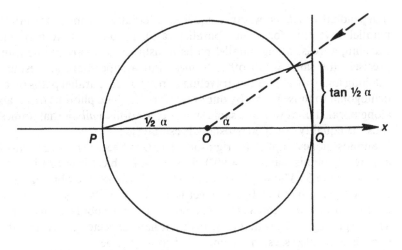

Figure 3.1

This shows that a moving sphere presents a circular outline to *all* observers *in spite* of length contraction! (Or rather: *because* of length contraction; for without length contraction the outline would be distorted.) By the same argument, moving straight lines (rods) will, in general, have the appearance of circular arcs, and flying saucers or fast-moving bicycle wheels are liable to look boomerang-shaped.

Another interesting though less realistic way of studying the visual appearance of moving objects is by use of what we may call "supersnapshots." These are life-size snapshots made by receiving *parallel* light from an object and catching it directly on a photographic plate held at right angles to the rays. One could, for example, make a supersnapshot of the *outline* of an object by arranging to have the sun behind it and letting it cast its shadow onto a plate. Moreover, what the eye sees (or the ordinary camera photographs) of a *small and distant* object approximates quite closely to a supersnapshot. Now, the surprising result (due to Terrell) is this: all supersnapshots that can be made of a uniformly moving object at a certain place and time by observers in any state of uniform motion are identical. In particular, they are all identical to the supersnapshot that can be made in the rest frame of the object.

To prove this result,[1] let us consider two photons P and Q traveling abreast along parallel straight paths a distance Δr apart, relative to some frame S. Let us consider two arbitrary events $\mathscr{P}$ and $\mathscr{Q}$ at P and Q, respectively. If $\mathscr{Q}$ occurs a time Δt after $\mathscr{P}$, then the space separation between $\mathscr{P}$ and $\mathscr{Q}$ is evidently $(\Delta r^2 + c^2\Delta t^2)^{1/2}$, and thus, by (**2.11**), the squared interval between $\mathscr{P}$ and $\mathscr{Q}$ is $-\Delta r^2$, independently of their time separation. But if, instead of traveling abreast, Q leads P by a distance Δl, then the space separation between $\mathscr{P}$ and $\mathscr{Q}$ would be $[\Delta r^2 + (c\Delta t + \Delta l)^2]^{1/2}$, and the squared interval would *not* be

---

[1] The proof given here was suggested by one used by V. F. Weisskopf in *Phys. Today*, Sept. 1960, p. 24.

independent of Δt. Now, since squared interval is an invariant (and since parallel rays transform into parallel rays), it follows that two photons traveling abreast along parallel paths a distance Δr apart in one frame do precisely the same in all other frames. But a supersnapshot results from catching an array of photons traveling abreast along parallel paths on a plate orthogonal to those paths. By our present result, these photons travel abreast along parallel lines with the same space separation in *all* inertial frames, and thus the equality of supersnapshots is established.

Suppose, for example, the origin-observer O in S sees at $t = 0$ a small object, apparently on his y axis ($\alpha = 90°$). Suppose this object is at rest in the usual second frame S'. The origin-observer O' in S' will see the object at an angle $\alpha' < 90°$, given by (**3.9**). If the object is a cube with its edges parallel to the coordinate axes of S and S', O' of course sees the cube not face-on but rotated. The surprising result is that O, who might have expected to see a contracted cube face-on, also sees an uncontracted rotated cube!

What is even more surprising is that it took something like fifty years from the advent of special relativity before these simple facts of vision were discovered (apart from the basic aberration formula, which, together with the Doppler formula, already appear in Einstein's 1905 paper). The interested reader may wish to think up various other visual problems, but we shall not pursue these matters further, since their intrinsic significance appears to be limited. Unlike the Doppler effect and the drag effect, aberration has not so far led to experiments with the express purpose of testing special relativity. On the other hand, it *does* provide another good example of the effectiveness of relativity theory. Before Einstein, it was complicated to obtain an aberration formula for two frames like S and S' with an "ether wind" blowing through S in an arbitrary direction—even if one assumed, as we have done so far, the corpuscular (photon) nature of light. On the basis of a wave theory of light, the Galilean transformation gave no aberration of the wave normal at all, and one had to introduce the concept of "rays," whose direction did not necessarily coincide with the wave normals. In relativity, the wave analysis is as straightforward as the corpuscular analysis of the problem (see Exercise 3.14). Thus it is wrong to think that relativity always "complicates things." In optics it is certainly quite the other way round.

CHAPTER 4

Spacetime and Four-Vectors

4.1 Spacetime

Einstein taught us to regard *events* as the basic data of physics. He showed us the full significance of inertial frames and inertial observers. It was he who found that every inertial observer has his own privately valid time, and, correspondingly, his own "instantaneous three-spaces" consisting of all events (x, y, z, t) with fixed time coordinate. But it was the mathematician Minkowski (in 1908) who taught us to think of the totality of events in the world as the "points" of an absolute four-dimensional manifold called "spacetime." Different inertial observers draw different sections through spacetime as their "instants." In fact, each inertial observer, with his standard x, y, z, and t, coordinatizes all of spacetime, just as a choice of x and y axes coordinatizes the Euclidean plane. Figure 2.3 represents spacetime with two dimensions suppressed, but that suffices to illustrate how two observers make different instantaneous sections (for example, the x and x' axes) and how they have different time axes. Certain lines in spacetime correspond to the history of material particles, and are aptly called "particle worldlines." They are straight if the particles are free. Extended bodies have "worldtubes" in spacetime.

Of course, in Newtonian physics, governed by the Galilean transformation, one can define spacetime just as well and use it to carry various graphical representations. Here all observers agree on how to slice spacetime into instants, but they do have different time axes (these being the worldlines of their origins). Free particles also have straight worldlines, and extended bodies have worldtubes.

But in the relativistic case, spacetime is very much *more* than a convenient scheme for drawing graphs. It is a four-dimensional *metric* space.[1] Analogously to "distance," we can define an "interval"

$$\Delta s = (c^2 \Delta t^2 - \Delta x^2 - \Delta y^2 - \Delta z^2)^{1/2} \tag{4.1}$$

for any two of its points. And this is *absolute*, i.e., it has the same value when evaluated by *any* inertial observer using standard coordinates [cf. Section 2.7(ii)]. As Minkowski demonstrated, the existence of the "metric" **(4.1)** of spacetime has a significant mathematical consequence: it leads to a vector calculus ("four-vectors") beautifully adapted to the needs of special relativity. Moreover, spacetime with this metric provides us with a new *absolute* (i.e., oberver- and coordinate-independent) background for our intuitive thought about the physical world, a background that had seemed lost forever with the abolition of absolute space and absolute time. Minkowski was so struck by his discovery that he exclaimed: "Henceforth space by itself, and time by itself, are doomed to fade away into mere shadows, and only a kind of union of the two will preserve an independent reality." And again: "In my opinion physical laws might find their most perfect expression as the mutual relations between worldlines." (In the study of general relativity we shall see how prophetic this statement was.) Minkowski may well be regarded as the father of the "fourth dimension."

No corresponding development is possible in Galilean spacetime, where there is only an absolute "temporal distance" Δt between events. The spatial distance $(\Delta x^2 + \Delta y^2 + \Delta z^2)^{1/2}$ depends on the motion of the observer, and no Galileo-invariant "metric" involving all of $\Delta x, \Delta y, \Delta z, \Delta t$ exists. Consequently, Galilean spacetime is nonmetric, and Galilean four-vectors are "unnormed," and thus not very useful. (But see Section 5.5.)

We may discuss here briefly the physical significance of Δs, or better, of Δs^2, though we shall have occasion to return to this point more than once. Evidently $\Delta s^2 = 0$ holds for two given events $\mathscr{P}$ and $\mathscr{Q}$ if and only if these events are connectible by a light signal. When $\Delta s^2 > 0$, then, in any inertial frame, $\Delta r^2/\Delta t^2 < c^2$ (where $\Delta r^2 = \Delta x^2 + \Delta y^2 + \Delta z^2$) and thus an observer moving with uniform velocity less than c can be sent from one of the events to the other; in *his* rest frame $\Delta r = 0$ and $\Delta s = c\Delta t$. Thus when $\Delta s^2 > 0$, Δs represents c times the time separation between $\mathscr{P}$ and $\mathscr{Q}$ measured in that inertial frame in which $\mathscr{P}$ and $\mathscr{Q}$ occur at the same point. Similarly, if $\Delta s^2 < 0$, then $|\Delta s|$ is the space separation between $\mathscr{P}$ and $\mathscr{Q}$ measured in an inertial frame in which they are simultaneous, and such a frame always exists. The first part of our assertion is obvious, and the second follows from the LT formula $\Delta t' = \gamma(\Delta t - v\Delta x/c^2)$. For if $\Delta t \neq 0$ in an arbitrarily chosen frame S, orienting the x axis along the locations of $\mathscr{P}$ and $\mathscr{Q}$ makes $\Delta y = \Delta z = 0$,

---

[1] Mathematicians actually call it "pseudometric," since the squared interval vanishes for certain pairs of points and becomes negative for others.

$\Delta x/\Delta t > c$, and thus $v = c^2/(\Delta x/\Delta t) < c$ leads to the required frame S' in which $\Delta t' = 0$.

Finally, note the identity

$$\Delta s^2 = \Delta t^2 \left(c^2 - \frac{\Delta r^2}{\Delta t^2} \right), \tag{4.2}$$

which very clearly relates the magnitude of $(\Delta r/\Delta t)$—the "signal velocity" from $\mathscr{P}$ to $\mathscr{Q}$—with the sign of Δs^2.

4.2 Three-Vectors

Before introducing four-vectors, it will be well to review the salient features of three-vectors, i.e., of "ordinary" vectors. Anyone who has done three-dimensional geometry or mechanics will be aware of the power of the vector calculus. Just what *is* that power? First, of course, there is power simply in abbreviation. A comparison of Newton's second law in scalar and in vector form makes this clear:

$$\left.\begin{array}{l} f_1 = ma_1 \\ f_2 = ma_2 \\ f_3 = ma_3 \end{array}\right\} \mathbf{f} = m\mathbf{a}.$$

This is only a very mild example. In looking through older books on physics or geometry, one often wonders how anyone could have seen the underlying physical reality through the triple maze of coordinate-dependent scalar equations. Yet abbreviation, though in itself often profoundly fruitful, is only one aspect of the matter. The other is the abolition of the coordinate-dependence just mentioned: vectors are absolute.

In studying the geometry and physics of three-dimensional Euclidean space, each "observer" can set up his standard coordinates x, y, z (right-handed orthogonal Cartesians) with any point as origin and with any orientation. Does this mean that there are as many spaces as there are coordinate systems? No: underlying all subjective "observations" there is a single space with *absolute* elements and properties, namely those on which all observers agree, such as specific points and specific straight lines, distances between specific points and lines, angles between lines, etc. Vector calculus treats these absolutes in a coordinate-free way that makes their absoluteness evident. All relations that can be expressed vectorially, such as $\mathbf{a} = \mathbf{b} + \mathbf{c}$, or $\mathbf{a} \cdot \mathbf{b} = 5$, are necessarily absolute. On the other hand, an observer's statement like $f_1 = f_2$ (about a force), which has no vector formulation, is of subjective interest only.

A (Cartesian) three-vector $\mathbf{a}$ can be defined as a number-triple (a_1, a_2, a_3) which depends on the choice of a Cartesian reference frame (x, y, z).[2] The

---

[2] Technically, $\mathbf{a}$ is a *function* from the set of Cartesian coordinate systems (x, y, z) to the space $\mathbf{R}^3$ of number-triples: $\mathbf{a}(x, y, z) = (a_1, a_2, a_3)$.

various vector operations can be defined via these "components"; e.g., $\mathbf{a} + \mathbf{b} = (a_1 + b_1, a_2 + b_2, a_3 + b_3)$, $\mathbf{a} \cdot \mathbf{b} = a_1 b_1 + a_2 b_2 + a_3 b_3$. But they can be *interpreted* absolutely (i.e., coordinate-independently): for example $\mathbf{a}$ itself as having a certain length and direction, $\mathbf{a} + \mathbf{b}$ by the parallelogram rule, etc. Only operations that have absolute significance are admissible in vector calculus. To check a vector equation, an observer could proceed *directly* by measuring absolutes like lengths and angles, but he would then really be a "superobserver." The observers we have in mind simply possess a standard coordinate lattice, and in fact they can be identified with such a lattice. Thus they can only read off components of all relevant vectors. To check a relation like $\mathbf{a} = \mathbf{b} + \mathbf{c}$, they would each obtain a set of three scalar equations $a_i = b_i + c_i$ ($i = 1, 2, 3$) which differ from observer to observer; but either all sets are false, or all are true. *A vector (component) equation that is true in one coordinate system is true in all*: this is the most salient feature of the vector calculus. Speaking technically, vector (component) equations are form-invariant under the rotations about the origin and the translations of axes which relate the different "observers" in Euclidean space and which in fact constitute the "relativity group" of Euclidean geometry. The reason for this form-invariance will appear presently.

The prototype of a three-vector is the displacement vector $\Delta \mathbf{r} = (\Delta x, \Delta y, \Delta z)$ joining two points in Euclidean space. Under a translation of axes its components remain unchanged, and under a rotation about the origin they suffer the same (linear, homogeneous) transformation as the coordinates themselves [cf. Section 2.7(i)], say

$$\Delta x' = \alpha_{11} \Delta x + \alpha_{12} \Delta y + \alpha_{13} \Delta z$$
$$\Delta y' = \alpha_{21} \Delta x + \alpha_{22} \Delta y + \alpha_{23} \Delta z \tag{4.3}$$
$$\Delta z' = \alpha_{31} \Delta x + \alpha_{32} \Delta y + \alpha_{33} \Delta z,$$

where the α's are certain functions of the angles specifying the rotation. *Any quantity having three components* (a_1, a_2, a_3) *which undergo exactly the same transformation* (4.3) *as* $(\Delta x, \Delta y, \Delta z)$ *under the comtemplated changes of coordinates (rotations and translations) is said to constitute a three-vector.* This property of three-vectors is often not stated explicitly in elementary texts; but it is implicit in the usual assumption that each three-vector can be *represented* by a displacement vector (a "directed line segment") in Euclidean space. Note that the position "vector" $\mathbf{r} = (x, y, z)$ of a point relative to an observer is a vector *only* under rotations and not under translations! The zero vector $\mathbf{0} = (0, 0, 0)$ is a three-vector according to (4.3); it is usually—if incorrectly—written as 0, as in $\mathbf{a} = 0$.

From (4.3) it follows that if the components of two three-vectors are equal in one coordinate system, they are equal in all coordinate systems; for both sets of new components are the same linear combination of the old components. If $\mathbf{a} = (a_1, a_2, a_3)$ and $\mathbf{b} = (b_1, b_2, b_3)$ separately transform like $(\Delta x, \Delta y, \Delta z)$, then so does $\mathbf{a} + \mathbf{b} = (a_1 + b_1, a_2 + b_2, a_3 + b_3)$, because of the linearity of (4.3); hence sums of vectors are vectors. Similarly, if k is a *scalar invariant* (often shortened to just "scalar" or "invariant"), i.e., a real number independent of the coordinate system, then clearly $k\mathbf{a}$, defined as (ka_1, ka_2, ka_3), is a vector, again from **(4.3)**.

If (x, y, z) is the current point of a curve in space, then the length of a chord, $\Delta l = (\Delta x^2 + \Delta y^2 + \Delta z^2)^{1/2}$, is evidently a scalar under the contemplated transformations. Dividing **(4.3)** by Δl and letting $\Delta l \to 0$, we find that the "unit tangent"

$$\mathbf{t} = \left(\frac{dx}{dl}, \frac{dy}{dl}, \frac{dz}{dl}\right) \tag{4.4}$$

is a vector. It is usually written as $d\mathbf{r}/dl$. Passing from geometry to Newtonian mechanics, consider a particle moving along this curve. The time interval Δt between two events at the particle is a scalar. Dividing **(4.3)** by Δt and letting $\Delta t \to 0$, we see that the velocity $\mathbf{u} = d\mathbf{r}/dt$ is a vector. One easily sees from **(4.3)**, written with a_1, a_2, a_3 in place of $\Delta x, \Delta y, \Delta z$, that the derivative of *any* vector $\mathbf{a}$ with respect to a scalar (defined by differentiating its components) is a vector. Thus the acceleration $\mathbf{a} = d\mathbf{u}/dt = (du_1/dt, du_2/dt, du_3/dt)$ is a vector. Multiplying $\mathbf{u}$ and $\mathbf{a}$ by the mass m (a scalar) yields two more vectors, the momentum $\mathbf{p} = m\mathbf{u}$ and the force $\mathbf{f} = m\mathbf{a}$. Note how the five basic vectors $\mathbf{t}, \mathbf{u}, \mathbf{a}, \mathbf{p}, \mathbf{f}$ all arise from the prototype $(\Delta x, \Delta y, \Delta z)$ by vector operations.

Associated with each three-vector $\mathbf{a} = (a_1, a_2, a_3)$ there is a very important scalar, its *norm* or *magnitude*, written $|\mathbf{a}|$ or simply a, and defined by

$$a^2 = a_1^2 + a_2^2 + a_3^2, \quad a \geq 0. \tag{4.5}$$

That this is invariant follows at once from the invariance of the squared norm $\Delta x^2 + \Delta y^2 + \Delta z^2$ of the prototype vector under rotations and translations; i.e., it follows from the existence of an invariant metric of Euclidean space. If $\mathbf{a}$ and $\mathbf{b}$ are vectors, then $\mathbf{a} + \mathbf{b}$ is a vector, whose norm must be invariant; but

$$|\mathbf{a} + \mathbf{b}|^2 = (a_1 + b_1)^2 + (a_2 + b_2)^2 + (a_3 + b_3)^2$$
$$= a^2 + b^2 + 2(a_1b_1 + a_2b_2 + a_3b_3),$$

and since a^2 and b^2 are invariant, it follows that the "scalar product"

$$\mathbf{a} \cdot \mathbf{b} = a_1b_1 + a_2b_2 + a_3b_3 \tag{4.6}$$

is invariant, i.e., coordinate-independent. If this were our first encounter with vectors, we would now look for the absolute (coordinate-independent) significance of $\mathbf{a} \cdot \mathbf{b}$, which *a priori* must exist; and by going to a specific coordinate system, we would soon discover it. The product $\mathbf{a} \cdot \mathbf{b}$, as defined by **(4.6)**, is easily seen to obey the commutative law $\mathbf{a} \cdot \mathbf{b} = \mathbf{b} \cdot \mathbf{a}$, the distributive law $\mathbf{a} \cdot (\mathbf{b} + \mathbf{c}) = \mathbf{a} \cdot \mathbf{b} + \mathbf{a} \cdot \mathbf{c}$, and the Leibniz differentiation law $d(\mathbf{a} \cdot \mathbf{b}) = d\mathbf{a} \cdot \mathbf{b} + \mathbf{a} \cdot d\mathbf{b}$. Also note that $\mathbf{a} \cdot \mathbf{a} = a^2$, which we may write as $\mathbf{a}^2 = a^2$.

4.3   Four-Vectors

We are now ready to develop the calculus of four-vectors by close analogy with three-vectors. And it is easy to guess what we are going to get: we get a vector calculus whose equations are form-invariant under *general* Lorentz transformations (see the final paragraph of Section 2.6), i.e., whose equations have precisely the property required by the relativity principle of all physical laws! This often enables us to recognize by its *form* alone whether a given or proposed law is Lorentz-invariant, and so assists us greatly in the construction of relativistic physics. However, let it be said at once that not *all* Lorentz-invariant laws are expressible as relations between four-vectors and scalars; some require "four-tensors." (But none require more: even the so-called "spinor laws" are expressible tensorially.) We shall touch briefly on four-tensors in Section 4.4.

The prototype of a four-vector is the displacement four-vector $\Delta \mathbf{R} =$ $(\Delta x, \Delta y, \Delta z, \Delta t)$ *between two events.*[3] Each four-vector can be *represented* by such a displacement. The admissible coordinate systems are the "standard" coordinates of inertial observers, and hence the relevant transformations are the general Lorentz transformations (compounded of translations, rotations about the spatial origin, and standard LT's). These give rise to a four-dimensional analog of (**4.3**) for any pair of inertial frames, with 16 constant α's instead of nine. We shall consistently use lower-case boldface letters to denote three-vectors, and boldface capitals to denote four-vectors. Under spatial and temporal translations, the components of $\Delta \mathbf{R}$ (and thus of *any* four-vector) are unchanged; under spatial rotations about the origin, the first three components of $\Delta \mathbf{R}$ (and thus of *any* four-vector) transform like a three-vector, while the last component is unchanged; and under a standard LT, the components of a four-vector transform precisely as in (**2.8**). The position four-"vector" $\mathbf{R} = (x, y, z, t)$ is a four-vector *only* under homogeneous general LT's, i.e., those that leave the coordinates of the event $(0, 0, 0, 0)$ unchanged. The *zero vector* $\mathbf{0} = (0, 0, 0, 0)$ is a true four-vector. Sums, scalar multiples, and scalar derivatives of four-vectors are defined by analogy with three-vectors and are recognized as four-vectors.

An important scalar under general LT's is the squared interval Δs^2 or, in differential form, ds^2. However, it is often convenient to work instead with the corresponding *proper time* interval $d\tau$, defined by

$$d\tau^2 = \frac{ds^2}{c^2} = dt^2 - \frac{dx^2 + dy^2 + dz^2}{c^2}. \tag{4.7}$$

This gets its name from the fact that, for a moving particle, $d\tau$ coincides with the dt that is measured by a clock attached to the particle; for in its instantaneous rest frame the particle satisfies $dx = dy = dz = 0$. We shall not be surprised, therefore, to find $d\tau$ appearing in many relativistic formulae where

---

[3] Other authors (e.g., RSR) take as the prototype $(\Delta x, \Delta y, \Delta z, c\Delta t)$, whose advantage is that in the resulting calculus all components of a four-vector (or four-tensor) always have equal physical dimensions.

in the classical analog there is a dt. For example, if (x, y, z, t) are the co-ordinates of a moving particle, we find by dividing the four-dimensional analog of (4.3) by $\Delta\tau$ and letting $\Delta\tau \to 0$, that

$$\mathbf{U} = \frac{d\mathbf{R}}{d\tau} = \left(\frac{dx}{d\tau}, \frac{dy}{d\tau}, \frac{dz}{d\tau}, \frac{dt}{d\tau}\right) \tag{4.8}$$

is a four-vector. This is called the *four-velocity* of the particle. It really is the analog of (4.4), and $\mathbf{U}$ can be regarded as the tangent vector of the worldline of the particle in spacetime. Now from (4.7),

$$\frac{d\tau^2}{dt^2} = 1 - \frac{dx^2 + dy^2 + dz^2}{c^2 dt^2} = 1 - \frac{u^2}{c^2},$$

u being the speed of the particle, whence (not surprisingly)

$$\frac{dt}{d\tau} = \gamma(u). \tag{4.9}$$

Since $dx/d\tau = (dx/dt)(dt/d\tau) = u_1\gamma(u)$, etc., we see that

$$\mathbf{U} = \gamma(u)(u_1, u_2, u_3, 1) = \gamma(u)(\mathbf{u}, 1). \tag{4.10}$$

We shall often recognize in the first three components of a four-vector the components of a familiar three-vector (or a multiple thereof), and in such cases we adopt the notation exemplified by (4.10).

As in three-vector theory, scalar derivatives of four-vectors (defined by differentiating the components) are themselves four-vectors. Thus, in particular,

$$\mathbf{A} = \frac{d\mathbf{U}}{d\tau} = \frac{d^2\mathbf{R}}{d\tau^2} \tag{4.11}$$

is a four-vector, called the *four-acceleration*. Its relation to the three-accelera-tion $\mathbf{a}$ is not quite as simple as that of $\mathbf{U}$ to $\mathbf{u}$; by (4.9), we have

$$\mathbf{A} = \gamma\frac{d\mathbf{U}}{dt} = \gamma\frac{d}{dt}(\gamma\mathbf{u}, \gamma) = \gamma\left(\frac{d\gamma}{dt}\mathbf{u} + \gamma\mathbf{a}, \frac{d\gamma}{dt}\right) \tag{4.12}$$

and it is seen from this that the components of $\mathbf{A}$ *in the instantaneous rest frame* of the particle ($u = 0$) are given by

$$\mathbf{A} = (\mathbf{a}, 0), \tag{4.13}$$

since the derivative of γ contains a factor u. Thus $\mathbf{A} = 0$ if and only if the *proper acceleration*—i.e., the norm of the three-acceleration in the rest frame —vanishes. The four-velocity $\mathbf{U}$, on the other hand, never vanishes.

The norm or magnitude $|\mathbf{V}|$ or V of a four-vector $\mathbf{V} = (V_1, V_2, V_3, V_4)$ is defined by

$$V^2 = -V_1^2 - V_2^2 - V_3^2 + c^2 V_4^2, \tag{4.14}$$

and its invariance follows from the invariance of the squared norm $-\Delta x^2 - \Delta y^2 - \Delta z^2 + c^2 \Delta t^2$ of the prototype vector. Precisely as for (4.6) we deduce from the invariance of $|\mathbf{V}|$, $|\mathbf{W}|$ and $|\mathbf{V} + \mathbf{W}|$ that the *scalar product* $\mathbf{V} \cdot \mathbf{W}$, defined by

$$\mathbf{V} \cdot \mathbf{W} = -V_1 W_1 - V_2 W_2 - V_3 W_3 + c^2 V_4 W_4, \qquad (4.15)$$

is invariant. As an example, if we consider $\mathbf{U}$ and $\mathbf{A}$ in the rest frame ($u = 0$), we see from (4.10) and (4.13) that

$$\mathbf{U} \cdot \mathbf{A} = 0 \qquad (4.16)$$

there; but since $\mathbf{U} \cdot \mathbf{A}$ is an invariant, the result if frame-independent. Thus the four-velocity is "orthogonal" to the four-acceleration.

If $\mathbf{A}$, $\mathbf{B}$, $\mathbf{C}$ are arbitrary four-vectors, one verifies at once from the definitions that

$$\mathbf{A} \cdot \mathbf{B} = \mathbf{B} \cdot \mathbf{A}, \quad \mathbf{A} \cdot (\mathbf{B} + \mathbf{C}) = \mathbf{A} \cdot \mathbf{B} + \mathbf{A} \cdot \mathbf{C}, \quad \mathbf{A} \cdot \mathbf{A} = A^2, \quad (4.17)$$

$$d(\mathbf{A} \cdot \mathbf{B}) = d\mathbf{A} \cdot \mathbf{B} + \mathbf{A} \cdot d\mathbf{B}. \qquad (4.18)$$

From (4.10) we have

$$U^2 = c^2. \qquad (4.19)$$

[The alert reader will have put $u = 0$ in (4.10) before calculating—why?] If we differentiate this equation with respect to τ—treating the left side as $\mathbf{U} \cdot \mathbf{U}$, we can reobtain (4.16). The norm of the four-acceleration is apparent from (4.13):

$$A^2 = -\alpha^2, \qquad (4.20)$$

where α is the proper acceleration.

We next inquire into the absolute significance of the scalar product. We do not expect this to be purely geometric, but rather kinematic, since LT's are essentially concerned with motion. Let $\mathbf{U}$ and $\mathbf{V}$ be the four-velocities of two particles, and let us look at $\mathbf{U} \cdot \mathbf{V}$ in the rest frame of the first, relative to which the second has velocity v, say; then, by (4.10) and (4.15),

$$\mathbf{U} \cdot \mathbf{V} = c^2 \gamma(v). \qquad (4.21)$$

Thus $\mathbf{U} \cdot \mathbf{V}$ is c^2 times the Lorentz factor of the *relative* velocity of the corresponding particles. Of course, four-velocities are not *arbitrary* four-vectors, if only because their norms are always c. A suggestive and slightly more general formula will be found in Exercise 4.12.

4.4 Four-Tensors

The typical transformation scheme (4.3) as applied to a three-vector $\mathbf{a}$ may be written in condensed form as

$$a_i' = \sum_{j=1}^{3} \alpha_{ij} a_j, \qquad (4.22)$$

while the analogous transformation of a four-vector **A** may be written

$$A'_\mu = \sum_{\nu=1}^{4} \alpha_{\mu\nu} A_\nu. \tag{4.23}$$

Greek indices are traditional when the values range from 1 to 4.

Just as four-*vectors* are defined by the behavior of their components under general LT's, so also are four-*tensors*. But the latter have more components than just four. For example, a four-tensor of *rank* 2 has 4^2 components $A_{\mu\nu}$, which obey the transformation rule

$$A'_{\mu\nu} = \sum_{\sigma,\tau=1}^{4} \alpha_{\mu\sigma} \alpha_{\nu\tau} A_{\sigma\tau}, \tag{4.24}$$

where the α's are the same as in (**4.23**). In general, a four-tensor of rank n has 4^n components, which are written with n indices, and they transform analogously to (**4.24**). Four-vectors, then, are simply four-tensors of rank 1. Many of their basic properties are typical for all four-tensors. E.g., four-tensor (component) equations are form-invariant under general LT's; and sums, scalar multiples, and scalar derivatives of four-tensors (defined by the corresponding operations on their components) are clearly four-tensors. One obvious way to construct a second-rank tensor is to form the "outer" (or "tensor") product $A_\mu B_\nu$ of two vectors **A** and **B** (sometimes written **A** ⊗ **B**). Certain combinations of the components of four-tensors, corresponding to the scalar product **A** · **B** of four-vectors, also play an important role in the theory.

Under linear coordinate transformations $x_\mu \to x'_\mu$, the Δx_μ transform like the dx_μ, namely

$$dx'_\mu = \sum \frac{\partial x'_\mu}{\partial x_\nu} dx_\nu. \tag{4.25}$$

Thus we recognize the coefficients $\alpha_{\mu\nu}$ in (**4.23**) as being in fact the partial derivatives $\partial x'_\mu/\partial x_\nu$. Tensors defined by use of these coefficients are called "contravariant." A dual class of tensors—"covariant" tensors—is defined with $\partial x_\mu/\partial x'_\nu$ for the α's in (**4.23**). A simple example of such a tensor is provided by the derivative of a scalar function of position φ: $\varphi_\mu = \partial\varphi/\partial x_\mu$.

We shall not pursue these matters here. What has been said will suffice for our applications in Sections 5.13 and 6.1 below. Tensor theory is taken up again in Section 8.1, though from a more general point of view. It should be noted that here, and right up to Section 8.1, we ignore a convention to which we shall adhere later, namely that only covariant tensor components are written with subscripts whereas contravariant components are written with superscripts.

4.5   The Three-Dimensional Minkowski Diagram

There is one particular in which the analogy between Euclidean three-space and spacetime (and between three-vectors and four-vectors) breaks down. Whereas the metric $\Delta x^2 + \Delta y^2 + \Delta z^2$ of three-space is always positive, the metric of spacetime can be positive, zero, or negative. (And so, correspondingly, can the squared norm of four-vectors.) From this it follows that spacetime is not isotropic, i.e., not all displacements away from a given event $\mathscr{P}$ are equivalent. Rather, they fall into three distinct classes according to the sign of their squared norm. Displacements with $\Delta s^2 > 0$ are separated from displacements with $\Delta s^2 < 0$ by those with $\Delta s^2 = 0$. These last, as we have noted before, correspond to events connectible with $\mathscr{P}$ by a light signal and evidently satisfy the equation

$$c^2 \Delta t^2 = \Delta x^2 + \Delta y^2 + \Delta z^2. \tag{4.26}$$

Under suppression of one spatial dimension, say the z dimension, this equation represents a right circular cone (see Figure 4.1) on a set of Cartesian axes Δx, Δy, $c\Delta t$ corresponding to any one particular observer. Nevertheless, this cone, called the *null cone* or *light cone* at $\mathscr{P}$, is absolute (i.e., independent of any particular observer), and so is the separation of events into three classes relative to $\mathscr{P}$. In full spacetime, the null cone at $\mathscr{P}$, regarded as a set of events, is the history of a spherical light front π which converges onto the event $\mathscr{P}$ and diverges again away from it. In Figure 4.1, π is reduced to a circle in the spatial xy plane collapsing to and expanding from the event $\mathscr{P}$, with the passage of time. Events whose displacement relative to $\mathscr{P}$ have positive squared norm lie inside the cone, i.e., where the time axis is, whereas those with negative squared norm lie outside. In full spacetime the former occur at points *within* the collapsing and re-expanding light sphere π. As we have seen in (4.2), Δs^2 between two events is positive if and only if the signal velocity

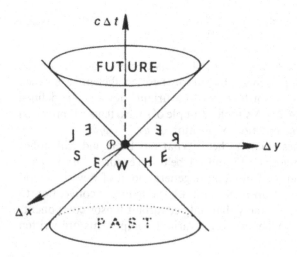

Figure 4.1

from one to the other is less than c. Thus all events that can be attended by a particle present at $\mathscr{P}$ lie inside the cone. Conversely, all vector displacements from $\mathscr{P}$ into the cone correspond to possible particle worldlines through $\mathscr{P}$, hence to possible inertial observers, and hence to possible t axes in spacetime; they are therefore called *timelike*. No particle can be present both at $\mathscr{P}$ and at an event outside the cone. Corresponding displacements are called *spacelike*. And for obvious reasons, displacements from $\mathscr{P}$ along the cone are called *lightlike* or *null*. According as their squared norm is positive, zero, or negative, one classifies *all* four-vectors as timelike, null, or spacelike. Thus, for example, the four-velocity U is timelike, while the four-acceleration A is spacelike. Note that a four-vector can be null without being zero, i.e., without having all its components zero.

We have already seen (in Section 2.7) that all observers agree on the time sequence of two events which are connectible in any frame by a signal with speed $\leq c$. Consequently, the events on and within the upper sheet of the null cone are regarded by *all* observers as occurring *after* $\mathscr{P}$, and they are said to constitute the *absolute future* of $\mathscr{P}$. Similarly, the events on and within the lower sheet of the cone constitute the *absolute past* of $\mathscr{P}$. Events in the remaining region form the *elsewhere*. Since no inertial observer can be present at $\mathscr{P}$ and one of the latter events, they can never be considered to happen at the same place as $\mathscr{P}$; hence the name. The time separation between $\mathscr{P}$ and an event in the "elsewhere," as measured by various inertial observers, varies all the way from minus to plus infinity [see Equation (2.13)].

The null cones at each event imprint a "grain" onto spacetime, which has no analog in isotropic Euclidean space, but is somewhat reminiscent of crystal structure. Light travels along the grain, and particle worldlines have to be within the null cone at each of their points.

The Minkowski diagram shown in Figure 2.3 is nothing but a spacetime diagram like that shown in Figure 4.1, with a specific origin and with yet another spatial dimension suppressed. Thus Figure 4.1 may be regarded as a three-dimensional Minkowski diagram and may be used, like Figure 2.3, to illustrate or solve SR problems. In fact, *it is a map of spacetime onto Euclidean space*. But it must be noted that the map distorts lengths and angles. Vectors that appear equally long in the diagram do not necessarily have the same "Minkowski length" $|A|$. For example, by rotating the hyperbolae of Figure 2.3 about the t axis we obtain hyperboloids of revolution in Figure 4.1 which are loci of displacements all having positive or negative unit square interval from $\mathscr{P}$. Also, vectors that appear orthogonal in the diagram are not necessarily "Minkowski-orthogonal" in the sense $A \cdot B = 0$, as exemplified by the axes of ξ and η in Figure 2.3. Conversely, the axes of x' and t' *are* Minkowski-orthogonal but do not appear orthogonal in that diagram. On the other hand, it is clear that vector sums in the diagram correspond to "Minkowski sums" $A + B$, that parallel vectors in the diagram correspond to "Minkowski-parallel" vectors (in the sense $A = kB$, k real), and that the "Minkowski ratio" k of such vectors is also the apparent ratio in the diagram. Consequently such theorems as the following can easily be read off from the diagram:

The sum of any number of future-pointing timelike or null vectors is itself timelike or null, and it is null only if all the summands are null and parallel.

Any vector $\mathbf{V} = (\mathbf{v}, V_4)$, which is timelike $(v^2 < c^2 V_4^2)$ and future-pointing $(V_4 > 0)$ can be expressed as a positive multiple of a four-velocity $\mathbf{U}$: $\mathbf{V} = (V/c)\mathbf{U}$. This is really obvious, since any vector with norm c can be interpreted as a four-velocity. Or consider the following reduction:

$$\mathbf{V} = V_4\left(\frac{\mathbf{v}}{V_4}, 1\right) = \frac{V}{c}\left(\frac{cV_4}{V}\right)\left(\frac{\mathbf{v}}{V_4}, 1\right) = \frac{V}{c}\gamma(u)(\mathbf{u}, 1) = \frac{V}{c}\mathbf{U}, \quad (4.27)$$

all components being relative to some specific frame S. Now consider a second frame S_0 which itself has four-velocity $\mathbf{U}$. In it, $\mathbf{U} = (0, 0, 0, 1)$ and so $\mathbf{V} = (0, 0, 0, Vc^{-1})$. The velocity of S_0 relative to S is clearly $\mathbf{u} = \mathbf{v}/V_4$. In effect, by choosing the time axis of S_0 parallel to $\mathbf{V}$, we have absorbed its spatial components. (If $\mathbf{V}$ is past-pointing, we can apply the same argument to $-\mathbf{V}$.) For *any* four-vector $\mathbf{W}$ in S we can absorb two of the spatial components (say W_2, W_3) by simply rotating the spatial axes of S. Then, as Figure 2.3 shows, unless $\mathbf{W}$ is null, we can choose the t' or x' axis along $\mathbf{W}$ and so absorb one more component: either W_1 (as in the preceding case), or W_4 if $\mathbf{W}$ is spacelike. These simplifications are often very useful.

4.6   Wave Motion

As a good example of the power of four-vectors, we shall here investigate the transformation properties of plane waves. Consider a series of plane "disturbances" or "wavecrests," of an unspecified nature, a wavelength λ apart, and progressing in a unit direction $\mathbf{n} = (l, m, n)$ at speed w relative to a frame S. A plane in S with normal $\mathbf{n}$ and at distance p from a point $P_0(x_0, y_0, z_0)$ satisfies the equation

$$l(x - x_0) + m(y - y_0) + n(z - z_0) = p.$$

If this plane propagates with speed w, the equation is

$$l\Delta x + m\Delta y + n\Delta z = w\Delta t,$$

where Δt measures time from when the plane crosses P_0, and $\Delta x = x - x_0$, etc. A whole set of such traveling planes, at distances λ apart, has the same equation with $N\lambda$ added to the right-hand side, N being any integer. And this can be written (after absorbing a minus sign into N) as

$$\mathbf{L} \cdot \Delta \mathbf{R} = N, \quad (N \text{ any integer}), \quad (4.28)$$

where

$$\mathbf{L} = \frac{1}{\lambda}\left(\mathbf{n}, \frac{w}{c^2}\right) = \nu\left(\frac{\mathbf{n}}{w}, \frac{1}{c^2}\right), \quad (4.29)$$

$\nu = w/\lambda$ being the frequency. Conversely, any equation of form (4.28) will be recognized in S as representing a moving set of equidistant planes, with λ, $\mathbf{n}$, and w determined by (4.29). We have written $\mathbf{L}$ *like* a four-vector, though we do not yet know that it is one. Suppose, however, we transform (4.28) directly to an arbitrary second frame S': since the components of $\Delta\mathbf{R}$ undergo a general LT, i.e., a linear homogeneous transformation, we shall get on the left a linear homogeneous polynomial in $\Delta x'$, $\Delta y'$, $\Delta z'$, $\Delta t'$, and on the right N as before:

$$\mathbf{L}' \cdot \Delta\mathbf{R}' = N,$$

where $\Delta\mathbf{R}'$ here stands for $(\Delta x', \ldots, \Delta t')$ and $\mathbf{L}'$ for the coefficients of these components divided by $-1, -1, -1, c^2$, respectively. Thus in S' too one has a moving set of equidistant planes of disturbance. Also, by the invariance of the scalar product, it certainly follows from (4.28) that

$$\mathbf{T}' \cdot \Delta\mathbf{R}' = N,$$

where $\mathbf{T}'$ stands for the vector transforms into S' of the components of $\mathbf{L}$. Each value of N determines a definite one in the set of traveling planes; hence, forming the difference of the last two equations, we see that

$$(\mathbf{L}' - \mathbf{T}') \cdot \Delta\mathbf{R}' = 0 \tag{4.30}$$

for any event on any of the planes. We can certainly find four linearly independent vectors $\Delta\mathbf{R}'$ corresponding to events on the planes, and thus satisfying (4.30), e.g., vectors of the form $(a, 0, 0, 0)$, $(0, b, 0, 0)$, $(0, 0, c, 0)$, $(0, 0, 0, d)$. But this implies that $(\mathbf{L}' - \mathbf{T}') = 0$, i.e., $\mathbf{L}' = \mathbf{T}'$, or, in words, that the components of $\mathbf{L}$ *do* transform as four-vector components. Hence $\mathbf{L}$ *is* a four-vector. It is called the *frequency four-vector*.

Now consider the usual two frames S and S' in standard configuration, and in S a train of plane waves with frequency v and velocity w in a direction $\mathbf{n} = -(\cos\alpha, \sin\alpha, 0)$. The components of the frequency vector in S are given by

$$\mathbf{L} = \left(\frac{-v\cos\alpha}{w}, \frac{-v\sin\alpha}{w}, 0, \frac{v}{c^2} \right). \tag{4.31}$$

Transforming these components by the scheme (2.7), we find for the components in S':

$$\frac{v'\cos\alpha'}{w'} = \frac{\gamma v(\cos\alpha + vw/c^2)}{w}, \tag{4.32}$$

$$\frac{v'\sin\alpha'}{w'} = \frac{v\sin\alpha}{w}, \tag{4.33}$$

$$v' = v\gamma\left(1 + \frac{v}{w}\cos\alpha \right). \tag{4.34}$$

The last equation expresses the Doppler effect for waves of all velocities, and, in particular, for light waves ($w = c$). In the latter case, it is seen to be equivalent to our previous Formula (3.5).

From (**4.32**) and (**4.33**), we obtain the general wave aberration formula

$$\tan \alpha' = \frac{\sin \alpha}{\gamma(\cos \alpha + vw/c^2)}. \tag{4.35}$$

In the particular case when $w = c$, this is seen to be equivalent to our previous Formulae (**3.7**) and (**3.8**)—in fact, it corresponds to their quotient.

Finally, to get the transformation of w, we *could* eliminate the irrelevant quantities from Equations (**4.32**)–(**4.34**), but it is simpler to make use of the invariance of $|\mathbf{L}|^2$. Writing this out in S and S', we obtain

$$v^2\left(1 - \frac{c^2}{w^2}\right) = v'^2\left(1 - \frac{c^2}{w'^2}\right), \tag{4.36}$$

whence, by use of (**4.34**),

$$1 - \frac{c^2}{w'^2} = \frac{(1 - c^2/w^2)(1 - v^2/c^2)}{(1 + v\cos\alpha/w)^2}. \tag{4.37}$$

This formula is *not* analogous to (**2.27**). The reason is that a particle riding the crest of a wave in the direction of the wave normal in one frame does not, in general, do so in another frame: there it rides the crest of the wave also, but not in the normal direction. The one exception is when $w = c$.

CHAPTER 5

Relativistic Particle Mechanics

5.1 Domain of Sufficient Validity of Newton's Laws

Newton's mechanics is not Lorentz-invariant. According to the program of special relativity, therefore, it was necessary to construct a new mechanics—even before any serious empirical deficiencies of the old mechanics had become apparent. The new mechanics is known as "relativistic" mechanics. This is not really a good name,[1] since, as we have seen, Newton's mechanics, too, is relativistic, but under the "wrong" (Galilean) transformation group. Newton's theory has excellently served astronomy (e.g., in foretelling eclipses and orbital motions in general), it has been used as the basic theory in the incredibly delicate operations of sending probes to the Moon and some of the planets, and it has proved itself reliable in countless terrestrial applications. Thus it cannot be *entirely* wrong. Before the twentieth century, in fact, only a single case of irreducible failure was known, namely the excessive advance of the perihelion of the planet Mercury, by about 43 seconds (!) of arc per century. Since the advent of modern particle accelerators, however, vast discrepancies with Newton's laws have been uncovered, whereas the new mechanics consistently gave correct descriptions. (Of course, Newton's mechanics has undergone *two* "corrections," one due to relativity and one due to quantum theory. We are here concerned exclusively with the former.) The new mechanics practically overlaps with the old in a large

---

[1] A better name is "Lorentz-invariant mechanics."

domain of applications (dealing with motions that are slow compared with the speed of light) and, in fact, it delineates the domain of sufficient validity of the old mechanics as a function of the desired accuracy. Roughly speaking, the old mechanics is in error to the extent that the γ-factors of the various motions involved exceed unity. In laboratory collisions of elementary particles γ-factors of the order of 10^4 are not uncommon, and γ-factors as high as 10^{11} have been observed in cosmic-ray protons. Applied to such situations, Newton's mechanics is not just *slightly* but *totally* wrong. Yet, within its acceptable slow-motion domain, Newton's theory will undoubtedly continue to be used for reasons of conceptual and technical convenience. And as a logical construct it will remain as perfect and inviolate as Euclid's geometry. Only as a model of nature it must not be stretched unduly.

5.2   Why Gravity Does Not Fit Naturally into Special Relativity

In this chapter we shall restrict ourselves to the mechanics of mass points *in the absence of gravity*. We shall construct a new collision mechanics within special relativity. Several attempts have been made to construct also new theories of gravitation within special relativity, but this can only be done at the heavy cost of abandoning either the equivalence principle or a "natural" interpretation of SR.

It is sometimes argued that, simply because freely falling cabins on opposite sides of the earth (for example) accelerate relative to each other, it is impossible for SR to hold in them, as required by the EP, *and also* in an extended frame at rest relative to the earth, as would be required by a special-relativistic theory of gravity. But why does this argument not damage *Newton's* theory, in which the (weak) EP also holds? In classical kinematics the Galilean transformation not only relates any two inertial frames, it relates *any* two rigid frames in uniform relative motion. So it relates the members of each freely falling local set and also the extended inertial frames; and no conflict arises from the fact that Newtonian gravity-free mechanics holds in the local freely falling frames. A similar duality is possible also in SR, and so *that* argument is inconclusive.

The first requirement of a special-relativistic theory of gravity is the existence of an inertial frame S large enough to contain systems of gravitational interest, e.g., a star together with its planets. According to SR, one would assume that the "real" geometry of S is Euclidean, though one might allow for the possibility that actual rulers will not conform to this. One would also assume the existence of a coordinate time with all the properties of SR time, and we already *know* that actual clocks will not conform to this: they will go slow near large masses, by the EP (cf. Section 1.21). Suppose that an actual clock fixed in S has a frequency v which is related to the coordinate frequency v_0 by $v = kv_0 (k \neq 1)$ at some event $\mathscr{P}$. Suppose, at first, that actual rulers agree with coordinate distances in S. Then the speed of light, being c as measured by actual clocks and rulers (by the EP and the clock and length hypotheses), must be kc in coordinate measure. Now let S' be a second inertial frame, moving with coordinate velocity v relative to S. According to SR, the coordinates of these two frames are related by a Lorentz transformation. Consider a light signal at $\mathscr{P}$, traveling in the direction of v in S. Since its coordinate speed is kc in S, and c is the only invariant coordinate speed [cf. (2.25)(i)], its coordinate speed in S' will *not* be kc but rather $k'c$, where $k' \neq k$. Also, according to the LT, the coordinate speed of S relative to S' is $-v$. Hence the velocities which observers at rest in these two frames assign to each other, as measured by actual clocks and rulers, are v/k and $-v/k'$, respectively, i.e., *unequal*. But this is absurd. (See third paragraph of Section 2.6.)

The only remedy is to assume that, in a gravitational field, rulers shrink by the same factor k by which clocks go slow. Then local coordinate speeds will agree with the actually measured speeds of all effects, in particular of light. It remains to specify field equations, which tell how the sources determine the field (and thus the factor k), and laws of motion, which tell how test particles move in the field. Light can be assumed to propagate rectilinearly in the coordinate space S. This still allows it to bend—as required by the EP—in the space determined by actual rulers.[2] *That* space will now be curved (cf. Section 7.2 below). Nordström's special-relativistic theory of gravity followed essentially this course.

However, there is something *unnatural* about regarding such theories as "special-relativistic." The underlying coordinate space S has no more physical significance than has a map. A more natural interpretation of such theories is that they (like GR) are curved-space theories. But then the restriction that there should exist a flat Minkowskian coordinate "map" with certain properties is seen to be rather artificial, and it is discarded in GR.

5.3 Relativistic Inertial Mass

Though there are many possible approaches to the new mechanics of "point-collisions" and particles in external fields, the result is always the same. If Newton's well-tested theory is to hold in the "slow-motion limit," and unnecessary complications are to be avoided, then only *one* Lorentz-invariant mechanics appears to be possible. Moreover, it is persuasively elegant, and far simpler than any conceivable alternative.

Here we begin by assuming tentatively that Newton's law of momentum

---

[2] Technically speaking, theories of this type have a "conformally flat" actual spacetime $\bar{S}$, because of the equal slowing of clocks and shrinking of rods. Null geodesics in $\bar{S}$ therefore correspond to null geodesics in S. The null-geodesic propagation of light is by far the most natural law to assume, since it satisfies the EP *and* the usual special-relativistic law in S. It is now easy to see why such theories lead to *zero* light deflection around a finite mass, between infinity and infinity. For at infinity S and $\bar{S}$ coincide, and the path is straight in S.

conservation can be salvaged in a suitably modified form, the momentum $\mathbf{p}$ of a particle being defined as inertial mass times velocity:

$$\mathbf{p} = m\mathbf{v}. \tag{5.1}$$

Inertial mass, which we need not expect to be invariant, is in turn defined by this law: ratios of m can be determined by subjecting, say, two particles to a collision and applying the law of momentum conservation to the observed velocities.

Consider now a very slight glancing collision of two equal spherical particles A and A', at rest, respectively, in the two usual inertial frames S and S', before collision. By symmetry, these spheres after collision will have equal and opposite transverse velocity components (at right angles to the x axes) in their respective frames, say of magnitude u. By (2.26)(ii), the transverse velocity of A' relative to S will therefore be $u/\gamma(v)(1 + u_1'v/c^2)$, where u_1' is the postcollision x' velocity of A' in S'. Transverse momentum conservation in S then implies

$$Mu = \frac{M'u}{\gamma(v)(1 + u_1'v/c^2)}, \tag{5.2}$$

where M and M' are the postcollision masses of A and A', respectively, *as measured in S*. Now let $u, u_1' \rightarrow 0$, i.e., consider ever more glancing collisions. Then $M \rightarrow m_0$, the "rest mass" of one particle, and $M' \rightarrow m$, the mass at velocity v of an identical particle. Proceeding to the limit in (5.2)—after dividing by u—we get

$$m = \gamma(v)m_0. \tag{5.3}$$

This conclusion is inevitable *if* momentum is conserved. However, it remains to be shown that *with* this definition of mass, momentum conservation in all collisions is indeed a Lorentz-invariant requirement: this will be done in Section 5.4. Note that, according to (5.3), the inertial mass of a particle increases with v from a minimum m_0 at $v = 0$ to infinity as $v \rightarrow c$. We should not be too surprised at this, since there must be *some* process in nature to prevent particles from being accelerated beyond the speed of light.

For accelerating particles, incidentally, we shall have to adopt a hypothesis, the *mass hypothesis* (analogous to our previous clock and length hypotheses), namely that the mass depends only on the instantaneous velocity according to (5.3) and *not* on the acceleration.

Equation (5.3) contains the germ of the famous mass–energy equivalence discovered by Einstein. For, expanding (5.3) by the binomial theorem, we find, to second order,

$$m = m_0\left(1 - \frac{v^2}{c^2}\right)^{-1/2} \approx m_0 + \frac{\tfrac{1}{2}m_0 v^2}{c^2}. \tag{5.4}$$

This shows that the inertial mass of a moving particle exceeds its rest mass by $1/c^2$ times its kinetic energy (assuming the approximate validity of the Newtonian expression for the latter). Consequently, energy appears to *contribute* to the mass in a way that would be consistent with Einstein's postulate of a general mass–energy equivalence according to the formula

$$E = mc^2. \tag{5.5}$$

Note that *if* this formula is correct, then the conservation of energy is equivalent to the conservation of inertial mass. It will, of course, be necessary to show (i) that the conservation of inertial mass is a Lorentz-invariant requirement and (ii) that it is compatible with the proposed law of momentum conservation.

5.4 Four-Vector Formulation of Relativistic Mechanics

The important questions of the Lorentz-invariance and consistency of the laws suggested in the last section are best discussed by making use of the powerful four-vector calculus. Our method consists essentially in *guessing* the relativistic laws in four-vector form. However, as in most such cases, we have at the back of our minds some concrete results or conjectures. In the present case, we have the tentative results of the preceding section.

It will be useful to begin by establishing an important lemma of vector theory: If a four-vector **V** has a particular one of its four components zero in *all* inertial frames, then the whole vector must be zero. For suppose first that one of the spatial components of **V** is always zero, and, without loss of generality, let it be the first. Then if there is a frame in which the y or z component is *nonzero*, apply a rotation to interchange the relevant space axis with the x axis, whereupon the x-component will become nonzero, contrary to hypothesis; if, on the other hand, there exists a frame in which the temporal component of **V** is nonzero, a LT (**2.7**) will produce a nonzero x-component, again contrary to hypothesis. Next, suppose **V** always has its temporal component equal to zero. Then if there exists a frame in which a spatial component is nonzero, without loss of generality let it be the first, and apply a LT: now a nonzero fourth component will appear, contrary to hypothesis. This establishes our result. We shall call it the *zero-component lemma*.

Now let us suppose that associated with each particle there is a scalar, characteristic of its internal state: the *rest mass* (or *proper mass*) m_0. (By the laws assumed for it, this will turn out to be identical with the Newtonian mass which the particle manifests in slow-motion experiments.) We then define for each particle, analogously to Newton's momentum $m_0 \mathbf{u}$, a *four-momentum*

$$\mathbf{P} = m_0 \mathbf{U} = m_0 \gamma(u)(\mathbf{u}, 1) = (\mathbf{p}, m), \tag{5.6}$$

where **U** is the four-velocity (**4.10**) of the particle, and **p** and m are defined by this equation, i.e.,

$$m = \gamma(u)m_0, \tag{5.7}$$

$$\mathbf{p} = m\mathbf{u}. \tag{5.8}$$

We call these quantities the *relativistic (inertial) mass* and the *relativistic momentum*, respectively. For the moment we shall assume $u < c$ and $m_0 \neq 0$. But, as will be shown in detail in Section 5.12, it is possible to include in the theory also particles of zero rest mass moving at speed c—e.g., photons— which nevertheless have finite m and $\mathbf{p}$. Most results about ordinary particles can be extended to zero rest mass particles also.

In any collision between two particles having four-momenta $\mathbf{P}_1$, $\mathbf{P}_2$ before collision and $\mathbf{P}_3$, $\mathbf{P}_4$ afterwards, the *conservation of four-momentum*,

$$\mathbf{P}_1 + \mathbf{P}_2 = \mathbf{P}_3 + \mathbf{P}_4, \tag{5.9}$$

would clearly be a Lorentz-invariant requirement, equivalent, because of (5.6), to the separate conservation laws [suffixes being used as in (5.9)]

$$\mathbf{p}_1 + \mathbf{p}_2 = \mathbf{p}_3 + \mathbf{p}_4, \tag{5.10}$$

$$m_1 + m_2 = m_3 + m_4. \tag{5.11}$$

(These equations clearly generalize to collisions in which more than two particles participate, or in which more, or fewer, or other particles emerge than went in.) Moreover, the truth of (5.10) in *all* inertial frames would imply the truth of (5.11) in all frames, and vice versa; this follows at once from our "zero component lemma", as applied to the vector $\mathbf{P}_1 + \mathbf{P}_2 - \mathbf{P}_3 - \mathbf{P}_4$

In the last section we proved that *if* a momentum of the form $m\mathbf{u}$ is conserved, then the mass m must be of the form (5.7), i.e., the momentum must be the "relativistic" momentum as defined in (5.8). We have now shown that the conservation of relativistic momentum *and* of relativistic mass together give a Lorentz-invariant law, (5.9), and that we cannot have the one without the other. Finally, in the case of slow motion, m reduces to m_0 and $\mathbf{p}$ to $m_0\mathbf{u}$. Thus if, for a certain collision, there exist frames in which all the motions are slow, the proposed relativistic momentum conservation law (5.10) reduces in such frames to the well-established Newtonian analog, and so does (5.11). (That identifies m_0 as the Newtonian mass.) In other situations there was no precedent to follow at the time when the new mechanics was first constructed, yet the formalism strongly suggested general adoption of (5.10) as a *postulate*, with its *corollary* (5.11) (or vice versa), i.e., equivalently, the adoption of (5.9). This has been thoroughly borne out by experiment in the sequel. It is now the basis of relativistic mechanics.

From (5.6) we obtain two alternative expressions for the squared norm of $\mathbf{P}$,

$$P^2 = c^2 m_0^2 = c^2 m^2 - p^2, \tag{5.12}$$

from which we deduce the important formula

$$p^2 = c^2(m^2 - m_0^2). \tag{5.13}$$

Furthermore, we have the following expressions for the scalar product of the four-momenta $\mathbf{P}_1$, $\mathbf{P}_2$ of two particles moving at relative speed v,

$$\mathbf{P}_1 \cdot \mathbf{P}_2 = c^2 m_{01} m_2 = c^2 m_1 m_{02} = c^2 m_{01} m_{02} \gamma(v), \qquad (5.14)$$

where, typically, m_{01} is the rest mass of the first particle and m_2 is the mass of the second in the rest frame of the first. To obtain these, we evaluate $\mathbf{P}_1 \cdot \mathbf{P}_2$ in the rest frame of either particle, using (5.6). [As reference to (5.44) below shows, (5.14)(i) remains valid even if the second "particle" is a photon; when both are photons, (5.14) becomes indeterminate and (5.45) below takes its place.]

Lastly, consider an *elastic* collision between two particles, i.e., a collision in which the individual rest masses are preserved. Writing $\mathbf{P}$, $\mathbf{Q}$ for the pre-collision momenta and $\mathbf{P}'$, $\mathbf{Q}'$ for the postcollision momenta, we have $\mathbf{P} + \mathbf{Q} = \mathbf{P}' + \mathbf{Q}'$, which, upon squaring, gives $\mathbf{P}^2 + \mathbf{Q}^2 + 2\mathbf{P} \cdot \mathbf{Q} = \mathbf{P}'^2 + \mathbf{Q}'^2 + 2\mathbf{P}' \cdot \mathbf{Q}'$. But, by hypothesis, $\mathbf{P}^2 = \mathbf{P}'^2$ and $\mathbf{Q}^2 = \mathbf{Q}'^2$. It follows that

$$\mathbf{P} \cdot \mathbf{Q} = \mathbf{P}' \cdot \mathbf{Q}'. \qquad (5.15)$$

This useful result we shall call the *elastic collision lemma*. (It remains valid even if one or both of the "particles" are photons: a rebounding photon cannot help but preserve its rest mass which is always zero.) By (5.14)(iii), Equation (5.15) is seen to be equivalent to the statement that the relative speed of the particles is the same before and after the elastic collision. In this guise the lemma is also true in Newton's theory.

5.5 A Note On Galilean Four-Vectors

In analogy to Minkowskian four-vectors, one can define "Galilean" four-vectors as those that transform like $(\Delta x, \Delta y, \Delta z, \Delta t)$ under a general Galilean transformation [i.e., (1.1) compounded with spatial rotations and spatial and temporal translations]. The vanishing of the spatial components of a Galilean four-vector in all inertial frames implies the vanishing also of its temporal component, since a Galilean transformation would otherwise produce a nonzero spatial component in some frame. Evidently, $m_0(\mathbf{u}, 1)$ constitutes a Galilean four-vector (multiply the prototype by the scalar $m_0/\Delta t$ and go to the limit), and thus, as in the relativistic case, the conservation of Newtonian momentum $m_0 \mathbf{u}$ under Galilean transformations implies the conservation of Newtonian mass m_0. But *not* vice versa!

5.6 Equivalence of Mass and Energy

Newtonian mass was often vaguely thought of as "quantity of matter," and so mass conservation was identified with the conservation of matter. But the conservation of relativistic mass, (5.11), expresses quite a different phenomenon. Since relativistic mass is velocity-dependent, its conservation is much more analogous to the classical conservation of kinetic energy in perfectly elastic collisions, except that it is postulated to hold in *all* collisions. Thus, independently of the considerations of Section 5.3, one can be led from a purely formal viewpoint to regard (**5.11**) as expressing the conservation of total *energy* and to postulate that the total energy of a particle is a multiple of its inertial mass. That this multiple must be c^2 is then clear from (**5.4**).

In fact, all through relativistic physics there occur indications that mass and energy are equivalent according to the formula

$$E = mc^2. \tag{5.16}$$

We have seen in (**5.4**) that kinetic energy *contributes* to the total mass of a particle according to this formula (at least, to lowest order). Hence *all* energy has mass, since all energy is exchangeable with kinetic energy. Suppose, for example, that two particles at room temperature collide inelastically and form a doublet at rest, which can give up ΔE units of heat energy in cooling to room temperature. By energy conservation, ΔE equals the original kinetic energy of the particles; but then, by mass conservation, the doublet just after impact had a mass $\Delta E/c^2$ units in excess of its room-temperature rest mass. Thus heat energy also contributes to mass according to (**5.16**).

We shall see in Section 5.10 that the work done on a particle by a relativistic force contributes to its mass exactly according to Formula (**5.16**). Further indications appear in Maxwell's theory: for example, Kelvin's energy density $(e^2 + h^2)/8\pi$ and Thomson's momentum density $\mathbf{e} \times \mathbf{h}/4\pi c$ reduce to $e^2/4\pi$ and $e^2/4\pi c$ in the case of radiation (when $\mathbf{e}$ and $\mathbf{h}$ are equal and orthogonal) so that (energy/c^2) × (velocity) = momentum. Even general relativity corroborates this equivalence (see Section 8.7). The ultimate reason for all this is that Lorentz-invariance severely restricts all possible conservation laws and that essentially there *can* be only one nondirectional velocity-dependent conserved quantity (as energy must be), namely m (see RSR, Section 27), and one directional one, namely $m\mathbf{u}$ as we have seen; and both or neither are conserved.

The theoretical indications available to Einstein when he proposed $E = mc^2$ suggested that energy *contributes* to mass (as in our example of the kinetic energy). To equate *all* mass with energy required an act of aesthetic faith, very characteristic of Einstein. For one of the prime attributes of energy is its transmutability, and implicit in (**5.16**) is the assertion that *all* the mass of a particle can be transmuted into available energy. It would be perfectly consistent with special relativity if the elementary particles were indestructible. Energy would then merely *contribute* to mass and the available energy of a compound particle of mass m would be $c^2(m - q)$, where q is the total rest mass of its constituent elementary particles. In free systems of particles, each of the sums $\sum c^2 m$, $\sum c^2 q$, $\sum c^2(m - q)$ would then be separately conserved, but only the last could justly be called energy. The bold hypothesis (**5.16**), however, was amply confirmed by later experience, especially by the observation of "pair annihilation" in which an elementary particle and its antiparticle annihilate each other and set free a corresponding amount of radiative energy; also by the spontaneous decay of neutral mesons into photon pairs; and by collisions in which different elementary particles emerge than went in, with different total rest mass.

We shall distinguish between the *kinetic* energy T, which a particle possesses by virtue of its motion,

$$T = c^2(m - m_0), \tag{5.17}$$

and its *internal* energy $c^2 m_0$. For an "ordinary" particle, this internal energy is vast: in each gram of mass there are 9.10^{20} ergs of energy, roughly the energy of the Hiroshima bomb (20 kilotons). A very small part of this energy resides in the thermal motions of the molecules constituting the particle, and can be given up as heat; a part resides in the intermolecular and inter-atomic cohesion forces, and in the latter form it can sometimes be given up in chemical explosions; another part may reside in excited atoms and escape in the form of radiation; much more resides in nuclear bonds and can also sometimes be set free, as in the atomic bomb. But by far the largest part of the energy (about 99 %) resides simply in the mass of the ultimate particles, and cannot be further explained. Nevertheless, it too can be liberated under suitable conditions, e.g., when matter and antimatter annihilate each other. Generally, therefore, *rest* mass will not be conserved.

One kind of energy that does *not* contribute to mass is *potential* energy of position. In classical mechanics, a particle moving in an electromagnetic (or gravitational) field is often said to possess potential energy, so that the sum of its kinetic and potential energies remains constant. This is a useful "book-keeping" device, but energy conservation can also be satisfied by debiting the *field* with an energy loss equal to the kinetic energy gained by the particle. In relativity there are good reasons for adopting the second alternative, though the first can be used as an occasional shortcut: the "real" location of any part of the energy is no longer a mere convention, since energy—as mass—gravitates.

According to Einstein's hypothesis, *every* form of energy has a mass equivalent: (i) If all mass exerts and suffers gravity, we would expect even (the energy of) an electromagnetic field to exert a gravitational attraction, and, conversely, light to bend under gravity (this we have already anticipated by a different line of reasoning). (ii) We shall expect a gravitational field *itself* to gravitate. (iii) The radiation which the sun pours into space is equivalent to more than four million tons of mass per second! Radiation, having mass and velocity, must also have momentum; accordingly, the radiation from the sun is a (small) contributing factor in the observed deflection of the tails of comets away from the sun. (The major factor is "solar wind.") (iv) An electric motor (with battery) at one end of a raft, driving a heavy flywheel at the other end by means of a belt, transfers energy and thus mass to the flywheel; in accordance with the law of momentum conservation, the raft must therefore accelerate in the opposite direction. (v) Stretched or compressed objects have (minutely) more mass by virtue of the stored elastic energy.

(vi) The total mass of the separate components of a stable atomic nucleus always exceeds the mass of the nucleus itself, since energy (i.e., mass) would have to be supplied in order to decompose the nucleus against the nuclear binding forces. This is the reason for the well-known "mass defect." Nevertheless, if a nucleus is split into two new nuclei, these parts may have greater *or* lesser mass than the whole. (With the lighter atoms, the parts usually exceed the whole, whereas with the heavier atoms the whole can exceed the parts.) In the first case, energy can be released by "fission," in the second, by "fusion."

5.7   The Center of Momentum Frame

Once again, it will be well to preface our discussion with a lemma from vector theory. We have seen in Section 4.5 that, given any timelike four-vector V, there exists a frame S_0 in which its components reduce to $(0, 0, 0, V_4)$. Moreover, the *sign* of the fourth component is invariant, since V either "points" into the absolute future ($V_4 > 0$) or into the absolute past ($V_4 < 0$). If W is a second timelike vector "isochronous" with V (i.e., $V_4 W_4 > 0$), then in S_0 we have

$$|V + W|^2 = c^2(V_4 + W_4)^2 - W_1^2 - W_2^2 - W_3^2$$
$$= V^2 + W^2 + 2c^2 V_4 W_4 > 0. \tag{5.18}$$

Hence the sum of two isochronous timelike vectors is another timelike vector, and clearly isochronous with the summands. By iteration we see that the same is true of any number of timelike isochronous vectors. That is our lemma.

Consider now in a given frame S a finite system of particles, subject to no forces except mutual collisions. We define its total mass $\bar{m}$, total momentum $\bar{p}$, and total four-momentum $\bar{P}$ as the sums of the respective quantities of the individual particles:

$$\bar{m} = \sum m, \quad \bar{p} = \sum p, \quad \bar{P} = \sum P = \sum (p, m) = (\bar{p}, \bar{m}) \tag{5.19}$$

[cf. (5.6)]. Because of the conservation laws, each of the barred quantities remains constant in time.

The quantity $\bar{P}$, being a sum of four-vectors, seems necessarily to be a four-vector itself. But, in fact, there is a subtlety involved here. If all observers agreed on which P's make up the sum $\sum P$, then $\sum P$ would *clearly* be a vector. But each observer forms this sum at one instant in *his* frame, which may result in different P's making up the $\sum P$ of different observers. A spacetime diagram such as Figure 4.1 is useful in proving that $\sum P$ is nevertheless a vector, and the suppression of the z dimension makes the argument a little more transparent without affecting its validity. A simultaneity in S corresponds to a "horizontal" plane π in the diagram and a simultaneity in a second frame S' corresponds to a "tilted" plane π'. In S, $\sum P$ is summed over planes like π, and in S' over planes like π'. However, we assert that in S' the same $\sum \mathbf{P}$ would result whether summed over π' or π. Imagine a continuous motion of π' into π. As π' is tilted, each individual $\mathbf{P}$ located on it remains constant (the particles move uniformly between collisions) except when π' sweeps over a collision; but then the sub-sum of $\sum \mathbf{P}$ which enters the collision remains constant, by momentum conservation. Thus, without affecting the value, each observer *could* sum his $\mathbf{P}$'s over the same plane π, and thus $\bar{\mathbf{P}}$ is indeed a four-vector.

Now, by (5.6), each particle's $\mathbf{P}$ is timelike (since $\mathbf{U}$ is) and future-pointing (since $m > 0$). Consequently, by the lemma introducing this section, $\bar{\mathbf{P}}$ is also timelike and future-pointing. By the final paragraph of Section 4.5, we can therefore find an inertial frame S_{CM} moving relative to S with velocity

$$\mathbf{u}_{CM} = \frac{\bar{\mathbf{p}}}{\bar{m}}, \qquad (5.20)$$

in which $\bar{\mathbf{P}}$ has no spatial components, i.e., in which $\bar{\mathbf{p}} = 0$. For this reason S_{CM} is called the CM or "center of momentum" frame. It is also the frame in which the center of mass of the system is at rest, though this needs some discussion. In relativistic mechanics, in contrast to Newtonian mechanics, the center of mass of a system varies from frame to frame. An example will make this clear: suppose two identical particles move with opposite velocities along parallel lines l_1 and l_2 in some frame S, and suppose l is the line midway between these. In S, the center of mass lies on l. But in the rest frame of either particle the other particle is the more massive, and thus the center of mass lies *beyond* l. The center of mass of a system is therefore frame-dependent. But it can be shown quite easily that *all* the centers of mass are at rest in the CM frame (see Exercise 5.7). It is useful to define *the* center of mass of the system as its center of mass in the CM frame.

Let us write $\mathbf{U}_{CM}$ for the four-velocity $\gamma(u_{CM})(\mathbf{u}_{CM}, 1)$ of S_{CM}. Then, from (5.19) and (5.20),

$$\bar{\mathbf{P}} = (\bar{\mathbf{p}}, \bar{m}) = \bar{m}(\mathbf{u}_{CM}, 1) = \bar{m}\gamma^{-1}(u_{CM})\mathbf{U}_{CM}. \qquad (5.21)$$

On taking norms of this equation, we see that $\bar{m}/\gamma(u_{CM})$ must be invariant, and thus equal to its value in S_{CM}:

$$\bar{m}/\gamma(u_{CM}) = \bar{m}_{CM}, \qquad (5.22)$$

where $\bar{m}_{CM}$ is the total mass in S_{CM}. Thus finally,

$$\bar{\mathbf{P}} = \bar{m}_{CM}\mathbf{U}_{CM}. \qquad (5.23)$$

This equation shows that $\bar{m}_{CM}$ and $\mathbf{U}_{CM}$ are for the system what m_0 and $\mathbf{U}$ are for a single particle. They are the quantities that would be recognized as the rest mass and four-velocity of the system if its composite nature were *not* recognized (as in the case of an "ordinary" particle—which is made up of possibly moving molecules). Since the kinetic energy in S_{CM} is given by $T_{CM} = c^2(\bar{m}_{CM} - \bar{m}_0)$, where $\bar{m}_0 = \sum m_0$, we see that the effective rest mass of the system exceeds the rest masses of its parts by the mass equivalent of the kinetic energy of the parts in the CM frame. This is precisely what one would expect.

5.8 Relativistic Billiards

As a first example on the new mechanics, let us consider the relativistic analog of the collision treated classically in Section 2.3. The situation in the CM frame is the same, and for the same reasons. Only the transformation to the rest frame of one of the particles (the "lab" frame) is different. We identify the lab and the CM frames with the usual frames S and S', respectively. Without loss of generality, we can write the postcollision velocities in S' as $\pm(u_1', u_2', 0)$, with

$$u_1'^2 + u_2'^2 = v^2. \tag{5.24}$$

By (2.26), the corresponding velocities in S are

$$\left(\frac{\pm u_1' + v}{1 \pm u_1' v/c^2}, \frac{\pm u_2'}{\gamma(1 \pm u_1' v/c^2)}, 0 \right). \tag{5.25}$$

Note that the first component is non-negative, because of (5.24). Thus, if θ and ϕ denote the magnitudes of the (acute) angles which the emerging particle paths make with the incident particle path (see Figure 5.1), we have, from (5.25) and (5.24),

$$\tan \theta \tan \phi = \left| \frac{u_2'}{\gamma(u_1' + v)} \cdot \frac{-u_2'}{\gamma(-u_1' + v)} \right| = \frac{1}{\gamma^2}. \tag{5.26}$$

This shows that in relativity the angle between the emerging particle paths is *less* than 90°: for if it were 90°, we would have $\tan \phi = \cot \theta$ and $\tan \theta \tan \phi = 1$. To express (5.26) in terms of the incident velocity V, say, (which is not $2v$ now), we apply (2.28) to the incident particle (setting $u_1' = u' = v, u = V$) and easily find $\gamma(V) = 2\gamma^2(v) - 1$, whence

$$\tan \theta \tan \phi = \frac{2}{\gamma(V) + 1}. \tag{5.27}$$

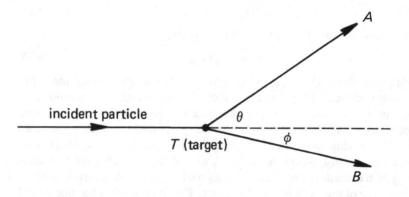

Figure 5.1

This result, though of limited interest to billiard players, is obviously important in connection with bubble-chamber and similar fast-particle collision experiments.

5.9 Threshold Energies

An important application of relativistic mechanics occurs in so-called threshold problems. Consider, for example, the case of a stationary proton (of rest mass M) being struck by a moving proton, such that not only the two protons but also a meson (of rest mass m) emerge. We shall ignore the electric interaction of the particles, which is negligible at the high energies of interest here. The question is, what is the minimum ("threshold") energy of the incident proton for this reaction to be possible? It is *not* simply $c^2(M + m)$; in other words, it is *not* enough to provide just the kinetic energy that will create the required extra rest mass. For, by momentum conservation, the end products cannot help but move, and thus have kinetic energy also.

In all such cases the least expenditure of energy occurs when all the end products are at rest in the CM frame. For consider—quite generally—two colliding particles: the "target" (at rest in the laboratory) and the "bullet," with respective precollision four-momenta $\mathbf{P}_1$ and $\mathbf{P}_2$. Let $\bar{\mathbf{P}}$ be the four-momentum of the entire outcoming system. Then $\mathbf{P}_1 + \mathbf{P}_2 = \bar{\mathbf{P}}$, and so

$$\mathbf{P}_1^2 + \mathbf{P}_2^2 + 2\mathbf{P}_1 \cdot \mathbf{P}_2 = \bar{\mathbf{P}}^2.$$

Substituting into this equation from (5.6), (5.14), and (5.23), we have, in the notation of (5.14),

$$m_{01}^2 + m_{02}^2 + 2m_{01}m_2 = \bar{m}_{CM}^2. \tag{5.28}$$

Now, m_{01} and m_{02} are given. It is clear, therefore, that c^2m_2, the energy of the "bullet" in the lab frame, is least when $\bar{m}_{CM}$ is least, and that evidently occurs when all outcoming particles are at rest in the CM frame, as we asserted. So to find a threshold energy we simply assume that the postcollision system travels in a lump, which means replacing $\bar{m}_{CM}$ in (5.28) by $\bar{m}_0$.

For the particular meson problem that we stated at the beginning, the general formula (5.28) in this way yields

$$2M^2 + 2M^2\gamma = (2M + m)^2,$$

where γ is the Lorentz factor of the threshold velocity. The solution for γ is immediate:

$$\gamma = 1 + \frac{2m}{M} + \frac{m^2}{2M^2}.$$

[Formula (5.28) with $\bar{m}_{CM} = \bar{m}_0$ even applies when the bullet is a photon that gets absorbed in the collision: in that case, as we shall see, $m_2 = h\nu/c^2$ and $m_{02} = 0$.]

5.10   Three-Force and Four-Force

Newton's second law can be written in either of the following two equivalent forms

$$\mathbf{f} = m_0 \mathbf{a}, \quad \mathbf{f} = \frac{d\mathbf{p}}{dt} = \frac{d(m_0 \mathbf{u})}{dt},$$

if m_0 is constant, as it usually is. When m_0 varies (as in the case of a rocket burning up its own substance, or a mothball evaporating isotropically), the second form is, in fact, inapplicable.[3] Nevertheless a generalization of the second form turns out to be the most useful definition of force in relativity. Thus we define the *relativistic* (*three-*) *force* $\mathbf{f}$ in terms of the relativistic momentum or mass as follows:

$$\mathbf{f} = \frac{d\mathbf{p}}{dt} = \frac{d(m\mathbf{u})}{dt}. \tag{5.29}$$

This definition has no physical content until other properties of force are specified, and the suitability of the definition will depend on these other properties.

Among them, we have Newton's third law, which asserts the equality of the forces of action and reaction. If an action and its reaction occur at different points (e.g., at opposite ends of a stretched spring), this law cannot be taken over directly into relativity because of the ambiguity of simultaneity. (This is what makes the discussion of systems of mutually interacting particles so difficult in relativity.) But if the law refers to the impact of two particles, it is well defined also in relativity, and in fact inevitable if we adopt (5.29), as we shall see presently. Secondly, (5.29) makes the usual definition of work compatible with Einstein's mass–energy equivalence, as we shall see in Equation (5.36). Further support comes from electrodynamics, where (5.29) is consistent with Lorentz's well-established law of force.

If $\mathbf{f}$ as defined above is useful in relativity, we would expect it to be part of a four-dimensional quantity. Let us therefore define the four-force $\mathbf{F}$ on a particle having four-momentum $\mathbf{P}$ by the relation

$$\mathbf{F} = \frac{d\mathbf{P}}{d\tau}, \tag{5.30}$$

which ensures the four-vector property of $\mathbf{F}$. From (5.6), (4.9), and (5.29), we then find

$$\mathbf{F} = \gamma(u) \frac{d}{dt}(\mathbf{p}, m) = \gamma(u)\left(\mathbf{f}, \frac{dm}{dt}\right). \tag{5.31}$$

The appearance of $\gamma(u)\mathbf{f}$ as the spatial part of a four-vector tells us the trans-

---

[3] A uniformly moving mothball would be subject to a force in all frames but its rest frame: force would not be invariant.

formation properties of **f**. Under a standard LT, for example, the two "trivial" transformation equations are $\gamma(u')f_2' = \gamma(u)f_2$ and $\gamma(u')f_3' = \gamma(u)f_3$. We note that they depend on the speed u of the particle on which **f** acts; hence **f**, unlike its Newtonian counterpart, is generally not invariant between frames.

We can now *prove* Newton's third law in the case of impact. In some particular frame S let the four-forces exerted by two colliding particles on each other be

$$\mathbf{F}_i = \gamma(u_i)\left(\mathbf{f}_i, \frac{dm_i}{dt}\right), \qquad (i = 1, 2).$$

Then, since the particles are in contact, $u_1 = u_2$, and, by energy conservation, $d(m_1 + m_2)/dt = 0$. Consequently the fourth component of the vector $\mathbf{F}_1 + \mathbf{F}_2$ vanishes in S, and, by the same argument, in all other inertial frames. But then, by the "zero-component" lemma of Section 5.4, it follows that $\mathbf{F}_1 + \mathbf{F}_2 = 0$. This, in turn, implies $\mathbf{f}_1 + \mathbf{f}_2 = 0$, and our assertion is established. We should not be too surprised that in our development Newton's third law appears as a theorem; after all, we took momentum conservation as an *axiom*, thus reversing Newton's logical sequence.

In the remainder of this section, we shall discuss particle motions with constant m_0. Then from (**5.30**), (**5.6**), and (**4.11**),

$$\mathbf{F} = m_0 \frac{d\mathbf{U}}{d\tau} = m_0 \mathbf{A}, \tag{5.32}$$

whence, by (**4.16**),

$$\mathbf{F} \cdot \mathbf{U} = 0. \tag{5.33}$$

Substituting from (**5.31**) and (**4.10**) into this equation, we get

$$\mathbf{f} \cdot \mathbf{u} = c^2 \frac{dm}{dt}, \tag{5.34}$$

and consequently, from (**5.31**),

$$\mathbf{F} = \gamma(u)(\mathbf{f}, \mathbf{f} \cdot \mathbf{u}/c^2). \tag{5.35}$$

If we adopt the classical definition of work, we now find from (**5.34**) that the work dW done by a force **f** in moving its point of application a distance $d\mathbf{r}$ is given by

$$dW = \mathbf{f} \cdot d\mathbf{r} = \mathbf{f} \cdot \mathbf{u}\,dt = c^2 dm. \tag{5.36}$$

This is another manifestation of Einstein's energy equation $E = mc^2$.

Now expand (**5.29**) and substitute from (**5.34**):

$$\mathbf{f} = m\mathbf{a} + \frac{dm}{dt}\mathbf{u} = \gamma m_0 \mathbf{a} + \frac{\mathbf{f} \cdot \mathbf{u}}{c^2}\mathbf{u}. \tag{5.37}$$

Thus, although **u**, **a**, and **f** are coplanar, the acceleration is not in general parallel to the force that causes it. Evidently there are just two cases when it *is*, namely when $\mathbf{f}$ is either parallel or orthogonal to $\mathbf{u}$. In particular, in the particle's rest frame, $\mathbf{f} = m_0\mathbf{a}$. It is instructive to consider the components of $\mathbf{f}$ and $\mathbf{a}$ along $\mathbf{u}$ and along a unit vector $\mathbf{n}$ orthogonal to $\mathbf{u}$ and coplanar with $\mathbf{u}, \mathbf{a}, \mathbf{f}$. Forming the scalar product of (5.37) with $\mathbf{u}/u$, we find

$$f_{\parallel} = \gamma m_0 a_{\parallel} + f_{\parallel} u^2/c^2, \quad \text{i.e.,} \quad f_{\parallel} = \gamma^3 m_0 a_{\parallel}, \tag{5.38}$$

where $f_{\parallel} = \mathbf{f} \cdot \mathbf{u}/u$ and $a_{\parallel} = \mathbf{a} \cdot \mathbf{u}/u$. Similarly, multiplying (5.37) with $\mathbf{n}$ gives

$$f_{\perp} = \gamma m_0 a_{\perp}, \tag{5.39}$$

where $f_{\perp} = \mathbf{f} \cdot \mathbf{n}$ and $a_{\perp} = \mathbf{a} \cdot \mathbf{n}$. It appears, therefore, that a moving particle offers different resistances to the same force, according as it is subjected to it longitudinally or transversely. In this way there arose the (now somewhat outdated but still often useful) concepts of "longitudinal" mass $\gamma^3 m_0$, and "transverse" mass γm_0. Since any force can be resolved into longitudinal and transverse components, Equations (5.38) and (5.39) provide one method of calculating the resultant acceleration in each case.

5.11 De Broglie Waves

In Section 3.3 we have already used the idea of photons, taking it for granted that light can be described alternatively as particles or as waves. This dualism, and its generalization to particles of nonzero rest mass, is an essential ingredient of quantum mechanics, but its very possibility depends on relativistic kinematics and dynamics. Guided by thermodynamic considerations, Planck in 1900 had made the momentous suggestion that radiant energy is *emitted* in definite "quanta" of energy

$$E = h\nu, \tag{5.40}$$

where ν is the frequency of the radiation and h a universal constant (Planck's constant) whose presently accepted value is 6.626×10^{-27} erg-sec. In 1905 Einstein, crystallizing the observed facts of the photoelectric effect, suggested that not only is radiant energy *emitted* thus, but that it also travels and is absorbed as quanta, which were later called photons. According to Einstein, a photon of frequency ν has energy $h\nu$, and thus a finite mass $h\nu/c^2$ and a finite momentum $h\nu/c$. It can be regarded as a limiting particle traveling at speed c, for which the mass γm_0 is finite, while $\gamma = \infty$ and $m_0 = 0$.

In 1923 de Broglie proposed a further extension of this idea, namely that associated with *any* particle of energy E there are waves of frequency E/h traveling in the same direction; however, as we have seen in Section 4.6, these waves cannot travel at the same speed as the particle (*unless that speed is c*), for this would not be a Lorentz-invariant situation. De Broglie found that, for consistency, the speeds u and w of the particle and its associated wave, respectively, must be related by the equation

$$uw = c^2. \tag{5.41}$$

Let us see why. If the particle has four-momentum $\mathbf{P}$, and the associated wave has frequency vector $\mathbf{L}$ [cf. (4.29)], de Broglie proposed the following relation, which is now known as *de Broglie's equation*:

$$\mathbf{P} = h\mathbf{L}, \quad \text{i.e.,} \quad m(\mathbf{u}, 1) = h\nu(\mathbf{n}/w, 1/c^2). \tag{5.42}$$

Being a four-vector equation, it is certainly consistent with special relativity. Moreover, if Planck's relation $h\nu = E(=mc^2)$ *does* apply to an arbitrary particle and its associated wave, then de Broglie's equation is in fact inevitable. For the vector $\mathbf{P} - h\mathbf{L}$ will then have its fourth component zero in all inertial frames; it must therefore vanish entirely, by the "zero-component" lemma of Section 5.4.

Equating components in (5.42), we find

$$\mathbf{u} \propto \mathbf{n}, \quad mu = \frac{h\nu}{w}, \quad m = \frac{h\nu}{c^2}. \tag{5.43}$$

The first of these equations locks the direction of motion of the particle to that of the wave-normal; the third is equivalent to (5.40); and the quotient of the last two is equivalent to (5.41).

For particles of nonzero rest mass, $u < c$ and thus $w > c$. The de Broglie wave speed then has an interesting and simple interpretation. Suppose a whole swarm of identical particles travel along parallel lines and something happens to all of them simultaneously in their rest frame: suppose, for example, they all "flash." Then this *flash* sweeps over the particles at the de Broglie velocity in any other frame. To see this, suppose the particles are at rest in the usual frame S', traveling at speed v relative to a frame S; setting $t' = 0$ for the flash, (2.7)(iv) yields $x/t = c^2/v$, and this is evidently the speed of the flash in S. Thus the de Broglie waves can be regarded as "waves of simultaneity."

De Broglie's idea proved to be of fundamental importance in the further development of quantum mechanics. It is one of the profound discoveries brought to light by special relativity, almost on a par with the famous relation $E = mc^2$. One of its first successes was in explaining the permissible electron orbits in the old Bohr model of the atom, as those containing a whole number of electron waves. A striking empirical verification of electron waves came later, when Davisson and Germer in 1927 observed the phenomenon of electron diffraction. And, of course, the superiority of the electron microscope hinges on de Broglie's relation, according to which electrons allow us to "see" with very much greater resolving power than photons since they have very much smaller wavelengths.

5.12 Photons. The Compton Effect

In the special case of light, (5.41) is satisfied by $u = w = c$ and, from (5.42), we then find for the four-momentum of a photon,

$$\mathbf{P} = \frac{h\nu}{c^2}(c\mathbf{n}, 1) \tag{5.44}$$

[This follows also from (5.6)(iii) and Einstein's values $h\nu/c^2$, $h\nu/c$ for the mass and momentum of a photon.] Clearly (5.44) is a *null* vector: $\mathbf{P}^2 = 0$.

We shall assume that, in collision problems, photons can be treated like any other particles, and that conservation of momentum and energy also applies to particle systems with photons. Systems consisting of photons only may or may not have a CM frame (a single photon certainly does *not*), but any system K containing at least one particle of nonzero rest mass *does* have a CM frame. For, if subscripts 1 and 2 refer to two arbitrary photons, we have from (5.44) that

$$\mathbf{P}_1 \cdot \mathbf{P}_2 = h^2 c^{-2} \nu_1 \nu_2 (1 - \cos\theta), \tag{5.45}$$

where θ is the angle between the paths of these photons. Consequently,

$$|\mathbf{P}_1 + \mathbf{P}_2|^2 = 2\mathbf{P}_1 \cdot \mathbf{P}_2 \geq 0,$$

and thus the sum of two photon four-momenta is null or timelike (null only if $\theta = 0$), and evidently it is future-pointing (fourth component positive) like the summands. On the other hand, the sum of a timelike vector and an isochronous null vector is always an isochronous timelike vector, as can be seen by adapting the sequence of Equations (5.18). By iteration of these results, it is seen that the total four-momentum of the system K is timelike and future-pointing, and that consequently a CM frame exists, as we asserted. (The reader may recall a heuristic derivation of this result in the penultimate paragraph of Section 4.5.)

It can also now be verified that the result (5.14)(i) still holds if the second particle is a photon, while (5.15) continues to hold even if both particles are photons. As an example, consider a stationary electron of rest mass m being struck by a photon of frequency ν, whereupon the photon is deflected through an angle θ and its frequency is changed to ν' (the *Compton effect*). The situation can be illustrated by Figure 5.1, with TA representing the path of the deflected photon. We wish to relate θ, ν, and ν' with m. If $\mathbf{P}$, $\mathbf{P}'$ are the four-momenta of the photon before and after collision, and $\mathbf{Q}$, $\mathbf{Q}'$ those of the electron, then $\mathbf{P} + \mathbf{Q} = \mathbf{P}' + \mathbf{Q}'$, by the conservation of four-momentum. Now, following a pleasant method of Lightman *et al.*,[4] we isolate the unwanted vector $\mathbf{Q}'$ on one side of the equation and square to get rid of it:

$$(\mathbf{P} + \mathbf{Q} - \mathbf{P}')^2 = \mathbf{Q}'^2.$$

Since $\mathbf{Q}^2 = \mathbf{Q}'^2$ and $\mathbf{P}^2 = \mathbf{P}'^2 = 0$, we are left with $\mathbf{P} \cdot \mathbf{Q} - \mathbf{P}' \cdot \mathbf{Q} - \mathbf{P} \cdot \mathbf{P}' = 0$, i.e.,

$$\mathbf{P} \cdot \mathbf{P}' = \mathbf{Q} \cdot (\mathbf{P} - \mathbf{P}'),$$

from which we find at once, by reference to (5.45) and (5.14), and canceling a factor h, the desired relation

$$hc^{-2}\nu\nu'(1 - \cos\theta) = m(\nu - \nu'). \tag{5.46}$$

[4] A. P. Lightman, W. H. Press, R. H. Price, S. A. Teukolsky, *Problem Book in Relativity and Gravitation*, Princeton University Press, 1975, p. 159.

In terms of the corresponding wave lengths λ, λ', and the half-angle $\theta/2$, this may be rewritten in the more standard form

$$\lambda' - \lambda = (2h/cm)\sin^2(\theta/2). \tag{5.47}$$

5.13   The Energy Tensor of Dust

After studying the mechanics of particles, it would be natural to turn our attention to the mechanics of continuous media (solids, fluids, and gases). And, indeed, special relativity has brought several interesting modifications to this subject, though these are perhaps of less direct interest to the general physicist than some of the other topics we have discussed. Yet there is one aspect of the relativistic mechanics of continua that is fundamental to the understanding of general relativity and cosmology, and that is the characterization of continua by their "energy tensor." A *particle* is dynamically characterized by its mass and momentum, which are conveniently exhibited in its four-momentum $\mathbf{P} = (\mathbf{p}, m)$. A moving *continuum*, on the other hand, not only has mass density and momentum density at each point, but also internal stresses, and it turns out that these are as inseparably tied to its density as the momentum of a particle is tied to its mass. In fact, the internal stresses, the momentum density, and the mass density of a continuum constitute the space-space, space-time, and time-time components, respectively, of a second-rank four-tensor $T_{\mu\nu}$. (Most four-tensors split in this way relative to a given frame, just as most four-vectors split into separately meaningful space and time parts. This is because spatial rotations and translations form a subgroup of the Lorentz transformations, and entities invariant under the former are those studied and named in classical physics. But it is also because space and time, though interrelated, are simply *not* equivalent.)

The analysis of the general continuum is not quite trivial (see, for example, RSR, Chapter 8), but for our purposes it will suffice to discuss the simplest of all continua, namely "dust." Technically, dust means a fluid entirely without pressure and viscosity. It can be imagined as a nearly continuous aggregate of incoherent—i.e., noninteracting—particles, moving at each point with common velocity (i.e., without random velocities). Its *proper* (mass) *density* ρ_0 at any event $\mathscr{P}$ is defined as the density measured by an observer O at rest relative to the dust at $\mathscr{P}$. To get the density ρ relative to an arbitrary frame in which the dust at $\mathscr{P}$ has velocity $\mathbf{u} = (u_1, u_2, u_3)$, we must allow for the foreshortening of what is considered a unit volume by O and also for the mass increase. Both effects involve a factor $\gamma(u)$, and thus

$$\rho = \rho_0 \gamma^2(u). \tag{5.48}$$

Because of the occurrence of *two* γ factors, the simplest four-tensor containing ρ is of rank 2, namely (cf. Section 4.4)

$$T_{\mu\nu} = \rho_0 U_\mu U_\nu, \tag{5.49}$$

where U_μ are the components of the four-velocity $\mathbf{U} = \gamma(\mathbf{u}, 1)$ of the dust. The density ρ is given by the component T_{44}. $T_{\mu\nu}$ is called the *energy tensor*. Its first and most obvious property is symmetry:

$$T_{\mu\nu} = T_{\nu\mu}. \tag{5.50}$$

Its second important property is that, in the absence of external forces, it satisfies the differential equations

$$\sum_{\nu=1}^{4} \partial_\nu T_{\mu\nu} = 0, \tag{5.51}$$

where the four symbols ∂_ν denote partial differentiation with respect to x, y, z, t, respectively. This we shall now prove.

Setting $\mu = 4$ in the left member of (5.51), we get the expression

$$\sum \partial_\nu T_{4\nu} = \partial_1(\rho u_1) + \partial_2(\rho u_2) + \partial_3(\rho u_3) + \partial_4 \rho, \tag{5.52}$$

whose vanishing constitutes the well-known "equation of continuity." For consider a small coordinate volume $dV = dx\,dy\,dz$. In time dt a mass $\rho u_1 dt\,dy\,dz$ enters one face perpendicular to the x axis, while a mass

$$\left(\rho u_1 + \frac{\partial(\rho u_1)}{\partial x}\, dx \right) dt\,dy\,dz$$

goes out through the other. Hence the net mass increase due to this pair of faces is $-\{\partial(\rho u_1)/\partial x\}dV dt$; the total increase per unit time and unit volume is $-\sum_{i=1}^{3} \partial_i(\rho u_i)$, therefore, and this must equal $\partial\rho/\partial t$.

Next, setting $\mu = i$ $(i = 1, 2, 3)$ in the left member of (5.51), we get

$$\begin{aligned} \sum \partial_\nu T_{i\nu} &= \partial_1(\rho u_i u_1) + \partial_2(\rho u_i u_2) + \partial_3(\rho u_i u_3) + \partial_4(\rho u_i) \\ &= u_i\{\partial_1(\rho u_1) + \partial_2(\rho u_2) + \partial_3(\rho u_3) + \partial_4 \rho\} \\ &\quad + \rho\{u_1\partial_1 u_i + u_2\partial_2 u_i + u_3\partial_3 u_i + \partial_4 u_i\}. \end{aligned} \tag{5.53}$$

The first brace vanishes, as we have shown above; the second brace vanishes because of the dynamics, being equal to the i-component of the acceleration. For if the velocity field of the dust is given by $\mathbf{u} = \mathbf{u}(x, y, z, t)$, and an element of dust follows the path $x = x(t)$, $y = y(t)$, $z = z(t)$, then its acceleration is given by

$$\frac{d\mathbf{u}}{dt} = \frac{\partial\mathbf{u}}{\partial x}\frac{dx}{dt} + \frac{\partial\mathbf{u}}{\partial y}\frac{dy}{dt} + \frac{\partial\mathbf{u}}{\partial z}\frac{dz}{dt} + \frac{\partial\mathbf{u}}{\partial t}.$$

This must vanish since there are no forces. Equation (5.51) is thus established.

A good example of a dust continuum without random motions is provided by a uniformly expanding ball of dust, such as Milne's cosmos (cf. Section 9.4).

The occurrence in the dust case of the density as the time-time component of a tensor makes it likely that this is a general state of affairs. And indeed it can be shown that all continua are characterized by tensors $T_{\mu\nu}$ which, though more complicated than (5.49), nevertheless have the mass density ρ (or the energy density $c^2\rho$—depending on conventions) as the time-time component.

This is the one component of $T_{\mu\nu}$ that *cannot* vanish when matter is present: hence the name "energy" tensor. All energy tensors satisfy Properties (5.50) and (5.51), with (5.51) embodying the laws of conservation of energy and momentum for the continuum.

In order to uphold the principles of the conservation of energy and momentum when a charged fluid interacts with an electromagnetic field, it is necessary also to ascribe to such *fields* an energy tensor $E_{\mu\nu}$ with the same general properties (5.50) and (5.51). Its components are fairly simple combinations of the electric and magnetic field strengths. In general relativity it is the sum of the energy tensors of matter and electromagnetic fields that acts directly as the source of the gravitational field.

CHAPTER 6

Relativity and Electrodynamics

6.1 Transformations of the Field Vectors

This chapter is deliberately not called "relativistic" electrodynamics, because there really has never been a "nonrelativistic" precursor. Maxwell's theory has always been relativistic in Einstein's sense, though no one before Einstein realized this. To be precise, the *formal* transformations (of the coordinates *and* of the electric and magnetic field vectors) leaving Maxwell's equations invariant had already been found by Lorentz and others (and this was the origin of the "Lorentz transformations"), but it simply was not understood that these transformations had a *physical* significance, namely that they related actual measurements made in different inertial frames. Since the physically relevant coordinate transformation was thought to be the Galilean transformation, Maxwell's theory was in fact thought to be strictly true in only *one* inertial frame—that of still ether.

Thus, special relativity found here a theory whose laws needed no amendment.[1] Nevertheless, recognizing the relativity of Maxell's theory made a considerable difference to our understanding of it, and also to the techniques of problem solving in it. Moreover, its basic formal simplicity became apparent now for the first time. Within the Galilean framework, Maxwell's theory was a rather unnatural and complicated construct. Within relativity, on the other hand, it is one of the two or three simplest possible theories of a field of

[1] This is true, at least, of Maxwell's theory in vacuum. On the other hand, Minkowski's extension of that theory to the interior of "ponderable" media (1908) was a purely relativistic development.

force. (The other candidates for this distinction are probably Yukawa's meson field theory and Nordström's attempt at a special-relativistic theory of gravity.) Any competent mathematician asked to produce formal Lorentz–invariant field theories without regard to empirical data would probably produce these three in short order. That two of them apply very accurately to certain natural phenomena is rather remarkable. Unfortunately, with the few rudiments of four-tensor calculus that we have developed so far, we cannot demonstrate this aspect of Maxwell's theory, and we must content ourselves with a more pedestrian approach here. (The fully four-dimensional approach is sketched out in Appendix II. See also RSR, Chapter 6.)

First we derive the transformation equations of the electric and magnetic field vectors, e and h. These vectors can be defined by their effect on a moving charge. We shall assume that the three-force f on a charge q moving with velocity u is given by

$$f = q\left(e + \frac{u \times h}{c}\right), \tag{6.1}$$

and that this is a Lorentz-invariant law. (Justification will come later.) It is, of course, the usual Lorentz force law, written in electrostatic units, except that we shall understand f to mean the *relativistic* force as defined in (5.29). This equation can serve as definition of e and h only if the velocity dependence of q is known. In Maxwell's theory q is, in fact, velocity-independent, and this agrees well with even the most modern experiments. We shall further assume that the action of the electromagnetic field on a charged test particle leaves its rest mass invariant. Then we can deduce the transformation properties of e and h from the known transformation properties of f. As we have seen in (5.35), f enters into the components of a four-vector F (the four-force) thus:

$$F = \gamma(u)\left(f, \frac{f \cdot u}{c^2}\right), \tag{6.2}$$

where u is the velocity of the particle on which f acts. Substituting (6.1) into (6.2), we get

$$F = q\gamma(u)\left(e + \frac{u \times h}{c}, \frac{e \cdot u}{c^2}\right), \tag{6.3}$$

and we suppose that the various quantities are measured in a definite inertial frame S. Now let us consider the components of this same four-vector in the usual second frame S'. We are interested only in the field e, h, and we shall clearly not restrict the generality of this field by specializing the test charge to $q = 1$ and its velocity in S' to zero. Then $u = (v, 0, 0)$ and $\gamma = \gamma(v)$. Thus in S' we have

$$F' = (e', 0), \tag{6.4}$$

and since the four components of **F** and **F′** must be related as the corresponding coordinates in (2.7), we find

$$e'_1 = \gamma\left(\gamma e_1 - \frac{v^2 \gamma e_1}{c^2}\right) = e_1, \tag{6.5}$$

$$e'_2 = \gamma\left(e_2 - \frac{v}{c}h_3\right), \tag{6.6}$$

$$e'_3 = \gamma\left(e_3 + \frac{v}{c}h_2\right). \tag{6.7}$$

To get the transforms of the h's, we remember that from any transformation formula between S and S' we can always get another (the inverse) by interchanging primed and unprimed quantities and replacing v by $-v$ (see Section 2.15). Let us do this to Equation (6.7), and then eliminate e'_3 from the resulting pair. This yields at once (if we recall that $\gamma^2 - 1 = \gamma^2 v^2/c^2$) the equation

$$h'_2 = \gamma\left(h_2 + \frac{v}{c}e_3\right), \tag{6.8}$$

and, similarly, (6.6) yields

$$h'_3 = \gamma\left(h_3 - \frac{v}{c}e_2\right). \tag{6.9}$$

We still lack the transformation of h_1. Let us give the unit charge in S' an *infinitesimal y'-velocity* u'_2. Then, instead of (6.4), we have

$$\mathbf{F}' = \left(\mathbf{e}' + \frac{\mathbf{u}' \times \mathbf{h}'}{c}, \frac{\mathbf{e}' \cdot \mathbf{u}'}{c^2}\right), \tag{6.10}$$

where $\mathbf{u}' = (0, u'_2, 0)$, and in (6.3) $\mathbf{u}$ is now $(v, u'_2/\gamma, 0)$ [cf. (2.26)]. Equating the z-components of (6.3) and (6.10) in accordance with (2.7), we find

$$e'_3 - \frac{u'_2 h'_1}{c} = \gamma\left(e_3 + \frac{vh_2}{c} - \frac{u'_2 h_1}{\gamma c}\right),$$

whence, by reference to (6.7), $h'_1 = h_1$. [The transformation for the x-component of (6.10) would reproduce (6.9); and similarly we could reobtain (6.8).] Collecting our results, we have therefore found

$$e'_1 = e_1, \qquad e'_2 = \gamma\left(e_2 - \frac{vh_3}{c}\right), \qquad e'_3 = \gamma\left(e_3 + \frac{vh_2}{c}\right), \tag{6.11}$$

$$h'_1 = h_1, \qquad h'_2 = \gamma\left(h_2 + \frac{ve_3}{c}\right), \qquad h'_3 = \gamma\left(h_3 - \frac{ve_2}{c}\right). \tag{6.12}$$

For consistency it would now be necessary to check that, *with* these transformations, the Lorentz four-force (6.3) is indeed a four-vector. In the present formalism this is a straightforward but tedious task, and we content ourselves by saying that it leads to the desired result.

The interested reader can verify that, because of Equations **(6.11)** and **(6.12)**, the components of **e** and **h** actually form a second-rank contravariant four-tensor, according to our definition of Section 4.4, in the following way:

$$A_{\mu\nu} = \begin{pmatrix} 0 & h_3 & -h_2 & -e_1/c \\ -h_3 & 0 & h_1 & -e_2/c \\ h_2 & -h_1 & 0 & -e_3/c \\ e_1/c & e_2/c & e_3/c & 0 \end{pmatrix}. \tag{6.13}$$

For this reason, the most convenient mathematical apparatus for theoretical electrodynamics is the calculus of four-tensors. By its use one can write the usual four Maxwell equations as two tensor equations involving the "field tensor" $A_{\mu\nu}$ and then their invariance under Lorentz transformations becomes self-evident (see Appendix II). This invariance can also be established laboriously by direct use of **(6.11)** and **(6.12)**, without, however, gaining any conceptual insights. Again, by tensor methods one can quite easily verify that the Lorentz four-force **(6.3)** is indeed a four-vector, being q/c times the "inner product" of $A_{\mu\nu}$ and U_μ.

Still, for many applications, Equations **(6.11)** and **(6.12)** and the four-vector calculus serve quite well. For example, it is not hard to verify by direct use of these equations that the following expressions are invariant under general Lorentz transformations:

$$\text{(i) } \mathbf{h}^2 - \mathbf{e}^2, \quad \text{(ii) } \mathbf{e} \cdot \mathbf{h}. \tag{6.14}$$

They are evidently invariant under spatial rotation and translation (by their three-vector form) and trivially so under time translation; only their invariance under a standard Lorentz transformation need be checked. It is less clear, from our present viewpoint, that these are essentially the *only* invariants of the field, whereas from the tensor theory this is easily seen.

One thing that is very clear from Equations **(6.11)** and **(6.12)** is the intermingling of **e** and **h** under a Lorentz transformation. For example, a "pure" **e** field, or a "pure" **h** field, in one frame, has *both* **e** and **h** components in another frame. (As Einstein had rightly presumed, the magnetic force deflecting a moving charge—even if there is no electric field, as around a current-carrying wire—is felt as an electric force in the rest frame of the charge.) Of course, pure **e** or **h** fields are untypical: the general electromagnetic field cannot be transformed into one of these by going to a suitable frame, for that would imply the vanishing of the invariant **e** · **h**.

On the other hand, our present analysis shows that *any* three-force **f** (whether of electrostatic origin or not) which in one frame S is independent of the velocity of the particles on which it acts and leaves their rest mass unchanged, must be a Lorentz-type force; and as such it will be velocity-dependent in any other frame S'—unless the field in S happens to be parallel to the motion of S'. For the four-force **(6.2)** in S can be identified with a four-force **(6.3)** specialized by $\mathbf{e} = \mathbf{f}$ and $\mathbf{h} = 0$; and **(6.3)**, being a four-vector, will then give the force in all frames. As evidenced by **(6.12)**, this force will have **h** components in S' unless $e_2 = e_3 = 0$: this is the exceptional case mentioned above.

6.2   Magnetic Deflection of Charged Particles

A free particle having rest mass m_0 and charge q is injected into a uniform pure magnetic field h at right angles to the lines of force; what happens? The exact answer to this question is important in bubble chamber magnetic analysis of elementary particles, and also in the design of accelerators. According to the Lorentz force law (6.1), the particle experiences a force

$$\mathbf{f} = \frac{q\mathbf{u} \times \mathbf{h}}{c} \tag{6.15}$$

at right angles to its velocity $\mathbf{u}$. Hence the acceleration $\mathbf{a}$ is in the direction of $\mathbf{f}$ [cf. (5.37)], and thus orthogonal to $\mathbf{u}$ and $\mathbf{h}$. Consequently, (i) the motion is in a plane orthogonal to $\mathbf{h}$, and (ii) u is constant (since $u\,du/dt = \mathbf{u} \cdot d\mathbf{u}/dt$).

Moreover, from (5.37), $a = q\dot{u}h/cm_0\gamma$. Consequently, the angular deflection per unit time is given by

$$\omega = \frac{d\theta}{dt} = \frac{|d\mathbf{u}|}{u\,dt} = \frac{a}{u} = \frac{qh}{cm_0\gamma}, \tag{6.16}$$

and thus the particle traces out a circle of radius

$$r = \frac{u}{\omega} = \frac{cm_0 u\gamma}{qh} \tag{6.17}$$

with angular velocity ω and therefore with period

$$T = \frac{2\pi}{\omega} = \frac{2\pi cm_0\gamma}{qh}. \tag{6.18}$$

According to classical mechanics, this formula would lack the γ factor and the period would be velocity-independent. At high speeds γ makes itself felt and this, for example, necessitated the development from the cyclotron to the synchrotron type of accelerator: the larger the orbit, the longer the period.

6.3   The Field of a Uniformly Moving Charge

As a good example of the power that special relativity brought to electromagnetic theory, we shall calculate the field of a uniformly moving charge q by that typically relativistic method of looking at the situation in a frame where everything is obvious, and then transforming to the general frame. In the present case, all is obvious in the rest frame S' of the charge, where there is a simple Coulomb field of form

$$\mathbf{e}' = (q/r'^3)(x', y', z'), \qquad \mathbf{h}' = 0, \qquad (r'^2 = x'^2 + y'^2 + z'^2), \tag{6.19}$$

if, as we shall assume, the charge is at the origin. Now we transform to the usual second frame S, in which the charge moves with velocity $\mathbf{v} = (v, 0, 0)$. We calculate the field in S at the instant $t = 0$ when the charge passes the origin. Reference to (6.12) (with $\mathbf{h}' = 0$), the inverse of (6.11), and (2.7) (with $t = 0$), shows

$$\mathbf{h} = (v/c)(0, -e_3, e_2), \quad \mathbf{e} = (e'_1, \gamma e'_2, \gamma e'_3) = (q\gamma/r'^3)(x, y, z)$$
$$r'^2 = \gamma^2 x^2 + y^2 + z^2 = \gamma^2 r^2 - (\gamma^2 - 1)(y^2 + z^2)$$
$$= \gamma^2 r^2 [1 - (v^2/c^2)\sin^2\theta], \tag{6.20}$$

where θ is the angle between the vector $\mathbf{r} = (x, y, z)$ and the x axis. Thus,

$$\mathbf{h} = \frac{1}{c}\,\mathbf{v} \times \mathbf{e}, \quad \mathbf{e} = \frac{q\mathbf{r}}{\gamma^2 r^3 [1 - (v^2/c^2)\sin^2\theta]^{3/2}}. \tag{6.21}$$

It is interesting to note that the electric field at $t = 0$ is directed away from the point where the charge is *at that instant*, though (because of the finite speed of propagation of all effects) it cannot be *due* to the position of the charge at that instant. Note that both the electric and magnetic field strengths in any fixed direction from the charge fall off as $1/r^2$. Also of interest is the angular dependence of the strength of the electric field: it is strongest in a plane at right angles to $\mathbf{v}$, and weakest fore and aft.

We can construct a surprising model in which the density of the lines of force represents field strength as usual. Consider a block of plexiglass at rest in S', into which the isotropic lines of force of q in S are "etched." The solid angle of a thin pencil of lines of x-cross sectional area dA at (x', y', z'), is given by $d\Omega' = dA \cos \theta'/r'^2 = dAx'/r'^3$ (see Figure 6.1a). In S the block moves with velocity v and all x-coordinates are shrunk by a factor γ^{-1}; the corresponding solid angle is given by $d\Omega = dAx/r^3$, whence, by reference to (6.20),

$$\frac{d\Omega'}{d\Omega} = \frac{x'r^3}{xr'^3} = \frac{\gamma r^3}{r'^3} = \frac{1}{\gamma^2 [1 - (v^2/c^2)\sin^2\theta]^{3/2}}. \tag{6.22}$$

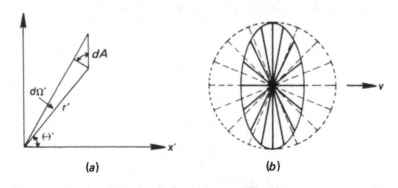

(a) (b)

Figure 6.1

Comparing this with **(6.21)**, we have for the electric field strength

$$\mathbf{e} = \frac{qd\Omega'}{r^2 d\Omega} = \frac{n}{d\Sigma},$$

where $n = qd\Omega'$ is the "number of lines of force" etched into the pencil in S' and $d\Sigma = r^2 d\Omega$ is the normal cross section of the pencil in S. Since in the model there are as many etched lines in $d\Omega$ as in $d\Omega'$, we see that the density of these lines represents the electric intensity in S as well as in S'. In other words, the lines of force transform like rigid wires attached to the charge! Figure 6.1b illustrates this.

Of course, this result is purely a consequence of the laws of electrodynamics and can be obtained without the explicit use of SR. Lorentz so obtained it, and thereon based an "explanation" of the length contraction of material bodies: if the electromagnetic fields of the fundamental charges "contract," then so must all matter, if it is made up of such charges. (Lorentz's argument, perforce, ignored the existence of the nuclear force fields, etc.)

To calculate the field of an *accelerating* charge, the above elementary method is inapplicable—unless one is willing to go to very great lengths and to make extra assumptions. The standard method involves use of the four-potential. (See, for example, RSR, page 141.) This is a four-vector of great importance in the full theory, but one which it is beyond our present scope to discuss.

6.4   The Field of an Infinite Straight Current

As a final example of relativistic reasoning in electromagnetic theory, we shall derive the field of an infinite straight current. We begin by calculating the field of an infinite, static, straight-line distribution of charge, of uniform line density σ_0, say (see Figure 6.2). Since the charge is static, there will be no magnetic field. By symmetry, the electric field at any point P will be along the perpendicular from P to the line; consequently, only field components in this direction need be considered. As reference to the figure shows, the contribu-

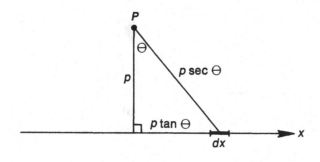

Figure 6.2

tion to the electric field in that direction from a line element dx is given by

$$de = \frac{\sigma_0 dx}{(p \sec \theta)^2} \cos \theta, \qquad (6.23)$$

where θ is the angle which the element subtends at P with the perpendicular p from P to the line. Since $x = p \tan \theta$, $dx = p \sec^2 \theta d\theta$ and thus, from (6.23),

$$e = \int_{-\pi/2}^{\pi/2} \frac{\sigma_0}{p} \cos \theta d\theta = \frac{2\sigma_0}{p}. \qquad (6.24)$$

Now suppose an infinite line charge with proper line density σ_0 moves with velocity v relative to a frame S. Because of length contraction, its line density σ in S is $\gamma\sigma_0$, and there it corresponds to a current $i = \gamma\sigma_0 v$. Let us identify the line charge with the x' axis of the usual frame S'. Then, by the above calculation, the only nonvanishing component of the field at the typical point $P(0, p, 0)$ in S' is given by

$$e_2' = \frac{2\sigma_0}{p}. \qquad (6.25)$$

Transforming this field to S by use of the inverse relations of (6.11) and (6.12), obtained by interchanging primed and unprimed symbols and writing $-v$ for v, we find

$$e_1 = 0, \quad e_2 = \frac{2\gamma\sigma_0}{p} = \frac{2\sigma}{p}, \quad e_3 = 0,$$

$$h_1 = 0, \quad h_2 = 0, \quad h_3 = \frac{2\gamma\sigma_0 v}{pc} = \frac{2i}{pc}. \qquad (6.26)$$

Note that the strength of the magnetic field is only a fraction v/c of that of the electric field, and another factor of order u/c reduces its effect, by comparison, on a charge moving with velocity u [see (6.1)]. Moreover, in a laboratory current of a few amperes, v is only about one millimeter per second. As C. W. Sherwin has said, it is hard to believe that this magnetic force, which has to suffer a denominator c^2, is the "work force" of electricity, responsible for the operations of motors and generators. And again, considering that this force arises from transforming a purely electric field to another frame having very small velocity relative to the first, A. P. French has remarked: who says that relativity is important only for velocities comparable to that of light? The reason is that an ordinary current moves a very big charge: there are something like 10^{23} free electrons per cubic centimeter of wire. Their *electric* force, if it were not neutralized, would be enormous—of the order of two million tons of weight on an equal cubic centimeter at a distance of 10 km.

But that force *is* neutralized in a "real" current flowing in a wire. Such a current corresponds to *two* superimposed linear charge distributions, one at rest and one in motion. The positive metal ions are at rest while the free electrons move, say, with velocity $-v$. Before the current is turned on, we can think of the ions schematically as a row of chairs on which the free electrons sit. When the current flows, the electrons play musical chairs, always moving to the next chair in unison since there can be no build-up of charge. But this means that the electrons are now as far apart *in motion* as are the stationary ions. Hence the respective line densities of ions and electrons are equal and opposite *in the lab frame*, say $\pm\sigma$, and the current is given by $i = \sigma v$. As can be seen from (**6.26**), the electric fields will cancel exactly, while the magnetic field is given as before by $2i/pc$.

Consider now a test charge moving with velocity **u** parallel to the wire. It experiences a force in the direction **u** $\times$ **h**, i.e., radially towards or away from the wire. In its rest frame, where it can be affected only by **e** fields, it sees two moving lines of positive and negative charge, respectively, but with slightly different line densities, numerically. For length contraction sees to it that if the line densities are equal and opposite in the lab frame, they cannot be so in any other frame moving parallelly to the wire. And it is this difference which provides the net **e** field that causes the charge to accelerate in its rest frame. This is about the closest we get to a direct manifestation of length contraction—a length contraction by a γ-factor of about $1 + 10^{-22}$!

Basic Ideas of General Relativity

7.1 Curved Surfaces

One of the most revolutionary features of general relativity is the essential use it makes of curved space (actually, of curved spacetime). Though everyone knows intuitively what a curved *surface* is, or rather, what it looks like, people are often puzzled how this idea can be generalized to three or even higher dimensions. This is mainly because they cannot visualize a *four*-space in which the three-space can *look* bent. So let us first of all try to understand what the curvature of a surface means *intrinsically*, i.e., without reference to the embedding space. Intrinsic properties of a surface are those that depend only on the measure relations *in* the surface; they are those that could be determined by an intelligent race of two-dimensional beings, entirely confined to the surface in their mobility and in their capacity to see and to measure. Intrinsically, for example, a flat sheet of paper and one bent almost into a cylinder or almost into a cone, are equivalent (see Figure 7.1). (If we closed up the cylinder or the cone, these surfaces would still be "locally" equivalent but not "globally.") In the same way a helicoid (spiral staircase) is equivalent to an almost closed catenoid (a surface generated by rotating the shape of a freely hanging chain); and so on. One can visualize intrinsic properties as those that are preserved when a surface is bent without stretching or tearing.

An important intrinsic feature of a surface is the totality of its *geodesics*. These are lines of minimal length between any two of their points, provided these points are "sufficiently" close together. (For example, on a sphere they

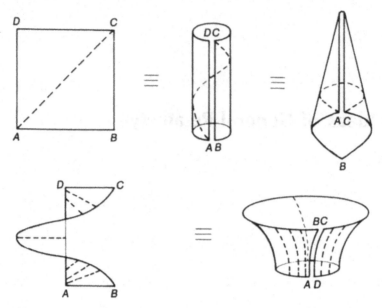

Figure 7.1

are the great circles—which are *not* minimal for paths longer than half a circumference.) Thus if, in Figure 7.2, A and B are two sufficiently close points on a geodesic g of a surface S, then all nearby lines joining A and B on S (e.g., l and l') would have greater length than the portion of g between A and B. Since geodesics depend only on distance measurements *in* the surface, they are intrinsic, i.e., they remain geodesics when the surface is bent. (See the dotted lines in Figure 7.1.)

Now let us see how the two-dimensional beings would discover the curvature of their world. We can assume that light in this world would travel along geodesics, in accordance with Fermat's principle. Then to each ob-

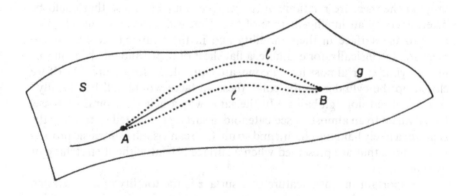

Figure 7.2

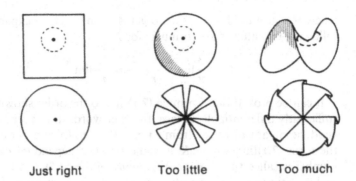

Figure 7.3	**Just right**	**Too little**	**Too much**

server his world would *look* flat, indistinguishable from its tangent plane. So he must do more than look: he must measure. As prototypes of three different kinds of surface regions, consider a plane, a sphere, and a saddle. On each of these draw a small geodesic circle of radius r, i.e., a locus of points which can be joined by geodesics of length r to some center (see Figure 7.3). In practice this could be done by using a taut string like a tether. Then we (or the flat people) can measure both the circumference C and the area A of these circles. In the plane we get the usual "Euclidean" values $C = 2\pi r$ and $A = \pi r^2$. On the sphere we get *smaller* values for C and A, and on the saddle we get *larger* values. This becomes evident when, for example, we cut out these circles and try to flatten them onto a plane: the spherical cap must tear (it has too little area), while the saddle cap will make folds (it has too much area).

For a quantitative result, consider Figure 7.4, where we have drawn two geodesics subtending a small angle θ at the north pole P of a sphere of radius a. By definition, we shall assign a curvature $K = 1/a^2$ to such a sphere. At distance r along the geodesics from P, let their perpendicular separation be η. Then, using elementary geometry and the Taylor series for the sine, we have

$$\eta = \theta\left(a \sin\frac{r}{a}\right) = \theta\left(r - \frac{r^3}{6a^2} + \cdots\right) = \theta\left(r - \frac{1}{6}Kr^3 + \cdots\right), \quad (7.1)$$

and consequently,

$$C = 2\pi\left(r - \tfrac{1}{6}Kr^3 + \cdots\right), \quad A = \pi\left(r^2 - \tfrac{1}{12}Kr^4 + \cdots\right), \quad (7.2)$$

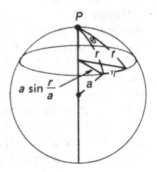

Figure 7.4

where we have used $A = \int C dr$ to get A from C. These expansions yield the following two alternative formulae for K:

$$K = \frac{3}{\pi} \lim_{r \to 0} \frac{2\pi r - C}{r^3} = \frac{12}{\pi} \lim_{r \to 0} \frac{\pi r^2 - A}{r^4}. \tag{7.3}$$

It can be proved that Formula (7.1)(iii), to the order shown, holds for *any* sufficiently differentiable surface. In other words, the spread of neighboring geodesics in any direction from a point P, up to third order in r, is always of the form (7.1)(iii), where K is some number depending only on P. This number is called the (*Gaussian*) *curvature* of the surface at P. Formulae (7.2) and (7.3) therefore apply quite generally. So the curvature of the saddle will be negative.

It can also be proved that a surface with $K = 0$ everywhere is necessarily intrinsically plane, and one with $K = 1/a^2$ everywhere is intrinsically a sphere of radius a (except for possible "topological identifications"—see Section 9.5).

7.2   Curved Spaces of Higher Dimensions

The ideas of the intrinsic geometry of surfaces can be extended to spaces of higher dimensions, such as, for example, the three-space of our experience. In particular, geodesics in all dimensions are defined exactly as in the two-dimensional case. For a sufficiently "well-behaved" space, two important theorems can then be proved: (i) there is a unique geodesic issuing from a given point in a given direction, and (ii) in a sufficiently small neighborhood of a given point P each other point can be connected to P by a unique geodesic.

A generalization of curvature from two to three dimensions might be to construct geodesic *spheres* (instead of circles) of radius r, and to compare their measured surface area or volume with the Euclidean values. It is logically quite conceivable that by very accurate measurements of this kind we would find our space to deviate slightly from flatness. The great Gauss himself made several experiments to determine its curvature, but with the available apparatus—then or now—none can be detected directly.

However, the direct generalization of Formulae (7.3) turns out to be too unrefined. More than one number is needed to characterize fully the curvature properties of spaces of higher dimensions. Consider all the geodesics issuing from a point P in the directions of a linear "pencil" $\lambda\mathbf{p} + \mu\mathbf{q}$ determined by *two* directions $\mathbf{p}$ and $\mathbf{q}$ at P. Such geodesics are said to generate a *geodesic plane* through P. Its curvature K at P is said to be the space curvature $K(\mathbf{p}, \mathbf{q})$ at P for the orientation $(\mathbf{p}, \mathbf{q})$. (In three dimensions K is completely known if it is known for 6 orientations, in four dimensions if it is known for 20.) A geodesic plane is the curved-space analog of a plane through a point, except that in general it satisfies its defining property only with respect to that one point. *If* light propagates along geodesics,[1] then each of the generators *looks* straight to the observer at P, and so the whole geodesic plane actually looks plane from there.

If the curvature K at P is independent of the orientation, we say P is an *isotropic point*. Then, indeed, all the curvature information is contained in knowing the surface area S or volume V of a small geodesic sphere of radius r (or suitable generalizations thereof in higher dimensions). In three dimensions S and V are easily seen to be given by the formulae

$$S = 4\pi(r^2 - \tfrac{1}{3}Kr^4 + \cdots), \quad V = \tfrac{4}{3}\pi(r^3 - \tfrac{1}{5}Kr^5 + \cdots). \tag{7.4}$$

The first follows from (7.1): the ratio of S to the Euclidean value $4\pi r^2$ must equal the square of the ratio of η to *its* Euclidean value θr. The second then follows from the relation $V = \int S\, dr$. If *all* points of a space are isotropic, it can be shown that the curvature at all of them must be the same (*Schur's theorem*), and then the space is said to be *of constant curvature*. A condensed statement of Schur's theorem would be "isotropy everywhere implies homogeneity."

Let us digress for a moment and as an example discuss the three-dimensional analog of a sphere, i.e., a three-space of constant positive curvature, say $1/a^2$. For such a *three-sphere* we can use (7.1)(i) to get the exact forms of (7.4):

$$S = 4\pi a^2 \sin^2 \frac{r}{a}, \quad V = 2\pi a^2 \left(r - \frac{a}{2} \sin \frac{2r}{a} \right). \tag{7.5}$$

Consider first what happens on an ordinary two-sphere of curvature $1/a^2$ as we draw circles about a given point: at first the circumferences get bigger as we increase the geodesic radius r, but after reaching a maximum they get smaller again and finally become zero when $r = \pi a$. (Note that there is nothing illogical in the fact that the later small circles *contain* early bigger ones.) If we lived in a three-sphere S_3 of curvature $1/a^2$ and drew concentric geodesic *spheres* around ourselves, their surface area would at first increase with increasing geodesic radius r (but not as fast as in the Euclidean case), reaching a maximum $4\pi a^2$, with included volume $\pi^2 a^3$, at $r = \tfrac{1}{2}\pi a$ [cf. (7.5)]. After that, successive spheres contract until finally the sphere at $r = \pi a$ has zero surface area and yet contains *all* our space: its surface is, in fact, a single point, our "antipode." The total volume of the three-sphere is finite, $2\pi^2 a^3$, and yet there is no boundary. Again, there is no center: every point is equivalent to every other. Every geodesic plane in S_3 is a 2-sphere. Its inside and outside are identical halves of S_3, just as a great circle divides a 2-sphere into identical halves. If I blow up a balloon in S_3, its surface will increase until it looks flat to me, then decrease and finally enclose me tightly! (An analogy is provided by a man "blowing up circles" on a sphere.) All this is not as

---

[1] This would be a natural assumption in classical physics. In GR, because the coordinate speed of light varies in the presence of gravitating matter, the quickest route for light is not generally a geodesic in the three-space, but always a geodesic in the four-dimensional spacetime.

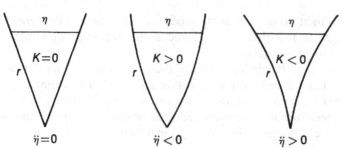

Figure 7.5

fantastic as it may seem. The very first cosmological model proposed by Einstein envisaged our three-space to be precisely of this kind.

Now we wish to introduce the concept of *geodesic deviation*, which is closely related to what we have already done. Differentiating the third equation in (7.1) twice with respect to r, we find, to first order,

$$\ddot{\eta} = -K\eta \quad (\cdot \equiv d/dr). \tag{7.6}$$

Thus we may use the second rate of spread (or "deviation") of two geodesics intersecting at a small angle as a very direct measure of the curvature of a surface (see Figure 7.5). [In fact, Formula (7.6) is valid for any two neighboring geodesics, even if they do not intersect at the location of interest. This can be seen by cutting obliquely across both with a third geodesic.] Suppose through some point P of an n-dimensional space V_n we draw two neighboring geodesics g_1 and g_2. They define a unique geodesic plane through P, of which they are two generators. Moreover, they are geodesics also relative to that geodesic plane. For if g_1, say, is the shortest path from A to B in V_n, there can be no shorter path between A and B on the geodesic plane. Thus the spread of g_1 and g_2 will measure the curvature of the geodesic plane and consequently the curvature of V_n at P for the orientation corresponding to g_1, g_2.

7.3   Riemannian Spaces

In the last two sections we have rather liberally quoted, without proof, theorems from a branch of mathematics called "Riemannian geometry," on the implicit assumption that the curved spaces under discussion were, in fact, *Riemannian*. This assumption we must now examine. On a curved surface we cannot set up Cartesian coordinates in the same way as in the plane (with "coordinate lines" forming a lattice of strict squares)—for if we could, we would *have* a plane, intrinsically. Certain surfaces by their symmetries suggest a "natural" coordinatization, like the plane, or the sphere (see Figure 7.6) on which one usually chooses co-latitude (x) and longitude (y) to specify points. On a general surface one can "paint" two arbitrary families of co-ordinate lines and label them $x = \ldots, -2, -1, 0, 1, 2, \ldots$ and $y = \ldots, -2, -1, 0, 1, 2, \ldots$, respectively, and one can further subdivide these as finely as one wishes. The resulting coordinate lattice may or may not be orthogonal.

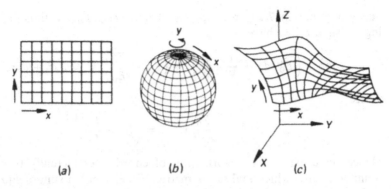

Figure 7.6

If we imagine the surface embedded in a Euclidean three-space with co-ordinates (X, Y, Z) (see Figure 7.6c), then it will satisfy equations of the form

$$X = X(x, y), \quad Y = Y(x, y), \quad Z = Z(x, y), \tag{7.7}$$

which we assume to be differentiable as often as required. For example, a sphere of radius a centered on the origin satisfies

$$X = a \sin x \cos y, \quad Y = a \sin x \sin y, \quad Z = a \cos x.$$

Since the distance between neighboring points in the Euclidean space is given by

$$d\sigma^2 = dX^2 + dY^2 + dZ^2, \tag{7.8}$$

distances in the surface are given by

$$d\sigma^2 = (X_1 dx + X_2 dy)^2 + (Y_1 dx + Y_2 dy)^2 + (Z_1 dx + Z_2 dy)^2, \tag{7.9}$$

where the subscripts 1 and 2 denote partial differentiation with respect to x and y, respectively. Evidently (**7.9**) is of the form

$$d\sigma^2 = E dx^2 + 2F dx dy + G dy^2, \tag{7.10}$$

where E, F, G are certain functions of x and y. (In the case of the sphere, this method yields $d\sigma^2 = a^2 dx^2 + a^2 \sin^2 x dy^2$—which can also be understood directly by elementary geometry.) Whenever the squared differential distance $d\sigma^2$ is given by a homogeneous quadratic differential form in the surface co-ordinates, as in (**7.10**), we say that $d\sigma^2$ is a *Riemannian metric*, and that the corresponding surface is *Riemannian*. It is, of course, not a foregone con-clusion that all metrics must be of this form: one could *define*, for example, a non-Riemannian metric $d\sigma^2 = (dx^4 + dy^4)^{1/2}$ for some two-dimensional space, and investigate the resulting geometry. (Such more general metrics give rise to "Finsler" geometry.) What distinguishes a Riemannian metric among all others is that it is *locally Euclidean*: At any given point P_0 the values of E, F, G in (7.10) are simply numbers, say E_0, F_0, G_0; thus, "completing the square," we have, *at P_0,*

$$d\sigma^2 = \left(E_0^{1/2} dx + \frac{F_0}{E_0^{1/2}} dy \right)^2 + \left(G_0 - \frac{F_0^2}{E_0} \right) dy^2 \doteq d\tilde{x}^2 + d\tilde{y}^2, \quad (7.11)$$

where

$$\tilde{x} = E_0^{1/2} x + \frac{F_0}{E_0^{1/2}} y, \quad \tilde{y} = \left(G_0 - \frac{F_0^2}{E_0} \right)^{1/2} y.$$

Hence there exists a transformation of coordinates (actually there exist infinitely many) which makes the metric "Euclidean" (a sum of squares of differentials) at any *one* preassigned point. Conversely, *if* there exist coordinates $\tilde{x}, \tilde{y}$ in terms of which the metric is Euclidean at a point P_0, then in general coordinates it must be Riemannian at P_0; for there must exist some transformation $\tilde{x} = \tilde{x}(x, y)$, $\tilde{y} = \tilde{y}(x, y)$ from the general to the special coordinates, and thus

$$d\tilde{x}^2 + d\tilde{y}^2 = (\tilde{x}_1 dx + \tilde{x}_2 dy)^2 + (\tilde{y}_1 dx + \tilde{y}_2 dy)^2,$$

which is Riemannian in x, y. Now we see that in order to predict the form of (7.10) we could have dispensed with the use of the embedding space. We could simply have postulated that the surface is locally Euclidean, i.e., that for any given point we can paint the coordinate lines so that $d\sigma^2 = dx^2 + dy^2$ *at* that point.

These ideas generalize directly from surfaces to spaces of higher dimensions. Such spaces, too, can (by definition) be coordinatized with arbitrary ("Gaussian") coordinates, just like surfaces. In three-space, for example, we can "paint" three families of coordinate *surfaces* and label them $x =$ constant, $y =$ constant, $z =$ constant. If there exists a metric analogous to (7.10), we say the space is Riemannian, and it *will* be so if and only if it is locally Euclidean.

The metric being locally Euclidean, in turn, ensures that the intrinsic *geometry* is locally Euclidean. For example, the circumference of a small geodesic circle is $2\pi r$ *to lowest order*, and thus the complete plane angle around a point is 2π; the complete solid angle around a point is 4π; the sum of the angles in a small geodesic triangle is 2π to lowest order; etc. On the other hand, a circle in the two-space with (Finsler) metric $(dx^4 + dy^4)^{1/2}$, e.g., the locus $(x^4 + y^4)^{1/2} = r^2$ for small r, would *not* have circumference $2\pi r$ even to lowest order.

A useful result to remember is that in 2 or 3 dimensions (but not in higher dimensions) it is always possible to find *orthogonal* coordinates, i.e., coordinates in terms of which the metric has no cross terms in the differentials. Thus, in the two-dimensional case, F in (7.10) can be made to vanish *globally*.

Now comes a formally slight—but for our purposes vital—generalization. It consists in admitting metrics that are not "positive definite": on a real surface we obviously have $d\sigma^2 > 0$ for all $dx, dy \neq 0$, but this condition is not essential for much of the theory. A "nondefinite" Riemannian metric corresponds locally to a "pseudo-Euclidean" metric (squares of differentials only, but some with negative sign), e.g., to one with "signature" $(+ + -)$: $dx^2 + dy^2 - dz^2$. It can be shown that the signature of a metric is invariant, i.e., no matter how a reduction to squares of differentials is achieved (at a given point—it usually cannot be done globally), there always results the same distribution of plus and minus signs, provided we adhere to real coordinates. The spacetime of special relativity (hereafter referred to as *Minkowski space, M_4*) with metric $ds^2 = c^2dt^2 - dx^2 - dy^2 - dz^2$ is an example of a Riemannian space of signature $(+ - - -)$. Of course, it is rather a special example, since it is not only Riemannian, but globally pseudo-Euclidean. The fact that ds here is not just a simple ruler distance in no way affects the mathematics.

As in the case of positive definite spaces, the differential equations of geodesics are obtained from the requirement of *stationary* length, $\delta \int ds = 0$. But in indefinite spaces the geodesics are not curves of *minimal* length. In general they have neighbors both longer and shorter than they. Only when the signature has a single positive [negative] sign, will there be geodesics of *maximal* length, namely those for which $ds^2 > 0$ $[< 0]$ (cf. Exercise 7.7). It can be shown that the sign of ds^2 must be constant along a geodesic, but it can be positive, negative, or zero. In all Euclidean or pseudo-Euclidean spaces geodesics correspond to linear equations in the Euclidean coordinates.

It turns out that the differential equation for geodesics can be reinterpreted thus: geodesics are "locally straight," i.e., they have zero curvature in the local Euclidean tangent space. This leads to the following empirical method of finding geodesics on surfaces. Cut a long thin strip of paper and draw a straight line down its middle. Then glue the strip carefully, bit after bit and without folds, to the surface. The drawn line will now be a geodesic on the surface.

In spaces of nondefinite metric, curvature is best visualized via geodesic deviation. The concepts of "isotropic point" and "constant curvature" apply to these spaces as to others; but the isotropy of an isotropic point is here restricted purely to curvature. The points of M_4, for example, are *not* isotropic in a general sense (because of the null cone), but nevertheless Equation (7.6) with $K = 0$ applies in *all* directions. We shall later meet nondefinite spaces of constant curvature in which (7.6) applies similarly in all directions with $K \neq 0$.

Once we know the metric, we know all distance relations in the space, and so we know all there *is* to know about the space intrinsically. The differential equation of geodesics, for example, involves only the coordinates and the metric coefficients [the generalizations of the E, F, G in (7.10)]; so does the formula for the curvature K in any given orientation; etc. Two spaces which are intrinsically equivalent clearly admit coordinates in terms of which the metrics are identical. (We need merely bring the spaces into coincidence and impress the coordinates of the one on the other.) Conversely, if the metrics can be made identical by a suitable choice of coordinates, the spaces are intrinsically equivalent. (We can apply the one to the other by matching corresponding points.) Hence instead of "intrinsically equivalent," one uses the shorter term "isometric."

This brings us to an important problem. Two spaces may be isometric and yet their metrics may *look* quite different. For example, of the four metrics

$$dx^2 + x^2 dy^2, \quad (4x^2 + y^2)dx^2 + (2xy - 4x)dxdy + (1 + x^2)dy^2,$$
$$y^2 dx^2 + x^2 dy^2, \quad ydx^2 + xdy^2,$$

the first three *all* represent the ordinary plane; the last does not. The first is actually the well-known polar metric $dr^2 + r^2 d\theta^2$ in unfamiliar notation, but still "recognizable"; the second results from the usual Euclidean coordinates $\tilde{x}, \tilde{y}$ by the undistinguished transformation

$$\tilde{x} = x^2 - y, \quad \tilde{y} = xy.$$

The reader will not guess in a hurry (and should not try) how the third arises, though it too results from transforming the Euclidean $\tilde{x}, \tilde{y}$. For spaces of constant curvature (like the plane), we have a powerful theorem to help us: *any two spaces of the same constant curvature, dimension, and signature, are isometric.* [As a corollary, all flat spaces ($K = 0$) must be isometric with pseudo-Euclidean space of the same signature.] But the general problem of deciding the equivalence of two arbitrary metrics (the so-called "equivalence problem" of quadratic differential forms) is very difficult, especially in practice.

7.4 A Plan for General Relativity

As we have seen in Section 5.2, the presence of gravitating matter precludes the existence of extended inertial frames, *if* we accept the EP. General spacetime will therefore not be the familiar flat M_4. (The EP provides another strong argument for curvature, by its prediction of curved light paths from the mere constancy of the light velocity—cf. the beginning of Section 1.21.) However, also by the EP, spacetime must be *locally* M_4, i.e., locally pseudo-Euclidean, and thus Riemannian! For, according to the EP, we can find at any event $\mathscr{P}$ (at least, in vacuum) a local inertial frame, i.e., a local coordinate system x, y, z, t with the property, among others, that the interval between $\mathscr{P}$ and *neighboring* events is given by

$$ds^2 = c^2 dt^2 - dx^2 - dy^2 - dz^2. \tag{7.12}$$

In its original form the EP does not apply inside matter, for example inside the earth, where we cannot have a material "freely falling lab"; nevertheless, by extension, we shall assume that even *inside* matter spacetime is Riemannian, with signature $(+ - - -)$. After all, matter is mostly vacuum anyway. To assign a conceptual meaning to ds^2 inside matter, we imagine a small cavity hollowed out at the point of interest, and assume that ds^2 is not altered appreciably thereby. This is analogous to the standard method of defining a Newtonian field inside matter.

The recognition of the Riemannian structure of the world led Einstein to his brilliant scheme for general relativity. Spacetime in the presence of gravitating masses may be curved. Perhaps it would then be true that free test particles trace out geodesics in this curved four-space and, by extension, that light in vacuum follows *null* geodesics (along which $ds^2 = 0$). In Newton's theory, absolute space provided the "rails" along which free particles moved in the absence of gravity (and of all other forces). In Einstein's theory, spacetime provides the rails in the absence of other forces; gravitational force no longer exists: it has become absorbed into the geometry.

But just *how* should spacetime be curved? Since the spacetime structure would now determine inertial *and* gravitational effects, Mach's principle suggests that it is the gravitating matter of the universe which alone should cause the structure of spacetime. The big task, and one that occupied even the genius of Einstein for many years, was to discover the *field equations*, i.e., the equations which predict quantitatively how the material contents of spacetime are related to its metric.

GR is thus a field theory, in which the geometry plays the part of a field. There is no "action at a distance" in GR to conflict with the SR speed limit. A particle does not "feel" the sources directly, but rather the field which they have built up in its neighborhood.

Bernhard Riemann, in his celebrated inaugural lecture of 1854, had already suggested that the differential geometry of our *three*-space might be determined by "external forces." It is fascinating to speculate on the course that physics might have taken if Riemann had lit upon gravity as the curver of space. A geodesic law of motion would not have been too far-fetched even then. Classical mechanicists knew, for example, that a particle confined without friction or external forces to a curved surface follows a geodesic on that surface. Of course, a direct generalization, namely that a particle in a gravitational field follows a geodesic in curved three-space, can be rejected at once. For a geodesic is uniquely determined by its initial direction, whereas a gravitational orbit depends on initial direction *and* velocity. Now in spacetime an initial direction *includes* velocity. This fact alone could suggest consideration of a four-dimensional manifold of space and time in which gravitational orbits are geodesics. Such a geodesic hypothesis incorporates *two* of Galileo's fundamental discoveries in mechanics: (i) that the path in space and time of a particle through a gravitational field is independent of the particle, and (ii) the law of inertia, according to which a free particle "goes straight in space and time": now it goes as straight as it is possible to go in curved spacetime. It totally eliminates from gravitational theory the concepts of inertial and passive-gravitational mass, and the puzzle of why they are equal: their equality in the Newtonian picture is both consequence and prerequisite of the geodesic law. As for the spacetime, its first version would undoubtedly have been positive-definite,[2] and who knows how long it

---

[2] Its metric, by comparison with Hamilton's principle (1834), would have been determined as approximately $(V^2 - 2\varphi)dt^2 + dx^2 + dy^2 + dz^2$—with φ the Newtonian potential and V a universal and rather large velocity—by arguments along the lines of our Section 7.7 below.

would have taken to arrive at the Minkowskian signature, and finally at SR as the local version of the theory?

But the kindling of Riemann's idea had to wait for Einstein. With Einstein's overall plan for GR in mind, let us now look at some of the details. The interval ds between neighboring events $\mathscr{P}$, $\mathscr{Q}$ in spacetime is a more complicated concept than that of distance between points on a surface or in three-space, but it is a perfectly definite physical quantity nevertheless. By definition, intervals from $\mathscr{P}$ can be determined by making the usual measurements in a small, freely falling, nonrotating, fully calibrated laboratory—in other words, in a LIF at $\mathscr{P}$. Since in the Gaussian coordinates of spacetime the metric has 10 coefficients [analogous to E, F, G in (7.10)] it will in general be sufficient to know 10 intervals at $\mathscr{P}$ in order to know the metric at $\mathscr{P}$ completely. (Not *all* of these 10 arbitrarily chosen intervals must be null: it can be shown that the null intervals define the metric only up to a factor. But none of them need be spacelike.) In the case of null and timelike intervals, we can in practice dispense with the LIF. If a light signal can be sent from $\mathscr{P}$ to $\mathscr{Q}$, or vice versa, then the corresponding ds^2 would be zero in any LIF, and therefore it is zero in *all* coordinate systems. If a freely falling clock can be sent from $\mathscr{P}$ to $\mathscr{Q}$, then in a LIF ds would be c times the proper time elapsed at the clock, and thus it is so in all coordinate systems. The direct measurement of spacelike intervals is more complicated. For example, we might arrange for a freely falling nonrotating ruler to have its ends present simultaneously at $\mathscr{P}$ and $\mathscr{Q}$: then its proper length would measure $|ds|$. But in order to ensure that it meets $\mathscr{P}$ and $\mathscr{Q}$ simultaneously, it would have to carry synchronized clocks, which is hardly an improvement over using a LIF. Methods can be devised for measuring spacelike intervals without rulers. E.g., among all observers freely falling through $\mathscr{P}$, there will be a subset of observers who can bounce a radar signal off $\mathscr{Q}$ in such a way that emission and retrieval of the signal occur at equal time intervals before and after $\mathscr{P}$, respectively; $|ds|$ is then equal to the radar distance of $\mathscr{Q}$ from these observers. (A Minkowski diagram, representing the LIF at $\mathscr{P}$, will make this clear.)

As we have seen, a freely falling clock (which, by hypothesis, follows a geodesic) will read $c^{-1} \int ds$ along its worldline. If that clock is pushed out of its free path, say by the action of a rocket engine or an electric field, and it is an "ideal" clock, then in each LIF along its path it still reads $c^{-1}ds$, by the clock hypothesis. Thus we arrive at the GR form of the clock hypothesis: an ideal clock, free or not, reads $c^{-1} \int ds$ along its worldline.

The worldlines of free particles in pseudo-Euclidean M_4 have linear equations and thus certainly satisfy the geodesic hypothesis. In M_4 the *maximal* property of these geodesics is also obvious: If a free particle A moves from event $\mathscr{P}$ to event $\mathscr{Q}$, then a clock attached to A reads $c^{-1} \int ds$ along this worldline. The clock on any other particle B present at $\mathscr{P}$ and $\mathscr{Q}$ must read *less*, since, in the rest frame of A, B can be regarded as the traveling "twin" of Section 2.14 who gets back *younger*.

It may be thought that the preceding argument can serve as *proof* of the geodesic hypothesis in general spacetime. Suppose, for example, a free test particle A, within a freely falling laboratory L, is in circular orbit around a mass center. Could not a neighboring particle B again be regarded as the twin in L? The flaw in this argument is that it assumes a LIF to exist over an extended time, whereas, in general, LIF's can be assumed to exist only in the immediate neighborhood of any given event. There is, in fact, no simple "proof" of the geodesic law of motion. However, it turns out that the field equations which Einstein eventually postulated contain that law implicitly.

The "grain" of special-relativistic spacetime, i.e., the existence of null cones at each event, and of three different kinds of displacement (timelike, spacelike, null), is impressed on the general-relativistic spacetime also through the LIF's. The cones will no longer be "parallel" to each other everywhere, and their generators (photon worldlines) will no longer be "straight." But still, a particle worldline will be *within* the cone at each of its points, and a photon will travel *along* the cones.

The reader may now be impatient to see whether indeed spacetimes exist that can be identified with known gravitational situations, and whether the geodesics in these spacetimes approximate to the Newtonian orbits. For we must not forget that Newton's theory of gravitation agrees almost perfectly with the observed phenomena throughout an enormous range of classical applications. *Any* alternative theory must yield the same predictions to within the errors of classical observations. In Sections 7.6 and 7.7 we shall compare Einstein's with Newton's theory in some simple situations. A useful tool for this work is developed in Section 7.5.

7.5   The Gravitational Doppler Effect

For the argument of our next section, we shall need a quantitative formulation of the gravitational Doppler effect which was already discussed qualitatively in Section 1.21. Suppose an elevator cabin of height dl is dropped from rest in a gravitational field of strength g, and at the same time a photon of frequency ν is emitted from its ceiling towards the floor. By the equivalence principle the signal takes a time $dt = dl/c$ to reach the floor, at which time this floor moves at speed $du = gdl/c$ relative to the elevator shaft. Also by the EP, no change in frequency is observed in the falling cabin. Hence an observer B at rest in the shaft a distance dl below the emission point (we neglect the small distance $\frac{1}{2}gdt^2$ moved by the floor in time dt) can be considered to move with speed du into a wave of frequency ν, thus observing a Doppler (blue) shift, given to first order by the classical formula

$$\frac{\nu + d\nu}{\nu} = \frac{c + du}{c} = 1 + \frac{gdl}{c^2}, \quad \text{or} \quad \frac{d\nu}{\nu} = \frac{gdl}{c^2}. \tag{7.13}$$

[see (3.5)]. Of course, B is not an inertial observer. However, here (and in all similar circumstances) we can assume that B makes the same measurements as a freely falling (inertial) observer B' momentarily at rest relative to B. This is justified by the clock and length hypotheses, according to which the readings of the clocks and rulers of B, though accelerated, will momentarily coincide with those of B'.

Formula (7.13) can be generalized for a light signal making an angle α with the field lines. We use the same cabin, but (3.5) now yields (7.13) with $du \cos \alpha$ in place of du, and thus with $gdl \cos \alpha$ in place of gdl, or

$$\frac{dv}{v} = \frac{\mathbf{g} \cdot d\mathbf{l}}{c^2} \tag{7.14}$$

in vectorial notation, where $d\mathbf{l}$ now measures the signal path.

If (as in Newtonian theory) the field $\mathbf{g}$ is derivable from a potential φ according to the equation

$$\mathbf{g} = -\mathbf{grad}\ \varphi, \tag{7.15}$$

then $\mathbf{g} \cdot d\mathbf{l} = -d\varphi$ for an infinitesimal path $d\mathbf{l}$. Substituting this into (7.14) we get

$$dv/v = -d\varphi/c^2, \tag{7.16}$$

or, integrating over a finite light path,

$$\frac{v}{v_0} = \exp\left(\frac{-\Delta\varphi}{c^2}\right), \tag{7.17}$$

where v_0 and v are the initial and final frequencies, respectively, and $\Delta\varphi$ is the total increment of potential over the path.

The theory of the gravitational Doppler effect can be applied to derive another interesting result: *gravitational time dilation*. If two standard clocks are fixed in an arbitrary static gravitational field at points whose potential differs by $\Delta\varphi$, then the clock at the lower potential goes slow by the Doppler factor $D = \exp(-\Delta\varphi/c^2)$ relative to the clock at the higher potential, as judged by mutual viewing. For suppose that the clock being viewed ticks in time with the wavecrests of the light whereby it is seen. The Doppler factor then tells precisely at what frequency these ticks are observed at the viewing clock, and our assertion is established.

If originally two standard clocks are adjacent and synchronized, and then one is taken to a place of lower potential and left there for a time, and finally brought back, that clock will clearly read slow by a factor D relative to the one that remained fixed—except for the error introduced by the two journeys. But whatever happens *during* the motions is independent of the total dilation at the lower potential, and can thus be dwarfed by it. Hence a "twin" at a lower potential stays younger than his twin at a higher potential. (At the surface of the earth the relevant dilation factor, compared to "infinity," is 1.000 000 000 8.) Indeed, it is cheaper to buy "youth" by going to live on a very dense planet than by fast cruising through space. For in the latter case the energy expended is directly proportional to the dilation factor attained (minus one), whereas in the former case it is evidently proportional to the logarithm of that factor.

It may be noted that gravitational time dilation and Doppler effect—which are essentially identical phenomena—are by their *nature* mixtures of the classical Doppler effect and the special-relativistic time dilation effect. This can be seen by translating two extreme special-relativistic situations via the EP into gravitational language. One is the rocket, corresponding to our argument leading to (7.13); it involves only the classical Doppler effect. The other is the turntable (cf. Exercise 7.13), which involves only SR time dilation (i.e., classically there would be *no* Doppler shift).

7.6 Metric of Static Fields

We now wish to explore the spacetime structure of a static gravitational field. Such a field might be caused, for example, by a large arbitrarily shaped massive body at rest. Let us coordinatize the static three-space of this field with a set of convenient coordinates x_1, x_2, x_3—which may not be Euclidean: it is very possible that this three-space is curved. In order to assign a time coordinate t to events, we place clocks at all the lattice points of a sufficiently fine grid of surfaces $x_i = $ constant $(i = 1, 2, 3)$.

The metric of the spacetime will be a quadratic form in the coordinate differentials,

$$ds^2 = Adt^2 + Bdx_1^2 + Cdx_2^2 + Ddx_3^2 + Edtdx_1 + \cdots, \qquad (7.18)$$

where in general the metric coefficients A, B, C etc., must be expected to be functions of all the coordinates. Can we assume that in the static case they are all time-independent? Not unless we choose t judiciously: even the time independence of the Minkowski metric (7.12) can be spoiled by introducing a "bad" time coordinate t', e.g., by the transformation $t = xt'$. (This leads to time-dependent terms in the metric since dt gets replaced by $xdt' + t'dx$.) A time coordinate which makes the metric time-independent will be called "good."

We shall now show that a necessary and sufficient condition for the metric coefficients to be time-independent is that light signals tracing out the same path always take the same coordinate time. Consider a light signal over some path

$$x_1 = x_1(\lambda), \quad x_2 = x_2(\lambda), \quad x_3 = x_3(\lambda), \qquad (7.19)$$

where λ is a parameter. The signal must satisfy $ds^2 = 0$ and thus, for an infinitesimal portion of the path, $0 = Adt^2 + B(dx_1/d\lambda)^2 d\lambda^2 + C(dx_2/d\lambda)^2 d\lambda^2 + \cdots$, i.e.,

$$Adt^2 + \alpha dt d\lambda + \beta d\lambda^2 = 0, \qquad (7.20)$$

where α, β involve time only through the metric coefficients. In general this equation yields two solutions for $dt/d\lambda$, corresponding to the two senses of traversing the path; and both solutions are time-independent *if* the metric coefficients are time-independent. Conversely, *all* the metric coefficients *must* be time-independent if (7.20) is to give time-independent solutions for all possible paths. (To see this, we first consider infinitesimal paths with two spatial coordinates constant, then paths with one spatial coordinate constant.) Since the coordinate time for traversing a finite path results from integrating a solution of (7.20), our condition is established.

To find a "good" time coordinate in practice, we first note that the condition "equal times for equal light paths" is equivalent to the condition that any two coordinate clocks go at the same rate as judged by mutual viewing. For if each sees the other "tick" at the same rate as itself, each light signal from the one clock to the other takes the same time as measured by these two clocks, and conversely. Now, as we have seen in Section 7.5, if two *standard* clocks, say C_1 and C_2, are fixed at potential φ_1 and φ_2, then the ratio of their rates (i.e., frequencies), as judged by mutual viewing, is given by

$$\frac{v_1}{v_2} = \exp\frac{\varphi_1 - \varphi_2}{c^2}. \tag{7.21}$$

Thus if we agree on some arbitrary zero point Z of the potential, a clock at potential φ will be rate-synchronized relative to the standard clock at $Z(\varphi = 0)$ if we readjust its natural rate v to Dv, where

$$D = \frac{v_Z}{v} = \exp\left(-\frac{\varphi}{c^2}\right). \tag{7.22}$$

Moreover, it follows from (7.21) that when all clocks are so synchronized with the standard clock at Z, *any* two clocks will be rate-synchronized with each other.

It remains to synchronize the *settings* of our coordinate clocks. To this end we must discuss the difference between *static* and *stationary* fields. Consider these two possible properties of fields: (i) time independence and (ii) time reversibility. By (i) we shall understand that the Doppler shift between any two fixed points in the field is constant in time. As we have seen, this allows us to find a "good" time coordinate in terms of which the metric coefficients are time-independent. (This process is independent of the existence of a potential, since it merely involves changing the natural rate of the coordinate clocks so as to neutralize their time dilation relative to some arbitrary point Z.) By (ii) we shall understand that for every possible motion of a photon, the time-reversed motion—in some "good" time—is also possible. (It then follows that the same property also holds for finite particles.) All fields which satisfy (i) are called *stationary*. They include, for example, the "gravitational" field on a turntable, which does *not* satisfy (ii). Fields which satisfy *both* (i) and (ii) are called *static*.

We shall now prove that staticness corresponds to the existence of a coordinate time t in terms of which the metric is not only time-independent, but also without time-space cross terms (such as $dtdx$). Assume that a "good" t has already been chosen. If we set $ds^2 = 0$ and $dx_2 = dx_3 = 0$ for a light signal in the spatial direction $(dx_1, 0, 0)$, then the metric (7.18) yields

$$Adt^2 + Bdx_1^2 + Edtdx_1 = 0.$$

This equation must give equal and opposite values of dt/dx_1 for the signal to be time-reversible, which implies $E = 0$. Similar arguments can be made for the vanishing of the coefficients of $dtdx_2$ and $dtdx_3$. Conversely, *if* all dt cross terms vanish, the α term in (7.20) is absent and *every* infinitesimal light signal is reversible. Since finite light signals result from integrating infinitesimal ones, our assertion is proved.

Lastly, we show how to achieve a synchronization of clock settings in practice, *if* the field is static. Assume that the clocks are already rate-synchronized. Now synchronize the settings of all clocks with that of an arbitrary clock C, by requiring that a signal from any clock A to C shall take the same coordinate time as a signal from C to A. Before such synchronization, a signal emitted at C at $t_C = 0$ might arrive at A at $t_A = \lambda$, whereas a signal emitted at A at $t_A = 0$ might arrive at C at $t_C = \mu$. To achieve equality of travel times it is merely necessary to reset the clock at A thus: $t_A \rightarrow t'_A = t_A + \frac{1}{2}(\mu - \lambda)$. Then the coordinate travel time in *either* direction becomes $\frac{1}{2}(\lambda + \mu)$. Now if all clocks are so synchronized with C, then any two of them are also setting-synchronized with each other. For suppose that the coordinate times for a photon to travel from C to B to A and back to C are α, γ, β, respectively (see Figure 7.7), so that the total time elapsed at C is $\alpha + \beta + \gamma$. By hypothesis, the motion is possible in reverse, with the same total elapsed time. Since the paths CA and BC will take β and α, AB must necessarily take γ, and our assertion is proved.

When we have synchronized the rates and settings of our coordinate clocks as indicated, the metric (7.18) reduces to

$$ds^2 = Adt^2 - d\sigma^2, \qquad (7.23)$$

where $d\sigma^2$ is a time-independent quadratic in the spatial differentials. So for $dx_1 = dx_2 = dx_3 = 0$ (the worldline of a coordinate clock) we get $ds^2 = Adt^2$. But, since we have deliberately altered the natural clock rates, we know

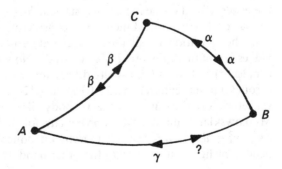

Figure 7.7

that $dt = D(c^{-1}ds)$, where D is given by (7.22). Consequently $A = D^{-2}c^2$, and so

$$ds^2 = \exp\left(\frac{2\varphi}{c^2}\right)c^2dt^2 - d\sigma^2. \tag{7.24}$$

For a light signal, $ds^2 = 0$, and thus $d\sigma = \pm D^{-1}cdt$, which shows that $d\sigma$ is "radar" distance in the three-space. But small rulers also measure radar distance, as can be seen by performing the experiment in their rest LIF and appealing to the clock and length hypotheses. Thus $d\sigma^2$ is just the spatial metric of the three-space of our field. To predict its exact form we need field equations.

We have now obtained in (7.24) *the general form of the spacetime metric of every static field*. The Doppler shift v_1/v_2 in the light between points with potential $\varphi = \varphi_1$ and $\varphi = \varphi_2$ is given *precisely* by (7.21). It may be thought that our apparently approximate method of obtaining Equation (7.21) in Section 7.5 makes (7.24) approximate also. But this is not so. Our method of arriving at (7.23) is exact, and we can evidently write (7.23) in the form (7.24). [As in the Newtonian case, φ is indeterminate up to an additive constant K; adding K to φ is equivalent to (i) shifting the arbitrary zero point Z, and (ii) rescaling the coordinate time t by a factor $\exp(-K/c^2)$.] The metric (7.24) *implies* that natural and coordinate clock frequencies at each point differ by a factor $\exp(\varphi/c^2)$. Consequently, since in this metric each clock sees every other clock tick at the same rate as itself, the Doppler shift between two points is given precisely by (7.21). We may regard this as the physical definition of the potential. Equation (7.21) implies (7.16), and thus (7.14) with (7.15), in which we recognize g as the field strength. In fact, it can be shown by tensor methods that the φ of (7.24) is related *precisely* by (7.15) to the field strength (which is now defined as minus the proper vector acceleration of a point fixed in the metric $d\sigma^2$), provided (7.15) is evaluated in the rest LIF of the point. (Cf. Exercise 8.6.)

7.7 Geodesics in Static Fields

The chief use of the spacetime metric occurs in calculating particle and light paths in the field as the geodesics of that metric. In the last section we obtained the general form of the metric of a static field, (7.24), up to uncertainty of its spatial part. However, dimensional considerations suggest that in a *weak* field the deviation of all the metric coefficients from their Minkowskian values are of the same order of magnitude. Now the coefficient of c^2dt^2, viz. $\exp(2\varphi/c^2) \approx 1 + 2\varphi/c^2$, usually differs very little from unity. For example, throughout the external field of the sun, $|2\varphi/c^2| < 0.5 \times 10^{-5}$. Thus the spacetime even around a massive body like the sun will be very nearly Minkowskian. Suppose we approximate for $d\sigma^2$ with a flat-space metric, e.g., $dx_1^2 + dx_2^2 + dx_3^2$. Do we thereby introduce a 50% inaccuracy? It depends: for light paths, yes. On the other hand, for "slow" orbits ($v \ll c$) the coefficient of $c^2 dt^2$ contributes vastly more than the spatial coefficients. If we approximate the former with unity, we essentially throw out gravity. If we approximate $d\sigma^2$ with flat space, we introduce only a small error. This can be seen as follows. A timelike geodesic is found as the longest of a bundle of neighboring worldlines connecting two events in spacetime. A slow-motion wordline in slightly deformed M_4 is almost parallel to the time axis. A deformation of the time dimension therefore has a first-order effect on the length of that line, whereas a deformation of the spatial dimensions has only a second-order effect. Quantitatively, consider a static metric $ds^2 = Ac^2 dt^2 - B d\sigma^2$ ($d\sigma^2$ flat). Inasmuch as they differ from unity, A and B measure deviations from M_4. The worldline of a particle moving with coordinate speed $v = d\sigma/dt$ has $ds^2 = dt^2(Ac^2 - Bv^2)$. Thus the space and time deviations are weighted in the ratio $v^2 : c^2$. For all the sun's planets, for example, $v < 50$ km/sec, so that $v^2 : c^2 < 3 \times 10^{-8}$, which illustrates the smallness of the spatial contribution. But for high-speed particles, and especially for light, this contribution can and does become significant, even in weak fields.

And now comes the big test. In the light of the above remarks, the "slow-motion" geodesics of the metric (7.24), with $d\sigma^2$ approximated by a flat 3-metric, must give, to high accuracy, the familair Newtonian orbits in a "weak" field. If they do not, the situation will not be saved by any field equations telling us the exact form of $d\sigma^2$, and the plan for GR will have to be abandoned.[3] Working to first order, let us approximate $\exp(2\varphi/c^2)$ by $1 + 2\varphi/c^2$, so that (7.24) reads

$$ds^2 = \left(1 + \frac{2\varphi}{c^2}\right)c^2 dt^2 - d\sigma^2, \qquad (7.25)$$

with $d\sigma^2$ assumed flat. For a particle worldline between events $\mathscr{P}_1$ and $\mathscr{P}_2$, at coordinate times t_1 and t_2, we have

$$\int_{\mathscr{P}_1}^{\mathscr{P}_2} ds = \int_{t_1}^{t_2} \frac{ds}{dt}\, dt = \int_{t_1}^{t_2} (c^2 + 2\varphi - v^2)^{1/2} dt, \qquad (7.26)$$

where $v = d\sigma/dt$ is the coordinate velocity of the particle. For the worldline to be geodesic, this integral must be a maximum. Now, if φ/c^2 and v^2/c^2 are both small compared to unity, we have

$$(c^2 + 2\varphi - v^2)^{1/2} = c[1 + (2\varphi - v^2)/c^2]^{1/2} \approx c[1 + (\varphi - \tfrac{1}{2}v^2)/c^2],$$

and thus

$$\int_{\mathscr{P}_1}^{\mathscr{P}_2} ds = \frac{1}{c} \int_{t_1}^{t_2} (c^2 + \varphi - \tfrac{1}{2}v^2) dt = \frac{1}{c} \int_{t_1}^{t_2} (U - T) dt, \qquad (7.27)$$

where we have written U for $c^2 + \varphi$ and T for $\tfrac{1}{2}v^2$. But the definition of Newtonian potential is arbitrary up to an additive constant, and so U is as good a potential as φ; and in T we recognize the Newtonian kinetic energy per unit mass of the particle. Hence the requirement that the last integral of

---

[3] It is hard to see what other geometric law of motion one could possibly fall back on.

(7.27) be maximal is recognized as Hamilton's principle, which requires that $\int (T - U)dt$ be minimal for the path of a particle. (In our formula the terms are reversed, since U contains the large constant c^2 which makes $U - T > 0$.) This shows that the slow-motion geodesics of **(7.25)** indeed coincide with the Newtonian orbits, in first approximation. It must have been a memorable moment when, by some such calculations as the above, Einstein first realized that his geodesic hypothesis "worked."

The present result well illustrates the "man-made" character of physical theories. It is really remarkable how the same empirically known orbits can be "explained" by two such utterly different models as Newton's universal gravitation and Einstein's geodesic law.

We can illustrate the geodesic law from an even more elementary point of view, in a very simple situation. Consider a spherically symmetric mass m far away from all other masses. Its Newtonian potential φ is $-mG/r$, G being the constant of gravitation and r the distance from the center of the mass. We adopt the usual polar coordinates r, θ, ϕ for $d\sigma^2$. Under these conditions, the metric **(7.25)** becomes

$$ds^2 = \left(1 - \frac{2Gm}{rc^2}\right)c^2dt^2 - dr^2 - r^2(d\theta^2 + \sin^2\theta d\phi^2). \qquad (7.28)$$

Let us look for circular orbits around the central mass. It is well known that such orbits are possible in classical mechanics, say at distance r from m, and with constant angular velocity ω: it is merely necessary that the centrifugal force $\omega^2 r$ should balance the gravitational force Gm/r^2, and so

$$\omega^2 = Gm/r^3. \qquad (7.29)$$

This, of course, is a special example of Kepler's third law.

Now, starting from Einstein's point of view, let us look for a circular orbit in the spacetime **(7.28)** which maximizes s. We shall assume that the orbit lies in the equatorial plane $\theta = \pi/2$, and that it has constant angular velocity $\omega = d\phi/dt$. Then, by **(7.28)**, we have for a complete revolution, say from time t_1 to time t_2,

$$s = \int_{t_1}^{t_2} ds = \int_{t_1}^{t_2} \left\{c^2\left(1 - \frac{2Gm}{rc^2}\right) - r^2\omega^2\right\}^{1/2} dt = \{\ \ \}^{1/2}\frac{2\pi}{\omega}. \qquad (7.30)$$

For a given ω, s will be maximal when the braced expression is maximal, and that occurs when

$$\frac{d}{dr}\{\ \ \} = \frac{2Gm}{r^2} - 2r\omega^2 = 0,$$

which is equivalent to **(7.29)**.

Actually, the maximizing procedure we applied to **(7.30)** is not quite the one relevant to geodesics. Geodesics have greater interval length than all neighboring curves between any two *fixed points*. The variation of r in **(7.30)** indeed produces neighboring curves, but curves that have *no* points (events) in common. Nevertheless, the fact that *a* maximum can be so obtained is an indication that it is probably the relevant one. (Our method is similar to that of tentatively finding geodesics on a sphere as the largest in a family of parallel circles.) And indeed, the rigorous calculation of the geodesics of the metric (7.28) yields (7.29) as a solution (see the end of Section 8.4).

What we have done so far is not yet GR. It merely shows the reasonableness of Einstein's plan for GR. In GR one does not have to fall back on Newtonian approximations in order to find a metric, but instead one uses the field equations. And one does not have to make do with approximate ways to work out geodesics or any other geometric features of the spacetimes under discussion: one has at one's disposal the elegant and fully developed theory of Riemannian geometry. This, incidentally, is one of the classic examples where a pure mathematician's flight of fancy (Riemann's n-dimensional geometry of 1854 and Ricci's tensor calculus for it) later became the physicists' bread and butter. We have put off this mathematics as long as possible, but now at last we must take a quick dip into it.

CHAPTER 8

Formal Development of General Relativity

8.1 Tensors in General Relativity

We shall not manipulate tensors very extensively in this book. Nevertheless tensors are indispensable for any genuine understanding of GR. Even the field equations cannot be expressed without them. In this section we sketch out the basic tensor theory as far as we shall need it. That includes the metric tensor, the geodesic equations, the absolute derivative, and the curvature tensor. The reader may skim it lightly at first, and refer back to it as need arises.

The four-tensors of SR are tied to "standard" coordinate systems (x, y, z, t), in the sense that it is sufficient to consider four-tensor components only relative to each such system. These systems are related to each other by general Lorentz transformations, and, as we have seen, there is a rule which tells how the components of a four-tensor transform when a LT is applied to the coordinates. In GR, as in Riemannian geometry, more general coordinate systems are forced on us, and it is therefore convenient to allow *fully* arbitrary ("Gaussian") coordinates. These need have no direct physical significance; often there is not even a preferred coordinate that can be regarded as *the* time. For example, even in M_4 (the flat spacetime of SR) we can go from a set of standard coordinates x, y, z, t to a set of Gaussian coordinates of no direct physical significance such as $x^1 = x + 2y, x^2 = 2x - y,$ $x^3 = \exp(z + 2t), x^4 = \exp(z - 2t).$

The tensors of GR, accordingly, have components relative to *arbitrary* coordinate systems. These tensors are necessarily localized, i.e., associated with points in spacetime. As we transform from one Gaussian coordinate system to another, the components of any tensor undergo a typical transformation, which generally depends on the point at which the tensor is located. The prototype of a *contravariant* first-rank tensor (a vector) A^μ is the coordinate *differential* dx^μ ($\mu = 1, 2, 3, 4$). Since contravariant tensors are traditionally written with superscripts (earlier in this book we ignored that tradition), the coordinates themselves are now also written with superscripts (x^μ) so as to make dx^μ look like what it is, a contravariant vector. The coordinate *differences* Δx^μ, however, behave as vector components *only* under linear transformations, and the coordinates x^μ themselves only under linear homogeneous transformations, for only then do they transform like dx^μ.

Consider now a transformation of coordinates from x^μ to $x^{\mu'}$. (It is becoming usual to denote a new coordinate system—and tensor components in the new system—by priming the indices, rather than the kernel letters, and μ, μ' are regarded as different indices, just like μ, ν. Earlier in the book we ignored this convention too.) By the chain rule, the coordinate differentials transform thus:

$$dx^{\mu'} = \sum_{\mu=1}^{4} p^{\mu'}_\mu dx^\mu, \quad p^{\mu'}_\mu = \frac{\partial x^{\mu'}}{\partial x^\mu}. \tag{8.1}$$

This will be the transformation pattern for *any* contravariant vector. We can write it without the $\sum$ sign,

$$dx^{\mu'} = p^{\mu'}_\mu dx^\mu, \tag{8.2}$$

if we adopt Einstein's *summation convention*, according to which any repeated index in one term, once up, once down, implies summation over all its values (e.g., $A^\mu_\mu = A^1_1 + A^2_2 + A^3_3 + A^4_4$, $A_{\mu\nu\sigma}B^{\mu\nu} = A_{11\sigma}B^{11} + A_{12\sigma}B^{12} + A_{21\sigma}B^{21} + \cdots$). We shall use this convention henceforth.

If we write

$$p^\mu_{\mu'} = \frac{\partial x^\mu}{\partial x^{\mu'}}, \tag{8.3}$$

and define other p's similarly, we have, by the chain rule,

$$p^\mu_{\mu'} p^{\mu'}_{\mu''} = p^\mu_{\mu''}, \tag{8.4}$$

where μ'' refers to a third coordinate system $x^{\mu''}$. In particular,

$$p^\mu_{\mu'} p^{\mu'}_\nu = \delta^\mu_\nu = \begin{cases} 1 & (\text{if } \mu = \nu) \\ 0 & (\text{if } \mu \neq \nu), \end{cases} \tag{8.5}$$

where δ^μ_ν is the *Kronecker delta*, defined by the last equation. This shows that the matrices $(p^\mu_{\mu'})$ and $(p^{\mu'}_\mu)$ are inverses of each other.

The two kinds of p's defined in (8.1) and (8.3) are the coefficients that enter into the definition of a general tensor. Usually they are not constant but depend on position. The prototype of a *covariant* first-rank tensor B_μ (also called a vector) is the "gradient" of a scalar function φ (i.e., a function whose value at each point is unchanged under a change of coordinates):

$$\varphi_{\mu'} := \frac{\partial \varphi}{\partial x^{\mu'}} = \left(\frac{\partial \varphi}{\partial x^{\mu}}\right) p^{\mu}_{\mu'} = \varphi_{\mu} p^{\mu}_{\mu'}. \tag{8.6}$$

We say that the numbers $A^{\mu \cdots}_{\nu \cdots}$ are the components of an nth-rank tensor, contravariant in the indices $\mu, \ldots$, and covariant in the indices $\nu, \ldots$, (n indices altogether) if under a coordinate transformation $x^{\mu} \to x^{\mu'}$ they transform according to the following (linear) scheme:

$$A^{\mu' \cdots}_{\nu' \cdots} = A^{\mu \cdots}_{\nu \cdots} p^{\mu'}_{\mu} \ldots p^{\nu}_{\nu'} \ldots. \tag{8.7}$$

There is one p for each index on $A^{\cdots}_{\cdots}$, and all the unprimed indices on the right are summed over. (It turns out that under the "orthogonal" coordinate transformations relevant to three-vectors, there is no distinction between covariance and contravariance, so that three-vectors can be written indifferently with superscripts or subscripts.) The most important property of general tensors, as of four-tensors, is that an identity between two sets of tensor components which is true in one coordinate system is true in all coordinate systems. The proof of this is immediate from (8.7). As a consequence, we can always be sure that tensor equations express physical or geometrical facts, i.e., facts transcending the coordinate system used to describe them. Another important property of the transformation (8.7) is the *group property* [similar to that of the LT's—cf. Section 2.7(iii)]. For a sample proof, multiply (8.6) by $p^{\mu'}_{\nu}$, which yields, by reference to (8.5),

$$\varphi_{\mu'} p^{\mu'}_{\nu} = \varphi_{\mu} \delta^{\mu}_{\nu} = \varphi_{\nu},$$

proving *symmetry*. Also, if $\varphi_{\mu''} = \varphi_{\mu'} p^{\mu'}_{\mu''}$, then, from (8.6) and (8.4),

$$\varphi_{\mu''} = \varphi_{\mu} p^{\mu}_{\mu'} p^{\mu'}_{\mu''} = \varphi_{\mu} p^{\mu}_{\mu''},$$

proving *transitivity*. As a result of the group properties, we can construct tensors by specifying their components arbitrarily in one coordinate system and using the tensor transformation law (8.7) to define their components in all other systems. The group properties ensure that *all* sets of components will then be related tensorially. For if φ_{μ}, say, is so related to $\varphi_{\mu'}$ and $\varphi_{\mu''}$, $\varphi_{\mu'}$ is related tensorially to φ_{μ} (by symmetry) and consequently to $\varphi_{\mu''}$ (by transitivity).

A *zero tensor* has all its components zero—in all coordinate systems, by (8.7). *Scalar invariants* (often called just scalars, or invariants) are numbers unaffected by coordinate transformations. They can be regarded as tensors of rank zero! Sums of tensors and products of scalars with tensors are defined by the corresponding operations on the tensor components and are clearly tensors. So are *outer products* of tensors, e.g., $A_{\mu\nu\sigma} B^{\rho}_{\tau}$. Another important tensor operation is *contraction*, the summing over a pair of indices, one up, one down, e.g., $A_{\mu\nu}{}^{\mu}$. It reduces the rank of a tensor by two. For example,

$$A_{\mu'\nu'}{}^{\sigma'} = A_{\mu\nu}{}^{\sigma} p^{\mu}_{\mu'} p^{\nu}_{\nu'} p^{\sigma'}_{\sigma}, \quad \text{so } A_{\mu'\nu'}{}^{\mu'} = A_{\mu\nu}{}^{\sigma} \delta^{\mu}_{\sigma} p^{\nu}_{\nu'} = A_{\mu\nu}{}^{\mu} p^{\nu}_{\nu'}.$$

When all indices are used up in a contraction—as in $A_{\mu\nu}B^\mu C^\nu$—the result is a scalar. The last algebraic tensor operation is *index permutation*. For example, if the tensor components $A_{\mu\nu}$ are exhibited as a square pattern, like a matrix, $A_{\nu\mu}$ denotes the components of the "transposed" pattern, and those form a tensor, as is evident from (8.7). As a result, we can form such sums as $A_{\mu\nu} + A_{\nu\mu}$, and such equations as $A_{\mu\nu} = A_{\nu\mu}$.

The metric of a general spacetime can be written in the form

$$ds^2 = g_{\mu\nu}dx^\mu dx^\nu, \quad (g_{\mu\nu} = g_{\nu\mu}), \tag{8.8}$$

where the $g_{\mu\nu}$ are certain functions of the coordinates. Written in full, Equation (8.8) reads

$$ds^2 = g_{11}(dx^1)^2 + g_{22}(dx^2)^2 + g_{33}(dx^3)^2 + g_{44}(dx^4)^2$$
$$+ 2g_{12}dx^1 dx^2 + 2g_{13}dx^1 dx^3 + \cdots.$$

Under a change of coordinates $x^\mu \to x^{\mu'}$ (8.8) becomes

$$ds^2 = g_{\mu\nu}dx^{\mu'}p^\mu_{\mu'}.dx^{\nu'}p^\nu_{\nu'} = g_{\mu'\nu'}dx^{\mu'}dx^{\nu'},$$

where

$$g_{\mu'\nu'} = g_{\mu\nu}p^\mu_{\mu'}p^\nu_{\nu'}, \tag{8.9}$$

which shows that $g_{\mu\nu}$ is a tensor. By analogy with Euclidean spaces one defines the *squared magnitude* of a vector A^μ by $A^2 = g_{\mu\nu}A^\mu A^\nu$ and the *orthogonality* of two vectors A^μ, B^μ by $g_{\mu\nu}A^\mu B^\nu = 0$. Note that this is consistent with our earlier definitions in the case of four-vectors, where $g_{\mu\nu} = \text{diag}(-1, -1, -1, c^2)$.

The Kronecker delta is a tensor, since $\delta^\mu_\nu p^{\mu'}_\mu p^\nu_{\nu'} = p^{\mu'}_\nu p^\nu_{\nu'} = \delta^{\mu'}_{\nu'}$. It follows that $g^{\mu\nu}$, the elements of the (symmetric) inverse matrix of $(g_{\mu\nu})$, are components of a contravariant tensor. For they are uniquely defined by the equation $g_{\mu\nu}g^{\nu\sigma} = \delta^\sigma_\mu$, and this equation is tensorial, and thus universal, *if* the $g^{\mu\nu}$ are tensorial. The $g_{\mu\nu}$ and $g^{\mu\nu}$ are used to define the operations of *raising and lowering indices*. E.g., given A_μ, we define $A^\mu := g^{\mu\nu}A_\nu$; and, given B^μ, we define $B_\mu := g_{\mu\nu}B^\nu$. Similarly, $C^\mu_{\ \nu} = g^{\mu\sigma}C_{\sigma\nu}$, and so on. These operations are often applied throughout an equation; e.g., each of $A_\mu + B_\mu = C_\mu$ and $A^\mu + B^\mu = C^\mu$ implies the other.

We recall from Section 7.4 that s measures proper time along a timelike path, if $c = 1$, as we shall usually assume. Now suppose a particle has a worldline whose parametric equation is

$$x^\mu = x^\mu(s), \tag{8.10}$$

which gives the coordinates at each moment of its proper time s. The derivative dx^μ/ds is evidently a vector since

$$\frac{dx^{\mu'}}{ds} = \frac{\partial x^{\mu'}}{\partial x^\mu}\frac{dx^\mu}{ds} = \frac{dx^\mu}{ds}p^{\mu'}_\mu. \tag{8.11}$$

But d^2x^μ/ds^2 is *not* a vector. For, if we differentiate (8.11), we get

$$\frac{d^2x^{\mu'}}{ds^2} = \frac{d^2x^\mu}{ds^2}p^{\mu'}_\mu + \frac{dx^\mu}{ds}\frac{d}{ds}(p^{\mu'}_\mu), \tag{8.12}$$

and the last term will not usually vanish. This argument shows quite generally that a scalar derivative of a tensor is usually not a tensor, the reason being that the p's are in general not constant (as they are in four-tensor theory).

There is, however, a way of defining certain derivatives of tensors which *are* themselves tensors, and which, moreover, reduce to the ordinary derivatives in the LIF. For this purpose one needs the so-called *Christoffel symbols* $\Gamma^\mu_{v\sigma}$, defined by the equation (summation convention!)

$$\Gamma^\mu_{v\sigma} = \Gamma^\mu_{\sigma v} = \tfrac{1}{2}g^{\mu\tau}(g_{\tau v,\sigma} + g_{\tau\sigma,v} - g_{v\sigma,\tau}), \tag{8.13}$$

where here and hereafter a comma denotes partial differentiation: $g_{\tau v,\sigma} = (\partial/\partial x^\sigma)g_{\tau v}$. These Γ's are *not* tensor components. It will turn out that they vanish at the origin of each LIF.

Now, if A^μ is any contravariant vector, it can be verified that the object DA^μ/ds, defined by the equation

$$\frac{D}{ds}A^\mu = \frac{d}{ds}A^\mu + \Gamma^\mu_{v\sigma}A^v\frac{dx^\sigma}{ds}, \tag{8.14}$$

is a tensor. It is called the *absolute derivative of A^μ* in the direction of dx^μ. [The absolute derivative of other tensors can be defined similarly, see **(8.24)**.] At the origin of each LIF it reduces to the ordinary derivative, if—as we shall prove—the Γ's vanish there. Thus, for example, $(D/ds)(dx^\mu/ds)$ will reduce to d^2x^μ/ds^2 in the LIF. But that is recognized as the four-acceleration of our particle. We shall therefore not be surprised to learn that the vanishing of this vector,

$$\frac{D}{ds}\left(\frac{dx^\mu}{ds}\right) = \frac{d^2x^\mu}{ds^2} + \Gamma^\mu_{v\sigma}\frac{dx^v}{ds}\frac{dx^\sigma}{ds} = 0, \tag{8.15}$$

turns out to be the differential equation for a geodesic, i.e., the rigorous solution of the problem of maximizing $\int ds$. Equation **(8.15)** is analogous to the classical equation $\mathbf{a} = \sum \mathbf{f}$ for the acceleration in a noninertial reference frame, where $\sum \mathbf{f}$ stands for the inertial forces (such as centrifugal force) per unit mass. Thus the Γ's are analogs of inertial forces, arising from the motion of the reference system, which explains why they vanish in the LIF.

We can test Equation **(8.15)** at once for the metric **(7.25)**, with Euclidean $d\sigma^2$. There, if $x^\mu = (x, y, z, t)$, $(g_{\mu v}) \approx \text{diag}(-1, -1, -1, c^2)$ and thus $(g^{\mu v}) \approx \text{diag}(-1, -1, -1, c^{-2})$. All $g_{\mu v}$ except g_{44} are constant. Hence, by **(8.13)**, of all $\Gamma^i_{\mu v}(i = 1, 2, 3)$ only

$$\Gamma^i_{44} = \frac{1}{2}\frac{\partial}{\partial x^i}g_{44} = \frac{\partial\varphi}{\partial x^i} \tag{8.16}$$

is nonzero. For a slowly moving particle, $ds \approx cdt$, and thus the first three of Equations **(8.15)** reduce to

$$\frac{d^2x^i}{dt^2} = -\frac{\partial\varphi}{\partial x^i}, \tag{8.17}$$

i.e., $\mathbf{a} = -\mathbf{grad}\ \varphi$, while the last turns out to be an identity, to the same approximation. This shows a little more directly (than via Hamilton's principle) the coincidence of slow-motion geodesics with Newtonian orbits.

A most important result for the development of GR is the following: Given any point P in a Riemannian n-space, we can construct a system of coordinates around P in which the Γ's vanish *at* P. First choose arbitrary coordinates y^μ. Consider any geodesic $y^\mu = y^\mu(s)$ through P. Suppose its "tangent vector" dy^μ/ds at P is a^μ. Now if Q lies on that geodesic at distance s from P, let it be labeled by new coordinates

$$x^\mu = a^\mu s. \qquad (8.18)$$

In the new coordinates, all geodesics through P have equations of the form (**8.18**), and therefore satisfy $d^2x^\mu/ds^2 = 0$. Comparison with (**8.15**) then shows that the Γ's vanish at P. *Any* coordinates in which the Γ's vanish at P are called "geodesic" at P. There are, in fact, infinitely many such geodesic coordinate systems at P. The relation between any two, x^μ and $x^{\mu'}$, is "locally linear," i.e., $p^{\mu'}_{\mu\nu} := \partial^2 x^{\mu'}/\partial x^\mu \partial x^\nu = 0$ at P; and any system so related to a geodesic system is itself geodesic. For, at P, the tensor (**8.15**) reduces to d^2x^μ/ds^2 in each geodesic system, thus the last term in (**8.12**) vanishes, i.e., $p^{\mu'}_{\mu\nu}(dx^\mu/ds)(dx^\nu/ds) = 0$ for all geodesics through P; hence $p^{\mu'}_{\mu\nu} = 0$. This argument is reversible, and so our assertion is established. We also note that the Γ's and the first derivatives of the g's vanish together. This follows from (**8.13**) and its "inverse"

$$g_{\mu\nu,\sigma} = g_{\mu\tau}\Gamma^\tau_{\nu\sigma} + g_{\nu\tau}\Gamma^\tau_{\mu\sigma}. \qquad (8.19)$$

The importance of the above result to GR lies in the fact that the geodesic coordinate systems at an event $\mathscr{P}$ in *spacetime* are, in fact, the LIF's at $\mathscr{P}$, plus all systems obtainable from them by locally linear transformations. For consider any geodesic system at $\mathscr{P}$. Complete squares at $\mathscr{P}$ as we did in (**7.11**)—a linear transformation—to produce locally Minkowskian coordinates x, y, z, t with origin at $\mathscr{P}$. Then consider the history of the set of points $x, y, z = $ constant. The distance between all neighboring such points near $\mathscr{P}$ is constant, since $\partial g_{\mu\nu}/\partial t = 0$ at $\mathscr{P}$. So we have a *rigid* frame. Moreover, each of its points traces out a geodesic: $x, y, z = $ constant, $t = s$, satisfies the "local" geodesic equation $d^2x^\mu/ds^2 = 0$. Hence the system "falls" freely, and our assertion is established. Note that LIF's are not just locally Minkowskian (i.e., locally orthogonal) systems: *these* could be produced from *any* coordinate system simply by completing squares at $\mathscr{P}$. LIF's (like all geodesic systems) are geometrically characterized by a coordinate net that is locally a net of geodesics, calibrated linearly. This can be illustrated by a curved 2-space, e.g., the sphere. The usual polar coordinates θ and ϕ (cf. Figure 7.6) are *not* geodesic except at points on the equator: only there are the coordinate lines locally (and in fact globally) geodesic.

Closely connected with the idea of geodesic coordinates in Riemannian spaces is the idea of *parallel transport*. In order to transport a vector A^μ parallelly along a prescribed curve, we go to a local geodesic system at a point P of the curve, and *in that system* define parallel transport at P by

$dA^\mu/ds = 0$. Reference to (8.14) shows that in *all* coordinate systems the tensor equation $DA^\mu/ds = 0$ will then be true. This is therefore the *equation of parallel transport*. On a curved surface it amounts to the following: Draw an arbitrary curve on a plane piece of paper, and a parallel vector field along that curve. Cut out a thin strip bounded on one side by this curve, and glue it to the surface. The vector field will then be parallel along the curve also on the surface. Note from (8.15) that a geodesic transports its tangent vector dx^μ/ds parallelly along itself, *and could be so defined*.

As an application of this idea to spacetime, consider the free fall of a test gyroscope (a spinning test particle). We would expect its axis to be parallelly transported along its geodesic path. This is one theoretical basis for the recently suggested gyroscopic test of GR, which involves sending a gyroscope into free orbit around the earth. (More satisfactory *dynamical* arguments lead to the same predictions.)

One tensor that plays a fundamental role in GR is the *Riemann curvature* tensor $R^\mu_{\nu\rho\sigma}$ defined as follows [for notation, see (8.13)]:

$$R^\mu_{\nu\rho\sigma} = \Gamma^\mu_{\nu\sigma,\rho} - \Gamma^\mu_{\nu\rho,\sigma} + \Gamma^\mu_{\tau\rho}\Gamma^\tau_{\nu\sigma} - \Gamma^\mu_{\tau\sigma}\Gamma^\tau_{\nu\rho}. \tag{8.20}$$

That this *is* a tensor of the type indicated by its four indices is not obvious, but it can be verified from the definition. It has $4^4 = 256$ components in four-dimensional spacetime; but in fact only 20 of these are independent, because $R^\mu_{\nu\rho\sigma}$ can be shown to possess certain symmetries. These are best exhibited by its fully covariant version $R_{\mu\nu\rho\sigma} = g_{\mu\tau}R^\tau_{\nu\rho\sigma}$:

$$R_{\mu\nu\rho\sigma} = R_{\rho\sigma\mu\nu} = -R_{\nu\mu\rho\sigma} = -R_{\mu\nu\sigma\rho}, \quad R_{\mu\nu\rho\sigma} + R_{\mu\rho\sigma\nu} + R_{\mu\sigma\nu\rho} = 0. \tag{8.21}$$

Note that, because of the structure of the Γ's [cf. (8.13)], $R^\mu_{\nu\rho\sigma}$ is entirely built up of the $g_{\mu\nu}$ and their first and second derivatives. Being a tensor, it cannot be made to vanish by a special choice of coordinates—for it would then vanish in all coordinates, by (8.7). Thus even in a LIF it will in general be nonzero. Of course, in *flat* space (e.g., M_4) there exist coordinates which make the g's constant and thus the Γ's zero everywhere; consequently, in flat space $R^\mu_{\nu\rho\sigma} \equiv 0$. The converse is also true: $R^\mu_{\nu\rho\sigma} \equiv 0$ is necessary *and* sufficient for a space to be flat (i.e., Euclidean or pseudo-Euclidean).

One way of seeing how $R^\mu_{\nu\rho\sigma}$ is connected with the geometric concept of curvature is to study "geodesic deviation." Consider two nearby and almost parallel geodesics, and let η^μ be a vector joining them and orthogonal to both. Then it can be shown from the geodesic equation (8.15) that

$$\frac{D^2\eta^\mu}{ds^2} : = \frac{D}{ds}\left(\frac{D\eta^\mu}{ds}\right) = (R^\mu_{\nu\rho\sigma}U^\nu U^\rho)\eta^\sigma, \tag{8.22}$$

where $U^\nu = dx^\nu/ds$ for one of the geodesics. Compare this with (7.6). [It is not difficult to deduce from (8.22) a formula for $K(U^\mu, \eta^\mu)$, the space curvature for the orientation (U^μ, η^μ); but we shall not need it.] Here, incidentally, we have a practical way of discovering the components of $R^\mu_{\nu\rho\sigma}$ in spacetime: Take a set of neighboring free test particles (four will generally be enough), and measure their four-velocities U^μ and 20 of their mutual acceleration components $D^2\eta^\mu/ds^2$; substitute in (8.22) and solve for the 20 independent components of the curvature tensor.

Finally, we must mention the *covariant derivative* of an arbitrary tensor $A^{\mu\cdots}_{\nu\cdots}$. It is defined thus,

$$A^{\mu\cdots}_{\nu\cdots;\sigma} = A^{\mu\cdots}_{\nu\cdots,\sigma} + \Gamma^\mu_{\tau\sigma}A^{\tau\cdots}_{\nu\cdots} + \cdots - \Gamma^\tau_{\nu\sigma}A^{\mu\cdots}_{\tau\cdots} - \cdots, \qquad (8.23)$$

where ",σ" stands for $\partial/\partial x^\sigma$ as in (8.13), and where we have a positive Γ term for each contravariant index of $A^{\mu\cdots}_{\nu\cdots}$ and a negative Γ term for each covariant index. The covariant derivative can be shown to be a tensor, with one covariant index more than $A^{\mu\cdots}_{\nu\cdots}$, namely σ. At the origin of geodesic coordinates (where $\Gamma^\mu_{\tau\sigma} = 0$) it reduces to the usual partial derivative. From this we may conclude that $g_{\mu\nu}, g^{\mu\nu}$, and δ^μ_ν have zero covariant derivative; and also that the operation " $;\sigma$ " is linear and satisfies the Leibniz rule. The *absolute derivative* is related to the covariant derivative as follows:

$$\frac{D}{ds}A^{\mu\cdots}_{\nu\cdots} = A^{\mu\cdots}_{\nu\cdots;\sigma}\left(\frac{dx^\sigma}{ds}\right). \qquad (8.24)$$

Thus it reduces to the ordinary derivative at the origin of geodesic coordinates [as we have already seen for the special case (8.14)]. Consequently it too is linear and satisfies the Leibniz rule. Second covariant derivatives, except of scalars, do not commute: one can prove that, for a scalar φ, $\varphi_{;\sigma;\tau} = \varphi_{;\tau;\sigma}$, but for a vector A^μ,

$$A^\mu_{;\sigma;\tau} - A^\mu_{;\tau;\sigma} = -A^\rho R^\mu_{\rho\sigma\tau}. \qquad (8.25)$$

8.2 The Vacuum Field Equations of General Relativity

In the neighborhood of any given event $\mathscr{P}_0$, the Newtonian gravitational field $\mathbf{f}$ can be split as follows:

$$\mathbf{f} = \mathbf{f}_0 + \Delta_0\mathbf{f},$$

where $\mathbf{f}_0$ is the field at $\mathscr{P}_0$ and $\Delta_0\mathbf{f}$ is defined by this equation. Since $\mathbf{f}_0$ can be "transformed away" by going to any (freely falling) local inertial frame S_0 at $\mathscr{P}_0$, all that is felt of $\mathbf{f}$ in S_0 is the so-called *tidal* field $\Delta_0\mathbf{f}$. For example, as the name implies, this is the kind of field that produces the tides on earth, since the earth (except for its rotation) constitutes a freely falling frame in the combined sun–moon gravitational field. Again, it is the tidal field in a freely falling elevator on earth that causes two free particles on a common horizontal to accelerate toward each other. Tidal forces always indicate the presence of an *intrinsic* gravitational field, i.e., one that cannot be ascribed to the choice of reference frame. Tidal forces are the only gravitational forces that can have tensorial representation, since the nontidal part can always be made to vanish in a LIF. The way to detect tidal forces is to observe a set of neighboring free test particles. If there are relative accelerations between them, then there are tidal forces. Hence tidal forces and curvature tensor occur together [cf. (8.22) *et seq.*] and the latter must be a measure of the former.

Now, every Newtonian gravitational field **g** is derivable from a potential φ in the usual way [cf. (7.15)],

$$g_i = -\partial\varphi/\partial x^i = -\varphi_i, \quad (i = 1, 2, 3), \tag{8.26}$$

where φ_i is defined by the last equation. (Similarly, φ_{ij} will denote $\partial^2\varphi/\partial x^i\partial x^j$.) The relative acceleration of two test particles separated by a small connecting three-vector η^i is therefore given by

$$d^2\eta^i/dt^2 = dg_i = -\varphi_{ij}\eta^j, \tag{8.27}$$

again using the summation convention. Thus it is the *second* derivatives of the potential that indicate an intrinsic field. We next recall that these second derivatives satisfy the so-called Poisson equation

$$\sum_{i=1}^{3} \varphi_{ii} = 4\pi G\rho, \tag{8.28}$$

which is the "field equation" of Newtonian gravitational theory, relating the sources of the field with the field itself. It is essentially the local version of the inverse square law. At first, however, we shall be interested in *vacuum* fields— such as the field around the sun—and then $\rho = 0$, and Poisson's equation reduces to Laplace's equation

$$\sum_{i=1}^{3} \varphi_{ii} = 0. \tag{8.29}$$

If we compare (8.27) with (8.22), we see that $R^\mu_{\nu\rho\sigma} U^\nu U^\rho$ corresponds to φ_{ij} and that the following is therefore an analog of Laplace's equation (8.29) in spacetime:

$$R^\mu_{\nu\rho\mu} U^\nu U^\rho = 0, \tag{8.30}$$

summation being implied again over the repeated index μ, as well as over the indices ν and ρ. If this is to hold independently of U^μ, which after all refers to a specific particle, we need

$$R^\mu_{\nu\rho\mu} =: R_{\nu\rho} = 0. \tag{8.31}$$

The tensor $R_{\nu\rho}$, defined by this equation, is called the *Ricci* tensor. (It *is* a tensor, since it arises from the "contraction" of another tensor, $R^\mu_{\nu\rho\sigma}$.) Because of (8.21) it possesses the symmetry

$$R_{\mu\nu} = R_{\nu\mu}, \tag{8.32}$$

and so the number of its independent components is 10. The vanishing of the 10 components of the Ricci tensor, then, is what the Newtonian analogy suggests as the vacuum field equations of GR. And this, indeed, was Einstein's proposal (1915). It has been strikingly vindicated: not only does GR, completed by these field equations (and their generalization to the interior of matter—see Section 8.8) reproduce within experimental error all those

Newtonian results that agree with observation, but where GR differs observably from Newton's theory, as in the original "three crucial effects" (the gravitational Doppler effect, the bending of light, and the advance of the perihelia of the planets), it is GR that is found to be correct. However, the gravitational Doppler effect as predicted by the EP (cf. Section 7.5) is not affected by the field equations. Hence, although it tests the EP—one of the cornerstones of GR—it cannot be regarded as a specific test for Einstein's field equations.

But how is it that instead of the *one* field equation (**8.29**) of Newtonian theory, there should be *ten* in GR? The reason is that the field equations must determine the whole metric, i.e., the $g_{\mu\nu}$. And there are just ten of these, since $g_{\mu\nu} = g_{\nu\mu}$. In fact, these ten $g_{\mu\nu}$ are the analogs of the *one* potential φ of Newton's theory. This analogy can be illustrated in many ways. The metric (**7.25**) already showed the close relation of g_{44} with the Newtonian φ in that particular coordinate system. An arbitrary change of coordinates would, of course, relate *all* the g's with φ. Again, let us recall the basic role of *any* potential: its first derivatives are directly related to the force (i.e., the acceleration), as in (**8.26**). In Maxwell's theory there is, instead of the Newtonian *scalar* potential φ, a *four-vector* potential Φ_μ, such that the four-acceleration is given by

$$\frac{d^2x^\mu}{ds^2} = \frac{q}{cm_0} \sum_{\tau,\nu} g^{\mu\tau}\left(\frac{\partial\Phi_\nu}{\partial x^\tau} - \frac{\partial\Phi_\tau}{\partial x^\nu}\right)\frac{dx^\nu}{ds} \tag{8.33}$$

(see Appendix II). The corresponding equation in GR is (**8.15**), which can be written in the form

$$\frac{d^2x^\mu}{ds^2} = -\frac{1}{2}\sum_{\tau,\nu,\sigma} g^{\mu\tau}\left(\frac{\partial g_{\tau\nu}}{\partial x^\sigma} + \frac{\partial g_{\tau\sigma}}{\partial x^\nu} - \frac{\partial g_{\nu\sigma}}{\partial x^\tau}\right)\frac{dx^\nu}{ds}\frac{dx^\sigma}{ds}. \tag{8.34}$$

This shows the g's in their role as potentials. GR, then, can be formally regarded as a gravitational field theory with a *tensor* potential $g_{\mu\nu}$.

The field equations (**8.31**), accordingly, are second-order differential equations in the potential (i.e., they involve second but no higher derivatives of the $g_{\mu\nu}$), as is clear from the definition of $R^\mu_{\nu\rho\sigma}$. And this, too, is analogous to the Newtonian (vacuum) case (**8.29**), and the Maxwellian (vacuum) case

$$\sum_{\mu,\nu} g^{\mu\nu}\frac{\partial^2\Phi_\sigma}{\partial x^\mu \partial x^\nu} = 0 \tag{8.35}$$

(see Appendix II). Unlike the Newtonian or Maxwellian field equations, however, Einstein's field equations are *nonlinear*: they contain products of the g's and their derivatives. But, as we have already remarked in Section 1.15, no linear theory (with the superposition principle enjoyed by its solutions) could take into account the gravitating effects of gravity itself.

And now a final point: do we not, after all, have too many field equations in (**8.31**)? Do not ten differential equations determine the ten unknowns $g_{\mu\nu}$

uniquely, given suitable boundary conditions? Yet surely we do not *want* to find the $g_{\mu\nu}$ uniquely since we ought to be at liberty to change coordinates in spacetime and so transform the metric into any equivalent metric. In fact, we would like to have four degrees of freedom in the determination of the $g_{\mu\nu}$, corresponding to the four arbitrary functions $x^{\mu'} = x^{\mu'}(x^\mu)$ which specify a change of coordinates. As it happens, however, the field equations satisfy four differential identities (see Section 8.10), and thus they effectively impose only six differential restrictions on the g's, which is precisely what is needed.

From every theoretical point of view, therefore, the field equations $R_{\mu\nu} = 0$ seem just right. Certainly there exist none that are simpler and still consistent with the fundamental ideas of GR. And it should be remembered that field equations are a matter of choice and not of proof. They belong among the axioms of a theory. The next logical step is to see whether they correctly predict verifiable results.

8.3 The Schwarzschild Solution

The first and most important exact solution of Einstein's field equations was found in 1916 by Schwarzschild. It is the metric for the spacetime around a spherically symmetric mass m, which itself may be surrounded, at some distance, by a spherically symmetric mass distribution. To obtain it, we first use the symmetry of the situation to narrow down the metric as much as possible. Since the configuration is static, our arguments of Section 7.6 apply, and the metric can be brought to the form (7.24) by a suitable choice of coordinate time. Imagine a sequence of reference spheres concentric with m, each made of a lattice of "weightless" (i.e., limitingly light) rulers, so that gravity does not compress them. Spherical symmetry implies that the three-space is a purely radial distortion of Euclidean space E_3. The reference spheres themselves will *not* be distorted, but the rate of increase of their areas with distance from m may differ from the Euclidean rate. Now, the metric of E_3 can be put in the form

$$d\sigma^2 = dr^2 + r^2(dr^2 + \sin^2\theta d\phi^2),$$

where r, θ, ϕ are the usual polar coordinates, measuring, respectively, distance from the origin, inclination from the z axis, and angle around the z axis. Imagine a sequence of ruler-made reference spheres $r = $ constant in E_3, and imagine each lattice point on the reference spheres marked with its coordinates r, θ, ϕ. Then consider a radial deformation of this space about the origin, while retaining the "painted on" coordinates: r will no longer necessarily measure ruler distance from the origin, but distances on the reference spheres will be unchanged, since they still coincide with rulers. Thus the meaning of the radial coordinate r is now *only* that a sphere $r = r_0$ has area $4\pi r_0^2$. The angular coordinates will retain their previous significance. Spatial geodesics through the origin ($\theta, \phi = $ constant) will remain geodesics, by symmetry, and also will remain possible tracks for light and particles. The spatial metric now differs from the above flat polar metric by having $e^B dr^2$ in place of dr^2, where B is some function of r. So we can write (7.24) in the form

$$ds^2 = e^A dt^2 - e^B dr^2 - r^2(d\theta^2 + \sin^2\theta d\phi^2), \qquad (8.36)$$

with A, B functions of r that must now be determined by the field equations. Clearly, we expect to find a metric not too different from our earlier approximation (7.28), but we shall let the field equations speak for themselves.

Since "orthogonal" metrics [i.e., metrics like (8.36) without mixed terms in the differentials, also called "diagonal" metrics] are of frequent occurrence, we have listed in Appendix I the Ricci tensor components for the most general such metric. It pays to calculate them once and for all, and then simply refer to the list as need arises. In particular, since every 3-space allows orthogonal coordinates, every static metric (7.24) allows orthogonal coordinates and so can be dealt with in this way.

If the indices 1, 2, 3, 4 refer to r, θ, ϕ, t, respectively, and if primes denote differentiation with respect to r, we find for the metric (8.36):

$$R_{11} = \tfrac{1}{2}A'' - \tfrac{1}{4}A'B' + \tfrac{1}{4}A'^2 - B'/r \qquad (8.37)$$

$$R_{22} = e^{-B}[1 + \tfrac{1}{2}r(A' - B')] - 1 \qquad (8.38)$$

$$R_{33} = R_{22} \sin^2\theta \qquad (8.39)$$

$$R_{44} = -e^{A-B}(\tfrac{1}{2}A'' - \tfrac{1}{4}A'B' + \tfrac{1}{4}A'^2 + A'/r) \qquad (8.40)$$

$$R_{\mu\nu} = 0 \quad \text{when } \mu \neq \nu. \qquad (8.41)$$

The vacuum field equations require $R_{\mu\nu} = 0$ for all indices. Thus, (8.37) and (8.40) yield

$$A' = -B', \qquad (8.42)$$

whence $A = -B + k$, where k is a constant. Reference to (8.36) shows that a simple change in the time scale $t \to e^{-k/2}t$ will absorb the k; let us suppose this done [it corresponds to adding a constant to the potential φ in (7.24)], and then

$$A = -B.$$

With that, (8.38) yields

$$e^A(1 + rA') = 1,$$

or, setting $e^A = \alpha$,

$$\alpha + r\alpha' = (r\alpha)' = 1.$$

This equation can at once be integrated, giving

$$\alpha = 1 - \frac{2m}{r},$$

where, at this stage, $-2m$ is simply a constant of integration. Since we have used the equations $R_{11} = 0$ and $R_{44} = 0$ only in combination, we must yet verify that our solution satisfies these equations separately. It is found to do so except at $r = 0$ and $r = 2m$, which are singularities.

Accordingly, we have found the following metric:

$$ds^2 = \left(1 - \frac{2m}{r}\right)dt^2 - \left(1 - \frac{2m}{r}\right)^{-1} dr^2 - r^2(d\theta^2 + \sin^2\theta d\phi^2). \quad (8.43)$$

Now compare this with (7.28) whose geodesics, as we already know, approximate to the Newtonian orbits round a mass m, independently of any (small) coefficient of dr^2. Consequently the m in (8.43) is identified with the mass of the central body, in units in which $G = c = 1$. (To restore conventional units we must write Gm/c^2 for m and ct for t.) In the present units the mass of the earth is 0.44 cm, and that of the sun 1.47 km.

Note that nowhere in our derivation did we have to *assume* that the spacetime is flat (i.e., Minkowskian) at infinity, yet the metric (8.43) possesses this property; it must therefore be a consequence purely of spherical symmetry, staticness, and the vacuum field equations.

The argument for (8.43) would go through even if the spherical mass m were surrounded by a spherically symmetric mass distribution beyond a sphere Σ of coordinate radius r_0: *between* m and Σ the metric (8.43) would apply. This, in turn, leads to the relativistic analog of the Newtonian theorem according to which there is *no* gravitational field inside such a sphere Σ if it is empty. For then we must put $m = 0$ in (8.43) in order to avoid a singularity, and the result is Minkowski space inside Σ, which indeed corresponds to the absence of gravity.

Birkhoff has shown (1923) that even the assumption of staticness is unnecessary in order to obtain the metric (8.43): spherical symmetry is all that is needed.[1] Thus even a spherically symmetric pulsating mass, surrounded by spherically symmetric pulsating matter beyond Σ, would give rise to the same metric between it and Σ.

Inspection of this metric now shows that, contrary to our simplicity assumption of Section 7.7, a mass *does* curve the three-space around it. Of course, the curvature is generally very small. For a "plane" of symmetry through the origin (e.g., the locus $\theta = \pi/2$) it turns out to be $-m/r^3$ at coordinate r. At the surface of the earth this is -2×10^{-27} cm^{-2}, and at the surface of the sun it is -4×10^{-28} cm^{-2}. It can be verified (see Exercise 8.9) that the intrinsic spatial geometry of such a "plane," with its metric $d\sigma^2 = dr^2/(1 - 2m/r) + r^2 d\phi^2$, is identical to that of the upper half of *Flamm's paraboloid*, generated by rotating the parabola

$$z^2 = 8m(y - 2m), \quad x = 0, \quad (8.44)$$

about the z axis in Euclidean three-space; r corresponds to $(x^2 + y^2)^{1/2}$ and ϕ to the angle about the z axis (see Figure 8.1). Of course, only that part of Flamm's paraboloid is relevant which corresponds to radii r greater than that of the central mass; *inside* that mass our *vacuum* metric does not apply.

[1] For a rigorous proof, see W. B. Bonnor's article in *Recent Developments in General Relativity*, p. 167, New York, Pergamon Press, Inc., 1962.

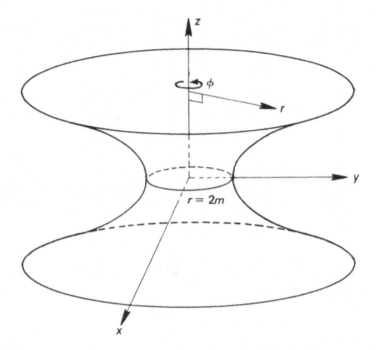

Figure 8.1

The reader may have noticed that something strange happens in **(8.43)** at the so-called *Schwarzschild radius* $r = 2m$, which corresponds to the waist-circle of the paraboloid. Remembering the significance of ds^2, one sees that standard clocks stand still and radial rulers shrink to zero coordinate length there. We shall discuss these phenomena later. At the moment we merely note that for "ordinary" bodies the Schwarzschild radius lies well inside them, where the vacuum solution **(8.43)** is inapplicable anyway; for example, for the sun it is 2.9 km, for the earth, 0.88 cm, and for a proton, 2.4×10^{-52} cm. (But see Sections 8.5 and 8.6.)

Minute though it is, the spatial curvature contributes significantly to two "post-Newtonian" effects of GR, namely the bending of light and the advance of the perihelia of the planets. If one calculates the geodesics of **(8.43)** with simply $-dr^2$ as the second term on the right, one gets only two-thirds of the advance of the perihelia, and one-half of the bending of light.

It is perhaps of some interest to understand directly how the space geometry contributes to these effects. Suppose that on the assumption of flat-space geometry an orbit is nearly circular, with mean radius a, and possibly with some perihelion advance, like the curve C in Figure 8.2a. To first approximation, the "plane" of the orbit is really the tangent cone to Flamm's paraboloid at radius $r = a \cos \psi$ (see Figure 8.2b), where the small angle ψ is given by

$$\psi \approx \frac{dz}{dy} = \frac{4m}{z},$$
(8.45)

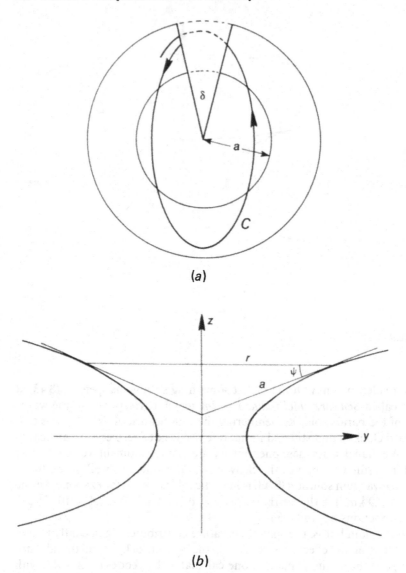

Figure 8.2

as we calculate easily from (8.44). To make the plane of the flat-space calculation into this cone, we must cut out of it a wedge of angle δ such that

$$a(2\pi - \delta) = 2\pi r = 2\pi a \cos \psi$$
$$\approx 2\pi a(1 - \tfrac{1}{2}\psi^2).$$

$$(8.46)$$

Clearly, δ will be the contribution of the spatial geometry to the perihelion advance. Solving (**8.46**) and substituting from (**8.45**) and (**8.44**) (with $y = r \gg m$), we get

$$\delta \approx \frac{2\pi m}{r}, \qquad (8.47)$$

which is one-third of the full perihelion advance, as we shall see in Section 8.4.

The contribution of the space geometry to the bending of light can be understood in the same kind of way. If a long, thin, rectangular strip of paper, with a straight line drawn down its middle (corresponding to a straight light path in flat space), is glued without wrinkles to the upper half of Flamm's paraboloid, and then viewed from the z axis at large z, the center line will appear bent: this is precisely the contribution of the space geometry to the bending of light. If the center line is already slightly bent relative to the strip, as implied by the EP for a light ray, then it will appear even more bent when applied to the paraboloid. The EP, as we have seen in Section 1.21, predicts the exact curvature of light in the static space *locally*, which we now recognize to mean: in the flat tangent space to the static 3-space. This corresponds to little tangent-plane elements on Flamm's paraboloid. We have needed the field equations to tell us how these tangent-plane elements are fitted together, namely into a portion of a paraboloid.

The angle δ, evaluated in (**8.47**) and illustrated in Figure 8.2, is also fairly obviously the contribution of the space geometry to the advance of the axis of a test gyroscope in circular orbit around a mass m at radius r, if the axis lies in the plane of the orbit (or of the projection of the axis onto this plane, otherwise). There is a second contribution to this advance, namely the so-called *Thomas precession*, which is a flat-space phenomenon: Imagine a set of inertial frames moving so that the spatial track of each origin is tangent to a circle of radius r, in such a way that we meet these origins at zero relative velocity as we go round the circle at constant angular speed ω; if any two successive such frames consider their axes to be oriented without relative rotation, then, as we complete the circle, the axes of the last frame are nevertheless rotated relative to those of the first by an angle $\pi r^2 \omega^2$, in the sense opposite to that of the orbit (see Exercise 2.29). In the present case, by (**7.29**), this amounts to $\pi m/r$, numerically. However, for a freely falling gyroscope the sense is reversed: it is the frame of the field that Thomas-precesses around the gyroscope, which itself is "free." The total effect, geometric and Thomas, gives the well-known *Fokker–de Sitter precession* of $3\pi m/r$, in the same sense as the orbit.

When using the Schwarzschild metric (**8.43**), especially in astronomy, one must, of course, understand the physical significance of the coordinates. That of the coordinate time t was already discussed in Section 7.6. And the meanings of θ and ϕ are clear enough: on each coordinate sphere $r = $ constant they are the usual "co-latitude" and longitude; if m were made of glass and we could observe from its center,[2] θ and ϕ would be the usual angular

---

[2] Though (**8.43**) does not extend *into* m, only the coefficients of dt^2 and dr^2 can have a different form there, and these do not affect the argument of the present paragraph.

measurements, from and about the z axis, respectively, made on the light coming from an event. (For by symmetry—and this can be checked by calculation—θ, ϕ = constant are possible light tracks.) From the same observation point, r would be "distance from apparent size," namely the known diameter of a distant object divided by the angle $d\alpha$ it subtends visually at the origin. For, from (8.43), the length of a ruler in the reference sphere r = constant, at coordinates differing by $d\theta$, $d\phi$, is $r(d\theta^2 + \sin^2\theta d\phi^2)^{1/2} = r d\alpha$.

However, this is only one possible way of measuring radial distance. Another consists of laying rulers end to end: this corresponds to distance along Flamm's paraboloid (cf. Figure 8.1). Such ruler distance is obtained by radially integrating the spatial part of (8.43). Since

$$\int\left(1 - \frac{2m}{r}\right)^{-1/2} dr = r\left(1 - \frac{2m}{r}\right)^{1/2} + 2m \log[(r - 2m)^{1/2} + r^{1/2}],$$

we obtain, for the ruler distance between coordinates r_1 and r_2,

$$\sigma \approx r_2 - r_1 + \log\frac{r_2}{r_1}, \tag{8.48}$$

approximately. The same result can be obtained directly by approximating for the integrand with $1 + m/r$. In the sun's field, the excess of ruler distance over coordinate distance from the sun's surface to the earth is thus about 8 km, a discrepancy of one part in 2×10^7.

Yet another way of measuring distance is by radar. Suppose we send a radio signal from coordinate r_1 to r_2 that is reflected at r_2 and returns to r_1 after a proper time $2T$ has elapsed there. The radar distance from r_1 to r_2 is then given by $R = cT$, or $R = T$ in units in which $c = 1$. For the signal we have $ds^2 = 0$ and thus, from (8.43),

$$dt = \pm \frac{dr}{1 - 2m/r} = \pm\left(1 + \frac{2m}{r - 2m}\right)dr. \tag{8.49}$$

Integrating in either direction along the path, we get the same elapsed coordinate time,

$$\int dt = \int_{r_1}^{r_2}\left(1 + \frac{2m}{r - 2m}\right)dr \tag{8.50}$$

This must yet be multiplied by the time dilation factor $(1 - 2m/r_1)^{1/2}$ at the observation point to be converted into proper time T. Thus we find

$$R = T = \left(1 - \frac{2m}{r_1}\right)^{1/2}\left[r_2 - r_1 + 2m \log\left(\frac{r_2 - 2m}{r_1 - 2m}\right)\right].$$

Other operational methods of defining distance, e.g., by parallax, by apparent brightness of a distant source, etc., can all be similarly related to coordinate distance. They are in general inequivalent, and this points to the need for caution when talking about distance in GR or about the relative speed of widely separated objects.

8.4   Rays and Orbits in Schwarzschild Space

An immediate result that can be simply read off from the Schwarzschild metric (**8.43**) is the gravitational Doppler shift in the light from a point at radial coordinate r_1 to a point at radial coordinate r_2. We need not assume that these points lie on the same radius vector. The relevant theory has been established in Section 7.6: In every static metric (**7.24**) the ratio of the frequency received (v_1) to the frequency of a similar source at the receiver (v_2) is given exactly by (**7.21**). That expression is the square root of the ratio of the coefficients of dt^2 at emitter and receiver, respectively. In the case of the metric (**8.43**), this yields

$$\frac{v_1}{v_2} = \left[\frac{1 - (2m/r_1)}{1 - (2m/r_2)}\right]^{1/2}. \tag{8.51}$$

This is exactly the same result as we can get from the metric (**7.25**) which was obtained from (**7.24**) by linearly approximating for $\exp(2\varphi/c^2)$, and then using the Newtonian potential as φ. The exactness of this agreement, however, is not only an accident of our earlier approximation, but also of our present choice of radial coordinate. For it is not clear which of several inequivalent distances in curved space correspond to the distance in Newton's potential. However, to first order in m/r (and this is the same for all distance definitions) Formula (**8.51**) can be predicted from the EP *without* the use of field equations. Its first-order verification—as provided by the terrestrial experiments mentioned in Section 1.21—consequently gives support to the EP but not specifically to the field equations of GR. Second-order verification, unfortunately, cannot be contemplated at present.

In order to obtain the exact light and particle paths in the Schwarzschild metric, we must solve the geodesic equations (**8.15**). They may be set up with the help of the Γ's listed in Appendix I. We shall omit some of the details.[3] It turns out that the equation with $\mu = 2$ implies that the paths are "plane" (as in Newton's theory); e.g., if a particle (or ray) initially moves in the plane $\theta = \frac{1}{2}\pi$, then it continues to do so; consequently we assume, without loss of generality, that the motion takes place in that plane. The remaining equations can then be reduced to the following most convenient pair (without approximation):

$$r^2 \frac{d\phi}{ds} = h, \quad (h = \text{constant}), \tag{8.52}$$

$$\frac{d^2u}{d\phi^2} + u = \frac{m}{h^2} + 3mu^2, \quad \left(u = \frac{1}{r}\right). \tag{8.53}$$

It may be of interest to note that for the flat-space metric (**7.28**) (with $G =$

---

[3] These may be found, for example, in A. S. Eddington, *The Mathematical Theory of Relativity*, Section 39, Cambridge University Press, 1924.

$c = 1$) Equation (8.52) is the same, whereas in Equation (8.53) there are two additional terms on the right, namely

$$m\left(\frac{du}{d\phi}\right)^2 + 2mu\,\frac{d^2u}{d\phi^2}. \tag{8.54}$$

On the other hand, in *Newton's theory*, angular momentum is conserved, i.e.,

$$r^2\,\frac{d\phi}{dt} = h, \tag{8.55}$$

and the inverse square law implies

$$\frac{d^2r}{dt^2} = -\frac{m}{r^2} + r\left(\frac{d\phi}{dt}\right)^2, \tag{8.56}$$

the last term being the centrifugal acceleration. By use of (8.55), Equation (8.56) can be converted to the form

$$\frac{d^2u}{d\phi^2} + u = \frac{m}{h^2}, \tag{8.57}$$

where again $u = 1/r$. Compare this with (8.53) and (8.52). We see that GR has added a small "correction" term to the Newtonian orbital equation, and has given a slightly different meaning to the conserved quantity h, as well as to the distance r. The latter we have already discussed. The smallness of the first two of these corrections can be judged, respectively, from the ratio $3mu^2 : m/h^2$ which for the earth's orbit, for example, is 0.000 000 03, and from the ratio ds/dt which for the earth is 0.999 999 995. (Of course, it is not clear that the Newtonian t is to be identified with GR coordinate time t rather than with proper time s, since Newtonian theory does not distinguish between these.)

Now a suitable solution of the Newtonian equation (8.57), if we are interested in planetary orbits, is given by

$$u = \frac{m}{h^2}\,(1 + e\cos\phi). \tag{8.58}$$

which represents an ellipse with focus at $r = 0$ and with eccentricity e. Substituting this into the right side of (8.53) as a first approximation, we get

$$\frac{d^2u}{d\phi^2} + u = \frac{m}{h^2} + \frac{3m^3}{h^4}\,(1 + 2e\cos\phi + e^2\cos^2\phi). \tag{8.59}$$

For the further solution of this equation, we need particular integrals of the following three types of equations,

$$\frac{d^2u}{d\phi^2} + u = A, \quad = A\cos\phi, \quad = A\cos^2\phi,$$

where each A is, in fact, a constant of order m^3/h^4. These must then be added to **(8.58)**. As can be verified easily, such particular integrals are, respectively,

$$u_1 = A, \quad = \tfrac{1}{2}A\phi \sin \phi, \quad = \tfrac{1}{2}A - \tfrac{1}{6}A \cos 2\phi. \tag{8.60}$$

Of these, the first simply adds a minute constant to u, while the third adds a minute constant and "wiggle," all quite unobservable. But the second cannot be neglected, since it has a "resonance" factor ϕ and thus produces a continually increasing and ultimately noticeable effect. Thus our second approximation is

$$u = \frac{m}{h^2} \left(1 + e \cos \phi + \frac{3m^2}{h^2} e\phi \sin \phi \right)$$

$$\approx \frac{m}{h^2} \left[1 + e \cos\left(1 - \frac{3m^2}{h^2} \right)\phi \right], \tag{8.61}$$

where we have used the formula $\cos(\alpha - \beta) = \cos \alpha \cos \beta + \sin \alpha \sin \beta$, and the approximations $\cos \beta \approx 1$, $\sin \beta \approx \beta$ for a small angle β. This equation shows u (and therefore r) to be a periodic function of ϕ with period

$$\frac{2\pi}{1 - 3m^2/h^2} > 2\pi.$$

Thus the values of r, which of course trace out an approximate ellipse, do not begin to repeat until somewhat *after* the radius vector has made a complete revolution. Hence the orbit can be regarded as an ellipse that rotates ("precesses") about one of its foci (see Figure 8.3) by an amount

$$\Delta = \frac{2\pi}{1 - 3m^2/h^2} - 2\pi \approx \frac{6\pi m^2}{h^2} \approx \frac{6\pi m}{a(1 - e^2)} \tag{8.62}$$

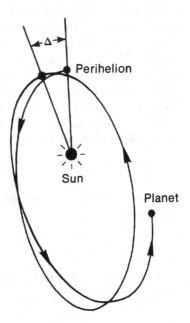

Figure 8.3

per revolution. Here we have used the Newtonian relation

$$\frac{2m}{h^2} = \frac{1}{r_1} + \frac{1}{r_2} = \frac{2}{a(1 - e^2)},$$ (8.63)

which follows from (8.58) on setting $\phi = 0, \pi$; a is the semi-major-axis, and r_1, r_2 are the maximum and minimum values $a(1 \pm e)$ of r. This Δ is the famous Einsteinian advance of the perihelion. For the flat-space metric (7.28) we have the two additional terms (8.54) on the right side of (8.53). They can be treated just as the term $3mu^2$: we substitute (8.58) and again use the particular integrals (8.60). The result is (8.61) with only two-thirds of the $\phi \sin \phi$ term, which leads to two-thirds of the full GR perihelion advance. The "missing" third comes from the spatial geometry, as we saw in (8.47)—at least for nearly circular orbits.

The reader may have questioned our justification for assuming ϕ small in (8.61) when, in fact, ϕ grows steadily. However, one can deduce from the *form* of (8.53) that, for initial conditions leading to approximate ellipses, r is a *strictly* periodic function of ϕ, so that our analysis gives a typical portion of the path. There are, on the other hand, solutions without Newtonian analog, such as orbits spiraling into the center.[4]

The relative accuracy with which planetary perihelia can be observed depends not only on the size of Δ, but also on the eccentricity and period of the orbit. The case of Mercury is by far the most favorable [period = 88 days; $a = 5.8 \times 10^{12}$ cm; $e = 0.2$; m(sun) $= 1.5 \times 10^5$ cm in present units]. Its actually observed perihelion advance, in *one hundred terrestrial years*, is $5599''.74 \pm 0''.41$ (seconds of arc). All but $43''$ of this total was explained by Newtonian perturbation theory as due to the remaining planets and other causes. The $43''$ discrepancy, however, had long been a notorious puzzle. It is precisely these missing $43''$ that were explained by GR in what must surely have been one of the most striking resolutions ever of a natural mystery. (Consider such pedestrian alternatives as changing Newton's $1/r^2$ law to $1/r^{2.000\,000\,16}$—as had been seriously proposed![5]) The advances of the perihelia of earth, Venus, and the asteroid Icarus have also been observed lately, and they agree to within the rather large observational uncertainties with the GR predictions of $3''.8, 8''.6$, and $10''.3$ per century, respectively.

Next we investigate the deflection of light. We can again start with Equations (8.52) and (8.53). But now $ds = 0$, whence $h = \infty$ and (8.53) reduces to

$$\frac{d^2u}{d\phi^2} + u = 3mu^2.$$ (8.64)

A suitable solution of this equation *without* the small right-hand term is given by the straight line

$$u = \frac{\sin \phi}{R},$$ (8.65)

---

[4] See, for example, H. P. Robertson and T. W. Noonan, *Relativity and Cosmology*, Section 9.6, W. B. Saunders Co., 1968.

[5] A. Hall, *Astron. J. 14*, 49 (1894).

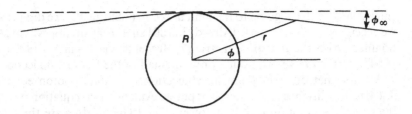

Figure 8.4

in which R can be regarded as the radius of the sun, if we are primarily interested in rays grazing the sun's edge (see Figure 8.4). Substituting **(8.65)** into the right side of **(8.64)** gives

$$\frac{d^2u}{d\phi^2} + u = \frac{3m}{R^2}(1 - \cos^2\phi),$$

of which a particular integral is [cf. **(8.60)**]

$$u_1 = \frac{3m}{2R^2}(1 + \tfrac{1}{3}\cos 2\phi).$$

Adding this to **(8.65)** yields the second approximation

$$u = \frac{\sin\phi}{R} + \frac{3m}{2R^2}(1 + \tfrac{1}{3}\cos 2\phi). \tag{8.66}$$

For large r, ϕ is evidently very small, and so $\sin\phi \approx \phi$, $\cos 2\phi \approx 1$. Going to the limit $u \to 0$ in **(8.66)**, we thus find $\phi \to \phi_\infty$ (see Figure 8.4), where

$$\phi_\infty = -\frac{2m}{R}.$$

Consequently the magnitude of the total deflection of the ray, by symmetry, is

$$\frac{4m}{R} \tag{8.67}$$

in circular measure.

For a ray grazing the sun, for example, this comes to $1''.75$. It has been tested by comparing the observed with the known positions of stars seen near the sun's limb during a total eclipse, and more recently by radio observations of quasars approximately in line with the sun (see Section 1.21). The agreement is very satisfactory, and certainly excludes the value $2m/R$ which can be predicted on the basis of Newtonian theory: Consider the orbit **(8.58)** with $\sin\phi$ in place of $\cos\phi$ (to conform with Figure 8.4). For a photon grazing the sun, $h = Rc$, and thus, setting $\phi = \tfrac{1}{2}\pi$ in **(8.58)** (and $c = 1$), we have

$$\frac{1}{R} = \frac{m}{R^2}(1 + e), \quad \text{or} \quad e = \frac{R}{m} - 1 \approx \frac{R}{m},$$

since $R/m \gg 1$. Then going to the limit $u \to 0$ in (8.58), just as we did in (8.66), we get $\phi_\infty = -1/e = -m/R$. (Already in 1801 such a calculation was made by Söldner.) Also the flat-space metric (7.28) yields, via Equation (8.53) with (8.54) and the above procedure, only one-half of the full GR deflection.

A somewhat different point in connection with the "photon equation" (8.64) is the existence of circular light paths: evidently that equation possesses the particular solution $u = 1/3m$, or $r = 3m$. [These orbits are three times larger than the corresponding "Newtonian" circular light orbits obtainable from (7.29) by setting $\omega = c/r = 1/r$ and $G = 1$; *that* yields $r = m$.] Since $r = 3m$ is *not* a geodesic track in the 3-space (as inspection of Figure 8.1 shows at once[6]), the present example illustrates that even in static spacetimes light does not in general follow a geodesic track in the 3-space. (If time were absolute, such a track would be expected on the basis of Fermat's principle, which, in fact, holds in GR.) But if the coefficient of dt^2 is constant in a static metric, light *does* follow a geodesic 3-track (see Exercise 8.5).

It is perhaps worth noting that the circular *particle* orbits of the Schwarzschild metric (8.43) *exactly* obey Kepler's law (7.29), namely

$$\omega^2 = \frac{m}{r^3}, \tag{8.68}$$

in present units, where $\omega = d\phi/dt$. Of course the exactness of this correspondence is physically meaningless, because there is no well-defined correspondence between Newtonian and GR coordinates. To obtain (8.68), we could use (8.53), but it is more straightforward to use the original geodesic equation with $\mu = 1$ (putting $\theta = \pi/2$). For the metric (8.36) it is (cf. Eddington *loc. cit.*):

$$r'' + \tfrac{1}{2}B'r'^2 - re^{-B}\phi'^2 + \tfrac{1}{2}e^{A-B}A't'^2 = 0, \tag{8.69}$$

where primes denote d/ds. Setting $r = $ constant and substituting for A, we get (8.68) at once. Note that this result is independent of B (which drops out) and therefore applies equally to the flat-space metric (7.28).

Finally, we consider the field strength **g** in the Schwarzschild metric (8.43). According to the remarks at the end of Section 7.6, we can calculate **g** at any point of the metric (7.24) as $\mathbf{g} = -\mathbf{grad}\,\varphi$. Comparing (8.43) with (7.24) we find that, in present units,

$$\varphi = \tfrac{1}{2}\log\left(1 - \frac{2m}{r}\right). \tag{8.70}$$

Hence, of course, the field is radial, and its magnitude is

$$g = \frac{d\varphi}{d\sigma} = \frac{d\varphi}{dr}\frac{dr}{d\sigma} = \frac{m}{r^2}\left(1 - \frac{2m}{r}\right)^{1/2}, \tag{8.71}$$

where σ denotes ruler distance and $dr/d\sigma$ can be read off from (8.43) as

---

[6] *Note:* if g is the shortest path between two points in the full space, there can be no shorter paths in any subspace; and $r = 3m$ is clearly *not* the shortest path between any two of its points on the paraboloid.

$(1 - 2m/r)^{1/2}$. Note how g becomes infinite at the Schwarzschild radius $r = 2m$. It follows that a "particle" at rest in the space at $r = 2m$ would have to be a photon [see the remark after (2.32)].

8.5 The Schwarzschild Horizon, Gravitational Collapse, and Black Holes

We shall now discuss a feature of the Schwarzschild field that was long misunderstood, namely the locus $r = 2m$—hereafter called the *horizon*—of the metric

$$ds^2 = \alpha dt^2 - \alpha^{-1}dr^2 - r^2(d\theta^2 + \sin^2\theta d\phi^2), \quad \left(\alpha = 1 - \frac{2m}{r}\right), \quad (8.72)$$

when it occurs outside the mass. We shall in fact study the metric (8.72) for all non-negative values of r, assuming that for $r > 0$ there is vacuum, and that we have a mass m confined entirely to $r = 0$. (We shall avoid calling it a "point"- mass, since the locus $r = 0$ will turn out to have a structure quite different from a point.) It can easily be seen, by looking at the steps of our original derivation, that the metric (8.72) satisfies the GR vacuum field equations for $r > 2m$ *and* for $r < 2m$.

However, *inside* the horizon the coordinates have a somewhat different significance. There the metric coefficient of dt^2 is negative, that of dr^2 positive. No particle or photon can have equation $r = $ constant, since that would imply $ds^2 < 0$, whereas for neighboring events on a particle ds^2 is proper time squared and thus positive, while for a photon it is zero. In fact, r is now a "time" coordinate in this sense: it cannot stand still for a particle or photon. While r, θ, ϕ inside the horizon have the same geometric significance as outside in terms of the reference spheres $r = $ constant, these spheres can no longer be realized by material lattices at rest. However, we must not explain this by saying that the force of gravity has become irresistible. For matter (and light) can move in the direction of increasing as well as decreasing r. In fact we must make a *choice* for the direction in which the "time" r shall run inside the horizon. Either r increases for all particles and photons, or it decreases for all. Anything else would lead to violations of causality. Note that the Schwarzschild metric inside the horizon is not static: the coefficients are time-dependent! We shall see that this is not due to a "bad" choice of coordinates, but rather to the intrinsically nonstationary nature of the inner spacetime. The significance of t inside the horizon will be discussed later.

For many years it was believed that there was a real singularity at $r = 2m$, in the sense that the local physics would become unusual there. (Indeed, as Bondi points out, if active gravitational mass were negative instead of positive, the Schwarzschild singularity would undoubtedly have been used as an argument to show that it *must* be negative, since a positive mass leads to a singularity...) However, in 1933 Lemaître found that the Schwarzschild singularity is not a *physical* singularity at all, but merely a *coordinate* singularity, i.e., one entirely due to the choice of the coordinate system: an observer in a small, freely falling cabin would pass through the sphere $r = 2m$ without noticing anything special at all.

It is very easy to *produce* coordinate singularities. Consider, for example, the Euclidean plane referred to the standard metric $d\sigma^2 = dx^2 + dy^2$; then simply introduce a new coordinate ξ by the equation

$$\xi = \tfrac{1}{3}x^3, \tag{8.73}$$

which gives a one-to-one relation between x and ξ. Thereupon the metric evidently becomes

$$d\sigma^2 = (3\xi)^{-4/3}d\xi^2 + dy^2,$$

and this now has a singularity at $\xi = 0$: a *coordinate* singularity! For it is fully removable by introducing a "new" variable x by Equation (8.73). Of course, in general it will not be so obvious how to remove a coordinate singularity, or even whether a given singularity is *due* to the coordinates or not. One way of deciding this last question is to calculate the invariants of the curvature tensor (14 coordinate-independent combinations of its components, which are constructed by standard tensor methods) and testing whether these remain finite as the singularity is approached: if they do, the singularity is probably not a physical one. (But it *could* be: for example, near the vertex of a cone all curvature components are zero!)

After Lemaître, a number of others rediscovered the coordinate nature of Schwarzschild's singularity, but the general appreciation of this fact came inexplicably slowly. It was finally helped by a widely read paper by M. D. Kruskal[7] containing a new coordinate system and, above all, a new topology for Schwarzschild space.

To study the Schwarzschild singularity, we first consider radial light signals in the metric (8.72). These satisfy $d\theta, d\phi = $ constant, and $ds^2 = 0$. For *ingoing* signals we choose the negative sign in (8.49) and then find, along the lines of (8.50),

$$t = -r - 2m \log|r - 2m| + v, \tag{8.74}$$

where v is a constant of integration, distinguishing one light signal from another. It is, in fact, that coordinate time t at which the signal passes the "checkpoint" $r = r_0$, where r_0 is the solution of the equation $r + 2m \log|r - 2m| = 0$ (a well-defined number between $2m$ and $2m + 1$). Equation (8.74) represents a light signal inside as well as outside the horizon. But since

---

[7] *Phys. Rev. 119*, 1743 (1960). Kruskal mistakenly seems to credit E. Kasner with the original discovery in 1921; Kasner's work is discussed and modified by C. Fronsdal in *Phys. Rev. 116*, 778 (1959). Actually, it was A. S. Eddington [*Nature 113*, 192 (1924)] who first transformed Schwarzschild's metric into a form not singular at $r = 2m$, but he seems not to have noticed this. [His paper, incidentally, contains a misleading misprint: Eq. (2) should have $r - 2m$ instead of $r - m$.] Eddington's transformation was rediscovered by D. Finkelstein [*Phys. Rev. 110*, 965 (1958)]. G. Lemaître's paper is in *Ann. Soc. Sci. Bruxelles A53*, 51 (1933). Kruskal's coordinates were discovered also by G. Szekeres at almost the same time: *Publ. Mat. Debrecen 7*, 285 (1960).

$t \to +\infty$ as $r \to 2m$, it looks as though the outer signal cannot penetrate the horizon. This will be seen to be the fault of the time coordinate. Even so, because of the coordinate discontinuity, it is not clear whether signals with the same v inside and outside are *one*.

Let us now use the above v as a new (pseudo-) time coordinate for events: each event shall be labeled by the v of the inward radial light signal it emits (or on which it lies). From (**8.74**) we then have

$$dt = dv - \alpha^{-1}dr, \tag{8.75}$$

which, when substituted into (**8.72**), converts that metric into the following ("Eddington–Finkelstein") form:

$$ds^2 = \alpha dv^2 - 2dvdr - r^2(d\theta^2 + \sin^2\theta d\phi^2). \tag{8.76}$$

This is regular for all $r > 0$. Since (**8.72**) satisfies the (tensor) GR vacuum field equations everywhere except at $r = 0$ and $r = 2m$, so must (**8.76**), since tensor equations remain valid under coordinate transformations. Moreover, the metric coefficients of (**8.76**) together with their first and second derivatives are continuous at $r = 2m$, and so (**8.76**) must satisfy the field equations *even there*.

Now whereas Equation (**8.74**) does not allow us to match light signals across the horizon, the signal $v = $ constant in (**8.76**) encounters no irregularity in traveling from $r = \infty$ to $r = 0$ and is clearly *one*. Nothing special happens at the horizon. We have, in fact, transformed the Schwarzschild singularity away by the transformation (**8.74**). (We called v a "pseudo"-time because for photons it *can* stand still.)

Note that our choice of *ingoing* signals determines the choice of time direction inside the horizon: r must decrease. Had we chosen *outgoing* signals instead, Equations (**8.74**) and (**8.76**) would read, respectively,

$$t = r + 2m \log|r - 2m| + u, \tag{8.77}$$

$$ds^2 = \alpha du^2 + 2dudr - r^2(d\theta^2 + \sin^2\theta d\phi^2). \tag{8.78}$$

The equation $u = $ constant corresponds to a light signal from $r = 0$ to $r = \infty$. The inner "time" r now increases.

But outside the horizon there are clearly signals traveling in *both* directions. The outgoing ones must have come from an inner space with r increasing, and the ingoing ones go to an inner space with r decreasing. Hence, to accommodate all light signals, we must join *two* inner spaces—isometric but with different time senses—to the outer space: one, as it were, in the past (before $t = -\infty$), one in the future (after $t = +\infty$). In fact, we also need a second *outer* space, with different time sense. For the metric (**8.76**) allows not only radial light signals satisfying $v = $ constant, but also the "locally opposite" ones satisfying

$$\frac{dv}{dr} = \frac{2}{\alpha}, \quad \text{i.e., } v = 2(r + 2m \log|r - 2m|) + \tilde{v}, \tag{8.79}$$

$\tilde{v}$ being a constant of integration here. Substituting from (8.74) we find that (8.79) is equivalent to

$$t = r + 2m \log|r - 2m| + \tilde{v}. \tag{8.80}$$

This is analogous to (8.77), except that we know it refers to a signal with *decreasing* r when inside the horizon. Hence it must come from an outer space with decreasing t! For future reference, let us denote by I_1, I_2 the inner spaces with increasing and decreasing r, respectively, and by O_1, O_2 the outer spaces with increasing and decreasing t, respectively. It would appear that the mass at $r = 0$ is divided: some of it in I_1, some in I_2. Full clarification of this apparently complicated topology of the Schwarzschild metric will come only with the introduction of Kruskal space.

We have already seen [after (8.71)] that a "particle" at rest on the horizon would have to be a photon. And from (8.76) and (8.78) we see that indeed $r = 2m$, θ, ϕ = constant, is a possible "radial" photon path. In fact, *the horizon can be regarded as a (potential) light front, standing still relative to the outer space*. Depending on which two of the four Schwarzschild regions it separates, this light front "points" outward or inward. For example, a photon going from O_1 to I_2 must meet the horizon *qua* outward-pointing light front, since one photon cannot overtake another. Similarly, a photon going from I_1 to O_1 must meet the horizon *qua* inward-pointing light front.

We shall now briefly discuss radial geodesics in Schwarzschild space. The geodesic equations (8.15) for the case of the Schwarzschild metric have already been cited, in part, in this and the preceding section. That for $\mu = 4$ reads

$$t'' + \left(\frac{dA}{dr}\right) r't' = 0 \quad \left(e^A = \alpha, \ ' \equiv \frac{d}{ds}\right). \tag{8.81}$$

It is worth remarking that the two "radial" equations (8.69) and (8.81) are unaffected by changing t into $-t$, which shows that every radial motion is t-reversible. Equation (8.81) has one immediate solution, t = constant, and that is a possible particle path inside the horizon. A general first integral is given by

$$t' = ke^{-A} = \frac{k}{\alpha}, \quad (k = \text{constant}). \tag{8.82}$$

When this is substituted into (8.69) (with $A = -B$ and $\phi' = 0$), we can solve to find

$$r' = \pm(k^2 - \alpha)^{1/2}. \tag{8.83}$$

For a particle dropped from rest at a great distance we have, from (8.72), $dt/ds \approx 1$ initially, and so, from (8.82), $k \approx 1$. Setting $k = 1$ in (8.82) and (8.83) yields

$$\frac{dt}{dr} = \frac{-r^{3/2}}{(r - 2m)\sqrt{(2m)}}, \quad \frac{ds}{dr} = -\frac{r^{1/2}}{\sqrt{(2m)}}. \tag{8.84}$$

Integration between two levels r_1 and $r_2(<r_1)$ then gives

$$\int dt = \frac{1}{\sqrt{(2m)}} \int_{r_2}^{r_1} \frac{r^{3/2}dr}{(r-2m)}, \quad \int ds = \frac{1}{\sqrt{(2m)}} \int_{r_2}^{r_1} r^{1/2}dr. \tag{8.85}$$

The first of these integrals diverges as $r_2 \to 2m$, and thus the particle crosses the horizon at $t = \infty$. The second integral, on the other hand, remains proper at $r_2 = 2m$, and finite right down to $r_2 = 0$. By continuity, it evidently represents *one* path. Our choice of $k = 1$ simplifies the analysis but does not affect the general result: *all* freely falling particles cross the horizon at $t = \infty$ but reach $r = 0$ in finite proper time. The time-reflected situation also holds: if the motion of any *outgoing* free particle is produced backwards, it is found that it left $r = 0$ a finite proper time earlier, but crossed the horizon at $t = -\infty$.

Now note the following extremely important fact. Once a particle has crossed into the horizon (i.e., into I_2: no particle can cross into I_1) it can never get out again, nor can it send out any signal. The reason for this is that the horizon bounding I_2 is an *outward*-pointing light front. Moreover, once the particle is inside, it will ultimately be annihilated by the infinite tidal forces (infinite curvature!) at $r = 0$. No amount of retro-jetting can save it from this fate. The *maximal* proper lifespan inside the horizon is πm (except for particles crossing from I_1 into I_2: theirs is $2\pi m$). For, looking at (8.72), while dr cannot vanish along the particle's worldline inside the horizon, *any* contribution from dt, $d\theta$, or $d\phi$ diminishes ds for a given dr, since their metric coefficients are negative. Consequently the maximum proper time is attained by a worldline t, θ, $\phi = $ constant, and that is

$$\int_0^{2m} \left(\frac{2m}{r} - 1\right)^{-1/2} dr = \pi m. \tag{8.86}$$

Such worldlines, as we have seen, are in fact geodesics, but not the continuations of those entering from O_1 or O_2.

Some of our findings can be illustrated on the "Schwarzschild diagram," Figure 8.5. This is a *map* of Schwarzschild r, t space, i.e., of events on a single radial direction θ, $\phi = $ constant. (If we rotate Figure 8.5 about the t axis and measure ϕ as angle around that axis, we can produce a three-dimensional map of the section $\theta = \pi/2$ of Schwarzschild space.) Note the deliberate break in the diagram at the horizon $r = 2m$. It is there to indicate that in these coordinates there is no continuity across this line, and also that it is a *branch* line bounding *two* inner spaces and *two* outer spaces. The arrows on the worldlines have been drawn on the assumption that we exhibit O_1 and I_2; the maps of O_2 and I_1 may be imagined beneath, with a cross-over at the horizon. Various light cones are drawn in the diagram, the shaded portions being in timelike relation to the vertex. Some possible worldlines are marked. Any particle crossing the horizon from an outer to an inner region (or vice versa) can do so only at $t = \pm\infty$, as we have seen. *All* events on the horizon are pushed to infinity in this diagram—except one: the entire *finite* portion of the line $r = 2m$ in Figure 8.5 represents a single event. This can be seen, for

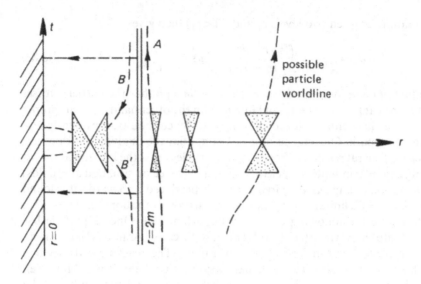

Figure 8.5

example, by considering the interval between any two lines $t =$ constant, which tends to zero as $r \to 2m$, since the metric coefficient of dt^2 does. The worldlines $t =$ constant inside the horizon represent free particles, which leave $r = 0$ in I_1, rise to the horizon, and there cross into I_2 to fall back to $r = 0$. A worldline such as B in I_2 may be the continuation of one like A in O_1. But the t-reversed worldline, B', which is also possible in I_2, cannot come from O_1! In fact, it comes from O_2, being the continuation of the t-reversal of A. Note also that the "mass locus" $r = 0$, though a genuine singularity and as such neither timelike nor spacelike, appears as the limit of the *spacelike* lines $r =$ constant ($<2m$)—quite unlike the *timelike* worldline one associates with an ordinary particle.

Fifteen years ago a discussion of the Schwarzschild horizon might have been no more than an intellectual exercise, of interest mainly because it brings out some quite unexpected properties of spacetime. Today, however, considerable practical importance attaches to it in astrophysics (and possibly even microphysics) in connection with *gravitational collapse* and *black holes*. Theoretically it is certainly possible for a mass to have its horizon outside itself. Consider, for example, a sphere of uniform density ρ and radius R. Its mass, naively (i.e., not allowing for the mass equivalent of gravitational binding energy, curvature, etc.) is $4\pi R^3 \rho/3$ and so its Schwarzschild radius is $8\pi R^3 \rho/3$. For this to exceed R we simply need

$$\tfrac{8}{3}\pi R^2 \rho > 1 \quad \text{or} \quad R > \left(\frac{3}{8\pi\rho}\right)^{1/2}. \tag{8.87}$$

Thus, spheres of any density can have their horizon showing, provided only R is big enough.

The Schwarzschild radius of a typical galaxy of mass $\sim 10^{45}$ gm is $\sim 10^{17}$ cm.

Since a typical galaxy has radius $\sim 10^{23}$ cm, a contraction by a linear factor of $\sim 10^6$ would bring it within its horizon. In a typical galaxy the stars are spaced like pinheads 50 km apart; the shrinkage would bring the pinhead-stars to within 5 cm of each other—still far from touching. Thus, unless a galaxy has some prevention device (like rotation), it *will* be sucked into the horizon by its own gravity. [Theoretically, intelligent beings could stop each star in a galaxy in its motion by ejecting a (small) part of its substance (see Exercise 5.19). The residual stars would then fall towards the center.] Once a galaxy is inside the horizon, its total collapse can no longer be halted by any means and proceeds very rapidly (in a few months). The galaxy can then no longer be seen from the outside, except by those photons which it emitted while still outside the horizon. The last of these will take an infinite time to reach us and will arrive infinitely red-shifted. Theoretically visible forever, such a galaxy would nevertheless soon fade from view: it has become a "black hole." On the other hand, its gravitational field will persist unchanged (by Birkhoff's theorem) and could lead to its detection. Some astronomers believe large numbers of collapsed or partly collapsed galaxies to be extant.

The collapse of the galactic core through its horizon began to be considered as a possible energy source for *quasars*—those puzzling, apparently hyperactive and most distant of all galaxies—soon after their discovery in 1963. When *pulsars* were discovered in 1968, the existence of black holes of *stellar* mass became a real possibility. It became apparent that under certain circumstances stellar matter can—by its own gravity—squeeze electrons into protons and pack the resulting neutrons tightly into a kind of super-nucleus of relatively small diameter (a few miles) but enormous density ($\sim 3 \times 10^{14}$ gm/cm^3). The result is a neutron star or "pulsar," so called because of the perfect radio pulse it emits by virtue of its rotation. The detailed process of formation is complicated and always involves thermonuclear reactions. When the mass of the parent star exceeds about eight solar masses, the Schwarzschild horizon will be crossed during the "squeeze," and instead of a pulsar we get a stellar black hole. Millions of these may exist in our galaxy alone. Hope for identification lies, for example, in the gravitational action that such a black hole may exert on a close (binary) stellar partner. Perhaps the first definite black-hole identification will come soon in the binary system Cygnus X-1, which has been studied intensively since 1971 for this very purpose. But though the circumstantial evidence mounts every year, conclusive proof is still lacking.

It may be noted that for a black hole resulting from gravitational collapse (as opposed to one created as a singular mass at $r = 0$), the Schwarzschild regions I_1 and O_2 do not exist. All outgoing signals in O_1 can be traced back to the body's surface before collapse, i.e., to events in O_1 rather than I_1. Similarly, "out-pointing" signals (**8.80**) in I_2 originate on the body's surface in I_2 rather than in O_2. The Eddington–Finkelstein metric (**8.76**) then covers the entire spacetime outside the collapsing matter.

It should be further noted that since most stars and galaxies have considerable angular momentum, collapse could be preceded by a stage of very fast rotation, to which the isotropic Schwarzschild metric would not be strictly applicable. There exists a famous generalization of this metric to *rotating* central masses, the *Kerr metric* (1963). However, that also has a horizon, and the qualitative facts of collapse, as outlined above, remain true under its regime also.

8.6   Kruskal Space and the Uniform Acceleration Field

In the last section we discovered certain unexpected complications of the Schwarzschild metric. What, for example, is the relation between the two inner regions I_1, I_2 and the two outer regions O_1, O_2? Why are there *two* horizons, one a light front pointing outward, the other a light front pointing inward? These questions and others found a complete answer with the introduction of Kruskal coordinates and Kruskal space (see footnote 7, page 150). However, before we discuss these, we shall consider a simpler situation which, by analogy, sheds much light on the above questions and on Kruskal's work. We propose to look at the familiar Minkowski space M_4 from an unfamiliar vantage point: an accelerating rocket.

Consider the following transformation from the usual coordinates x, y, z, t of M_4 to new coordinates X, Y, Z, T:

$$t = X \sinh T, \quad x = X \cosh T, \quad y = Y, \quad z = Z. \tag{8.88}$$

This implies the relations

$$x^2 - t^2 = X^2, \qquad t/x = \tanh T \tag{8.89}$$

and

$$\begin{aligned} ds^2 &= dt^2 - dx^2 - dy^2 - dz^2 \\ &= X^2 dT^2 - dX^2 - dY^2 - dZ^2. \end{aligned} \tag{8.90}$$

Comparing the first of (**8.89**) with (**2.33**), one sees that any point having constant X, Y, Z performs hyperbolic motion parallel to the x axis, with proper acceleration $1/X(c = 1)$. The metric (**8.90**)(ii) is a static metric of the general form (**7.24**), with Euclidean space part. It is, in fact, a metric adapted to the static field existing in the hyperbolically moving rocket-skyscraper of Section 2.16. Figure 8.6 is a Minkowski diagram for this rocket. It is basically the same as Figure 2.5, but with two additions. The straight lines issuing from the origin are lines of constant $T(t/x = \text{constant})$, and thus represent moments in rocket time. Also we have added a second rocket, the mirror image of the first, accelerating in the direction of the *negative* x axis and corresponding to negative values of X. Equations (**8.88**)–(**8.90**) apply to both rockets. In the second rocket, however, coordinate time T proceeds *backward*, i.e., from $+\infty$ to $-\infty$.

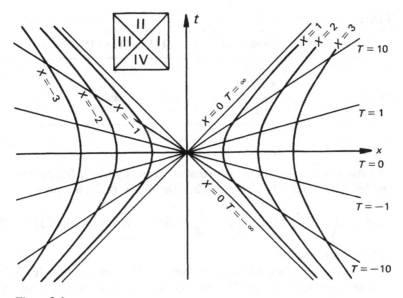

Figure 8.6

In either rocket, the "bottom floor" ($X = 0$) is made of photons. Although these "stick" to the rockets (except at $t = 0$, when the two rockets exchange bottom floors), we can distinguish between bottom photons pointing *into* or *out of* the rockets. Before $t = 0$, they point out: particles and signals *entering* the rockets from region IV in the diagram encounter these photons traveling *against* them. After $t = 0$, they point in: particles and signals entering region II *from* the rockets encounter these photons traveling against them. Note that when a particle has entered region II, it cannot get back into a rocket, nor send a signal back. Indeed, the origin light cone $t = \pm x$ is the *horizon* for the two-rocket system: the past horizon ($t < 0$) is penetrable *into* the rockets, the future horizon ($t > 0$) is penetrable *out of* the rockets.

It will not have escaped the reader's notice that an analogy exists between these rockets and the Schwarzschild case. Regions I and III in Figure 8.6 correspond to the outer Schwarzschild regions, O_1 and O_2, respectively. Region II corresponds to I_2, *into* which signals can be sent from O_1 and O_2. Region IV corresponds to I_1, *from* which signals can be sent into O_1 and O_2, and from which also particles can be sent directly into I_2 without going outside the horizon. Imagine a set of thin skyscrapers in outer Schwarzschild space O_1, standing perpendicularly on the horizon like spikes on a chestnut. Each of these corresponds to the rocket in region I. Of course, in the rocket-skyscraper the "field strength" varies as $1/X$ whereas in the Schwarzschild skyscaper it is given by (**8.71**). But otherwise the analogy is good. We shall presently make it even better.

First, however, we simplify the Schwarzschild metric (**8.72**) a little more, to facilitate the analogy. We have already chosen our units so that $c = G = 1$. But since we have *three* units at our disposal (length, time, and mass), we can,

Table 8.1

	I ($X > 0$) and III ($X < 0$)	II ($X > 0$) and IV ($X < 0$)
$t =$	$X \sinh T$	$X \cosh T$
$x =$	$X \cosh T$	$X \sinh T$
$t/x =$	$\tanh T$	$\coth T$
$x^2 - t^2 =$	X^2	$-X^2$

for a given mass m, so choose them that additionally $m = \frac{1}{4}$. Suppose this done, and let us also write R, T for r, t. Then the metric (**8.72**) becomes

$$ds^2 = \left(1 - \frac{1}{2R}\right)dT^2 - \left(1 - \frac{1}{2R}\right)^{-1} dR^2 - R^2(d\theta^2 + \sin^2\theta d\phi^2). \quad (8.91)$$

Its horizon is now at $R = \frac{1}{2}$.

Returning to Figure 8.6, observe that—so far—the coordinates X, T are undefined in regions II and IV. However, we can define them in those regions also, and so extend their use to all of Minkowski space. Table 8.1 summarizes the relations between the old and the new coordinates in all four regions.

In regions II and IV this leads to

$$ds^2 = -X^2 dT^2 + dX^2 - dY^2 - dZ^2, \quad (8.92)$$

a perfectly good metric, but form-distinct from (**8.90**)(ii). To remedy this defect, we make a change in the X coordinate. If R is defined by the relation

$$2R - 1 = x^2 - t^2 = \begin{cases} X^2 & \text{(I, III)} \\ -X^2 & \text{(II, IV)}, \end{cases} \quad (8.93)$$

then the Minkowski metric (**8.90**)(i) becomes

$$ds^2 = (2R - 1)dT^2 - (2R - 1)^{-1}dR^2 - dY^2 - dZ^2 \quad (8.94)$$

everywhere except on the horizon, where (**8.94**) has a coordinate singularity. Particles with constant $R > \frac{1}{2}$ are at rest in one of the rockets. (Unlike X, R does not by its sign distinguish between the rockets.) In regions II and IV, R is the "time" coordinate, proceeding backward in II and forward in IV. These regions are not filled with rocket history. But we can, if we wish, fill them with "ballistic" history. We can regard regions II and IV as generated by all the free paths through the origin; T measures their rapidity (cf. Exercise 2.22), and $1 - 2R$ measures the square of their proper time from the origin.

The Schwarzschild metric (**8.91**) has a real singularity at $R = 0$. Not so the metric (**8.94**). But we can give it an artificial one: let us mutilate Minkowski space by cutting out of it the two regions corresponding to $R < 0$, marked by shading in Figure 8.7. The boundaries of these regions (i.e., the two branches of the hyperbola $t^2 - x^2 = 1$) we shall call the "edges" E_1 and E_2. Mutilated Minkowski space is, in a sense, a universe of finite duration. Free paths all begin on E_1 and end on E_2: E_1 corresponds to the birth, E_2 to the death, of

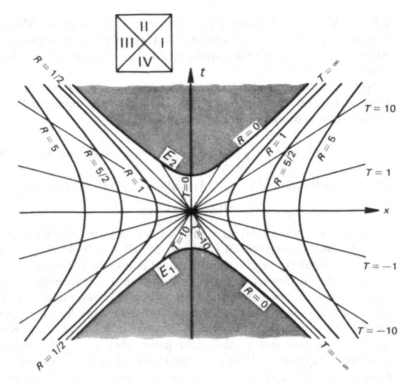

Figure 8.7

this universe. Though no limit exists to the total proper times of free paths, all are finite. Infinite life can be had only outside the horizon, and then only by ceaselessly accelerating away from it.

Simple though this universe is, it is rather difficult to understand from the vantage point of the rocket inhabitants with their metric (**8.94**). First, they live in an apparently changeless world, and the finiteness of their universe is hard for them to picture. Second, all free particles dropped by them from rest at level $R = R_0$ reach the edge E_2 in the same proper time $(2R_0)^{1/2}$ — as can be easily calculated (see Exercise 8.20). The rocket people therefore (quite mistakenly) tend to think of E_2 as a plane parallel to their bottom floor and at a constant distance below them. Even worse: since all free particles that rise in the rocket and momentarily come to rest at $R = R_0$ left E_1 also a proper time $(2R_0)^{1/2}$ earlier, the rocket people tend to identify E_1 with E_2. This leads to difficulties. For example, bottom photons must be exchanged between the rockets: where are the "planes" E_1 and E_2 when the rockets momentarily touch? However, we should sympathize with the rocket people. After all, *we* tend to think of the Schwarzschild locus $R = 0$ (from the vantage point of the static outer region) as a single spatial point, the permanent center of the horizon sphere, the beginning or end of every radial signal. And this is false! The rocket people's confusion evaporates when they learn about

Minkowski space, Minkowski coordinates x, y, z, t, and the Minkowski diagram, Figure 8.7. We, the "Schwarzschild" people, are similarly enlightened when we learn about Kruskal space, Kruskal coordinates, and the Kruskal diagram.

What is Kruskal space? Figure 8.7 can serve as a map for *it*, i.e., as a Kruskal diagram! If x, t are now Kruskal coordinates (a more usual notation for these is u, v) the metric of Kruskal space K_4 is[8]

$$ds^2 = \left[\frac{1}{2Re^{2R}}\right](dt^2 - dx^2) - R^2(d\theta^2 + \sin^2\theta d\phi^2), \qquad (8.95)$$

where R is a function of $x^2 - t^2$, defined implicitly by the relation

$$[e^{2R}](2R - 1) = x^2 - t^2, \qquad (8.96)$$

which unfortunately allows no explicit solution in terms of elementary functions. Kruskal's metric (8.95) is regular for all $R > 0$. At $R = 0$ it has a curvature singularity, and thus an intrinsic boundary. It satisfies Einstein's vacuum field equations, as we shall presently see. It has spherical symmetry at any "moment" $t = $ constant. (We must regard x as a *radial* coordinate.) The Kruskal diagram illustrates a single radial direction θ, $\phi = $ constant in full Kruskal space. Note that the $\pm 45°$ lines in Figure 8.7, *qua* Kruskal diagram, still represent (radial) light paths: they have $ds^2 = 0$ and are geodesics by symmetry. Note also that Kruskal's metric is invariant under homogeneous Lorentz transformations of x and t, since these preserve $x^2 - t^2$ and $dx^2 - dt^2$. This shows, for example, that the portions of the lines marked $T = $ constant cut off by E_1 and E_2 (in regions II and IV) all have the same proper-time "length," since they can be Lorentz transformed into each other. It also shows that the hyperbolas $R = $ constant $(R > \frac{1}{2})$ correspond to motions with constant proper acceleration (since each of their points can be Lorentz transformed into the vertex), and that all sections $T = $ constant in any one quadrant are identical.

Via Table 8.1 one can define a coordinate T throughout Kruskal space. The R defined in (8.96) serves as another. It is then straightforward to show that, in terms of these, the Kruskal metric (8.95) is equivalent to the Schwarzschild metric (8.91), i.e., to

$$ds^2 = \left(\frac{2R - 1}{[2R]}\right)dT^2 - \left(\frac{2R - 1}{[2R]}\right)^{-1} dR^2 - R^2(d\theta^2 - \sin^2\theta d\phi^2) \qquad (8.97)$$

everywhere except, of course, on the horizon $R = \frac{1}{2}$. This shows that Kruskal's metric (8.95) satisfies the Einstein vacuum field equations everywhere except at $R = \frac{1}{2}$; that it satisfies them even there then follows from continuity [cf. after (8.76)]. It also shows that Kruskal space is the *complete* space (in the sense that every geodesic in it is either extendable to infinity or ends on a singularity) which incorporates the four Schwarzschild regions O_1, I_2, O_2, I_1 in its four quadrants I, II, III, IV, respectively. The "Schwarz-

---

[8] The brackets here and below are inserted for later convenience.

schild skyscrapers" (which we likened to spikes on a chestnut) can now be regarded as rockets accelerating through Kruskal space.

Note the *formal* parallelism with the SR rocket situation: Table 8.1 and Figure 8.7 are shared by K_4 and mutilated M_4. Equations (**8.95**), (**8.96**), and (**8.97**), with Table 8.1, give the relation between K_4 and the Schwarzschild metric. The same equations, *without* the bracketed terms (and with the θ, ϕ terms suppressed), give the relation between mutilated M_4 and the rocket metric (with the Y, Z terms suppressed).

To gain an insight into the *topology* of Kruskal space, we shall cut it by a sequence of "constant time" surfaces. There is, however, much arbitrariness in deciding what "time" to hold constant. Rather than Kruskal time t, we prefer for this purpose a time τ adapted to the finite duration of K_4. Such a time can be defined, for example, by using the set of hyperbolas confocal with E_1, E_2 in the Kruskal diagram as its level surfaces (see Figure 8.8a). We may label the surface $t = 0$ as $\tau = 0$, and the others arbitrarily. Each point (R, T) in the Kruskal diagram represents an entire 2-sphere of radius R in the full Kruskal space. The interval along a curve $\tau = $ constant tells us how far these spheres are apart. In our graphical representation we must necessarily suppress one spatial dimension: we can get *typical* sections by putting $\theta = \pi/2$. Each point in the Kruskal diagram then represents a *circle* of radius R. Following any line $\tau = $ constant, we see that these circles start from a certain minimum radius (on the t axis) and become infinite in either direction. Each section will be a kind of Flamm paraboloid (cf. Figure 8.1). In fact, the section $\tau = 0$ will be *precisely* a Flamm paraboloid, since it corresponds to a section $T = 0$ of the Schwarzschild metric. All other sections $\tau = $ constant have a thinner waist. Figure 8.8b shows the complete sequence. The universe suddenly appears out of nowhere as an infinite line E_1. This immediately flares open into a long drawn out double trumpet, reaches maximum girth and maximum flare as a Flamm paraboloid, whereupon the entire sequence is reversed, back to a line E_2 momentarily, and then again nothing. As the sections open, observe how the two horizons (dashed circles) rush towards each other, running over the sections at the speed of light. While regions I and III (outside the horizons) grow, region IV (between the horizons) shrinks. As the horizons cross, IV disappears altogether and II appears. An observer at constant R is similarly running along the sections.

The following alternative slicing of Kruskal space is also of interest. Take the same confocal hyperbolas as in Figure 8.8, but where they intersect a given hyperbola $R = \frac{1}{2} + \varepsilon$ *very* close to the horizon, continue them with $T = $ constant (heavy line in Figure 8.9a). Each such slice consists, essentially, of two Flamm-paraboloid halves (corresponding to $T = $ constant, $R > \frac{1}{2} + \varepsilon$) joined by the portion of the double trumpet of Figure 8.8 that lies between the horizons (strictly, between the horizons "plus epsilon"). If we exaggerate the Flamm-paraboloid halves into planes, our sections look as shown in Figure 8.9b. The planes with their horizon "hole" remain unaltered, while the "neck" between them changes from infinite length and zero girth, to zero length (at which time the horizons cross over), and back again. An observer

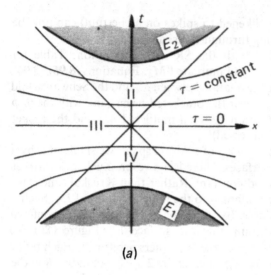

(a)

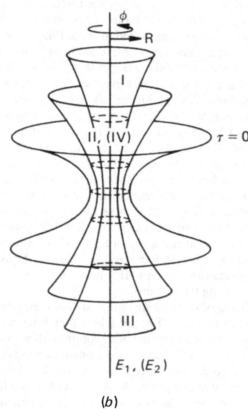

(b)

Figure 8.8

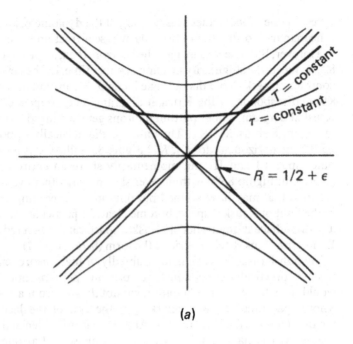

(a)

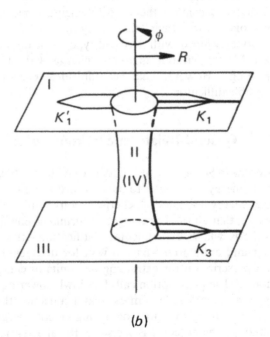

(b)

Figure 8.9

at rest in one of the planes sees nothing of the dynamic behavior of the neck. If he manages to stay out of the hole, whose attraction he must resist, he can live forever, But if he falls in, past the horizon, then, as we have seen in Section 8.5, he meets the singularity in a finite proper time, i.e., he gets squeezed in the contracting neck. Note that a single locus ϕ = constant (heavy line in Figure 8.9b) corresponds to the Kruskal diagram. The "opposite" rockets of the Kruskal diagram, whose bottom photons get exchanged at $t = 0$, are shown in Figure 8.9b as K_1, K_3. They are *not* diametrically opposite each other across the horizon, as are K_1, K_1'! Figure 8.9b illustrates nicely how the two outer spaces I and III can momentarily share an entire horizon 2-sphere (a circle in Figure 8.9b) without ever sharing *any* other event or signal.

Much that may have seemed puzzling in the preceding section will now, in the light of Kruskal space, become clear. In particular, it will be seen that any *two* adjacent quadrants of Kruskal space can be covered regularly by an Eddington–Finkelstein metric of the form (**8.76**) or (**8.77**).

Though Kruskal's work is undoubtedly of high theoretical interest, does it have practical applications? At present, perhaps not. Kruskal space would have to be *created in toto*: it cannot *develop* from a collapsing object, whose spacetime—as we remarked at the end of the last section—lacks regions O_2 and I_1, i.e., III and IV. And there is no evidence that full Kruskal spaces exist in nature. John Wheeler and his school at one time hoped to construct a *geometric* theory of elementary particles, in which Kruskal spaces together with their "electrically charged" generalizations, and a space-time honeycombed with Kruskal-type "wormholes," would play a basic role. ("Matter without matter," "charge without charge," "geometry is everything.") However, that beautiful idea seems to have run into unsurmountable difficulties.

8.7   A General-Relativistic "Proof" of $E = mc^2$

As we saw in Section 5.6, it is easy enough to establish by the methods of SR that all energy *contributes* to mass according to the relation $E = mc^2$ (since kinetic energy does, and all energy is convertible to kinetic energy). But the converse, that all mass is energy, i.e., *available* energy, was originally a pure hypothesis whose experimental justification—the mutual annihilation of matter and antimatter—had to wait for decades. In GR one can actually give a prescription for extracting μc^2 units of energy from a mass μ, continuously. The prescription calls for slowly lowering the mass μ down to the horizon of a black hole of mass m, and collecting the resulting energy. Why "slowly?" Simply so as to avoid losing kinetic energy to the black hole.

Although the outcome is the same, the analysis is more transparent if we distribute the mass μ uniformly over a thin spherical shell Σ concentric with the black hole, say at coordinate $r = r_0$. To say that Σ has mass μ shall mean that the force m exerts on an area-element dS of Σ is $\mu dS/4\pi r_0$ times the g of (**8.71**); in principle, observers on Σ could measure this by cutting out dS.

Now allow Σ to shrink to $r = r_1$. As it shrinks, the field does work on it, which can be stored as elastic energy in Σ. That energy in turn can be converted into photons and sent back radially from Σ to stationary observers at $r = r_0$, where it can be stored or reconstituted into matter. Let us assume this process to occur continuously, so that Σ always retains its mass μ. During a contraction from r to $r - dr$ the field does work dW on Σ which is given by [cf. (8.71), (8.70)]

$$dW = -\mu g d\sigma = -\mu d\varphi = -\frac{\mu m dr}{\alpha(r) r^2}, \quad \alpha(r) = 1 - \frac{2m}{r}. \tag{8.98}$$

When converted into photons and sent to r_0, the corresponding energy dE received is diminished by a Doppler factor (8.51), since the energy carried by each photon is proportional to its frequency:

$$dE = -\frac{\mu m dr}{\alpha(r)^{1/2}\alpha(r_0)^{1/2} r^2}. \tag{8.99}$$

Consequently the total energy received at r_0 is given by

$$E = \mu m \alpha(r_0)^{-1/2} \int_{r_1}^{r_0} \frac{dr}{\alpha(r)^{1/2} r^2} = \tfrac{1}{2}\mu\alpha(r_0)^{-1/2} \int_{\alpha(r_1)}^{\alpha(r_0)} \frac{d\alpha}{\alpha(r)^{1/2}}$$

$$= \mu\left[1 - \frac{\alpha(r_1)^{1/2}}{\alpha(r_0)^{1/2}}\right]. \tag{8.100}$$

This becomes precisely μ (or μc^2 in ordinary units) as $r_1 \to 2m$. When Σ falls through the horizon, the process necessarily stops. We shall therefore collect at r_0—as asserted—the energy equivalent μc^2 of Σ, while Σ itself is lost through the horizon. Admittedly, the process will take an infinite time, since the last photons from Σ reach r_0 at time $t = \infty$.

Now a possible objection arises. Have we really extracted the energy from Σ or have we extracted it from the field? This objection is best answered by showing that the field is unchanged when the process is completed. The field, as it were, serves merely as a catalyst: while it *does* work on Σ, it has work done *on* it by the outgoing photons. To prove that the field is unchanged, we note first that the field *outside* r_0 is certainly unchanged, by Birkhoff's theorem (cf. Section 8.3). The *jump* in field strength across the ultimately created energy shell at r_0 cannot be very different from the corresponding (small) classical value. Yet, if our procedure caused *any* effective mass increase of the black hole, and thereby a change of the Schwarzschild field between the horizon and r_0, that increase could be amplified indefinitely by repeating the process, thus creating a *large* jump across r_0. We conclude that the field, in fact, is totally unchanged. Without changing anything else, therefore, we have converted the entire mass of Σ into available energy.

As a corollary, we note that the mass of the black hole is unchanged in spite of it having absorbed Σ "from rest at the horizon." This is what is meant by saying that the number of elementary particles inside a black hole is indeterminate.

8.8   A Plane-Fronted Gravity Wave

One might well expect that curvature disturbances in spacetime would propagate at the speed of light, and so give rise to "gravitational radiation." That this is in fact the case has been shown by many theoretical investigations, using mainly various approximation methods. However, there exist certain exact solutions of Einstein's vacuum field equations that clearly represent gravity waves, and we shall here examine one of the simplest of these, an infinite "plane-fronted sandwich wave." It is not the kind of wave that could be generated by any reasonable source; rather, it would have to be created *in toto*. Nevertheless it exhibits some interesting properties which might be expected to apply also in more general radiative situations. Note that a gravity wave must satisfy the *vacuum* field equations. Although it may carry gravitational energy, such energy is *not* an explicit source term in Einstein's theory [see (iv) of the paragraph following (**8.138**)]; and we shall assume that there are no other forms of energy (like mass, electromagnetic fields, etc.) in its path.

Any scalar wave profile $p = f(x)$ propagating in the x direction at speed c will, after time t, be given by the equation $p = f(x - ct)$. We shall here take $c = 1$ and also, following tradition, write $p = g(t - x)$, where $g(x) = f(-x)$. Guided by the electromagnetic analogy, we might expect gravitational waves to be transverse, i.e., to distort spacetime only in directions at right angles to the wave normal. In this way we are led to consider metrics of the form

$$ds^2 = dt^2 - dx^2 - p^2 dy^2 - q^2 dz^2, \qquad (8.101)$$

where p and q are functions only of

$$u = t - x.$$

The "disturbance" at any time t is in the y, z plane only, and it propagates at the speed of light. We have introduced a further restriction by assuming that there is no term in $dydz$. As we shall see presently, this amounts—in the simplest case—to the choice of the y and z axes in the "planes of polarization" of the wave.

By reference to Appendix I, it is straightforward to write down the Ricci tensor $R_{\mu\nu}$ and the Riemann tensor $R_{\mu\nu\rho\sigma}$ for the metric (**8.101**). In this way we find that the vacuum field equations $R_{\mu\nu} = 0$ will be satisfied if and only if

$$\frac{p''}{p} + \frac{q''}{q} = 0, \qquad (8.102)$$

where primes denote differentiation with respect to u; and that the metric is flat ($R_{\mu\nu\rho\sigma} = 0$) if and only if

$$p'' = q'' = 0. \qquad (8.103)$$

Note that this latter condition is equivalent to p and q being linear functions of u.

The simplest nontrivial way to satisfy Equation (**8.102**) is to set

$$p = \cos ku, \qquad q = \cosh ku. \qquad (8.104)$$

As we shall presently see, this is also *physically* the most fundamental choice. However, we shall not assume (**8.104**) to hold *everywhere*. We wish to construct a *sandwich* wave, i.e., a zone of curvature, bounded by two parallel planes, which propagates at the speed of light in the direction of the normal to the planes; on either side of the sandwich there shall be flat spacetime, if possible. Figure 8.10a shows a three-dimensional sketch of such a wave at one instant $t = $ constant, while Figure 8.10b shows its "Minkowski diagram." (In both diagrams care must be exercised not to regard the spaces on either side of the wave zone as simple continuations of each other—as though the wave were absent.)

As we shall have to "patch" three solutions of Einstein's vacuum field equations together (two outside and one inside the sandwich), we must first clarify our "junction conditions." The reader will be familiar with the requirement in both Newton's and Maxwell's theories that the potential *and* its first derivatives be continuous, unless a surface concentration of sources is crossed. Physically, this means that the field strengths cannot suddenly jump. In GR one similarly requires the continuity of the metric coefficients $g_{\mu\nu}$ and their first derivatives, unless a surface concentration of mass is crossed. The physical meaning of this can be seen from the geodesic equations (**8.15**): the coordinate acceleration of test particles (a measure of force relative to the coordinate system) must not suddenly jump. In our metric (**8.101**) we therefore require the continuity of p, q, and p', q'.

If we assume that the front and back planes of the wave (**8.101**) have equations $u = 0$ and $u = a^2$, respectively, and that $p = q = 1$ before its passage, while $p = \cos(u/a)$, $q = \cosh(u/a)$ within the sandwich, we shall certainly satisfy the junction conditions at $u = 0$. If the postwave spacetime is to be flat, p and q must be linear there; the precise forms of p and q are then determined by the values they and their derivatives take at $u = a^2$ inside the wave, since these values must be continuous across the junction. In this way we find (cf. Figure 8.11a):

$$u \le 0: \quad p = 1, \; q = 1,$$

$$0 \le u \le a^2: \quad p = \cos\left(\frac{u}{a}\right), \; q = \cosh\left(\frac{u}{a}\right),$$

$$a^2 \le u: \quad p = (\cos a + a \sin a) - \left(\frac{u}{a}\right)\sin a,$$

$$q = (\cosh a - a \sinh a) + \left(\frac{u}{a}\right)\sinh a. \qquad (8.105)$$

We now ask: how are test particles affected by the passage of the wave? To

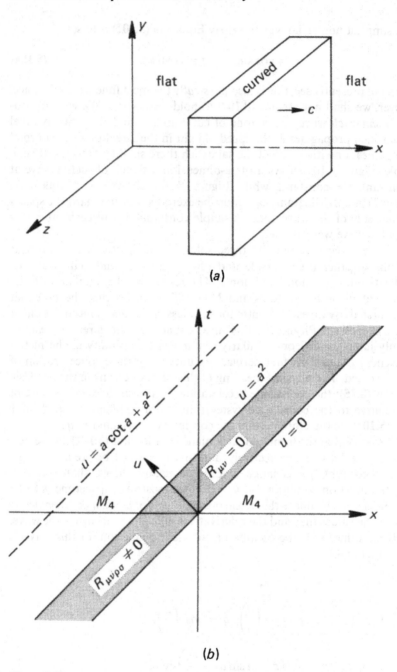

(a)

(b)

Figure 8.10

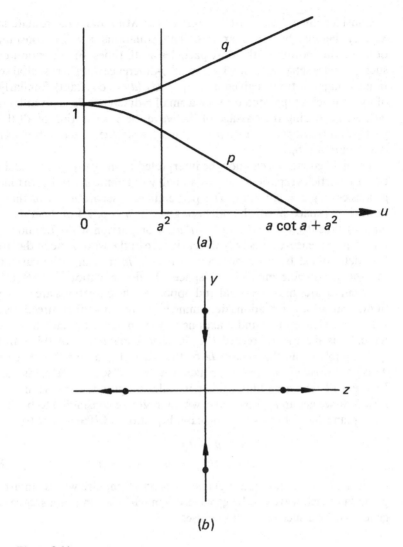

Figure 8.11

answer this question, we shall need a simple lemma. It states that the equations

$$x, y, z = \text{constant} \qquad (8.106)$$

represent timelike geodesics (i.e., possible particle worldlines) in the metric
(8.101), or, more generally, in *any* metric having no cross terms in dt and a
unit (or constant) coefficient of dt^2. For we find from (8.13) (with $x, y, z, t =$
x^1, x^2, x^3, x^4) that, for such metrics, $\Gamma^\mu_{44} = 0$. Equations (8.106) imply
$ds^2 = dt^2$, whence $dt/ds = 1$ and $d^2t/ds^2 = 0$. All of Equations (8.15) are
therefore satisfied. And the timelike property is obvious.

Consider, then, "test-dust" at rest in the Minkowski coordinate system x, y, z, t before the wave arrives. The equations $x, y, z = $ constant will describe the motion of the dust particles at all times. Fix attention on two such particles separated only by a small y-difference dy. Their spatial separation throughout time is given by $|ds| = p\,dy$ ($dy = $ constant). Similarly, that of two particles separated only by a small z-difference dz is given by $|ds| = q\,dz$. Hence during the passage of the wave the y-separation of all the dust particles *decreases* [$p = \cos(u/a)$ decreases], while their z-separation *increases* (see Figure 8.11b).

The field equations can now be interpreted geometrically. Consider again two test particles separated only by a small y-difference dy, and both following geodesics $x, y, z = $ constant. The perpendicular separation η of these geodesics in y, t space, in which both lie, is given by $\eta = p\,dy$. For them, $ds = dt = du$, and thus $d^2\eta/ds^2 = d^2\eta/du^2 = p''dy$. Comparison with (7.5) now shows that $-p''/p$ measures precisely the curvature of the wave zone in the orientation determined by y, t space. Similarly, $-q''/q$ measures the curvature in the orientation determined by z, t space. The field equations (8.102) state that these curvatures must be equal and opposite. The curvatures are clearly less, numerically, for orientations determined by t and any other directions in the y, z plane. Hence the y and z axes lie in the "planes of polarization" of the wave. It is natural to regard the numerical maximum of this curvature, $|p''/p| = |q''/q|$, as the *amplitude* of the wave. Equations (8.105) therefore describe a wave of *constant* amplitude, $1/a^2$, i.e., a "square wave," as shown in Figure 8.12. The area under this particular amplitude curve is unity. If we let $a \to 0$, we get an *impulse* wave, which might be considered to be the most basic plane wave. In this limiting case, Equations (8.105) reduce to

$$u \leq 0: \quad p = q = 1$$
$$u \geq 0: \quad p = 1 - u, \quad q = 1 + u. \tag{8.107}$$

By fitting together (integrating) a succession of impulse waves, an arbitrary plane sandwich wave can be generated—much as a continuous force can be generated by a succession of impulses.

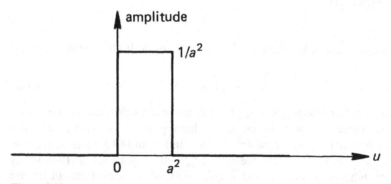

Figure 8.12

After the wave [(**8.105**) or (**8.107**)] has passed, p goes to zero linearly. It would therefore seem that *all* the dust particles on any line x, z = constant get "focused" together when $u = a \cot a + a^2$. Three troublesome questions now arise: (i) Where on the line x, z = constant do they all coalesce—and why there rather than elsewhere on that line? (ii) How do the distant particles get to the focus event without traveling faster than light? And (iii), does not the vanishing of p imply a singularity of the metric? Must we perhaps reject waves in which p or q have zeros as unphysical? No: first, we can show that at least *one* of p, q *must* attain a zero; secondly, that the apparent physical difficulties of the focusing effect are only apparent; and thirdly, that the metric singularity is only a coordinate singularity.

For these (and some other) purposes, it is useful to express p and q in terms of two new functions L and β of u, which are defined as follows:

$$p = Le^{\beta}, \quad q = Le^{-\beta}, \tag{8.108}$$

whence

$$pq = L^2, \quad \frac{p}{q} = e^{2\beta}. \tag{8.109}$$

Note that, like p and q, L and β together with their first derivatives must be continuous to fulfil the junction conditions. From (**8.108**) we find

$$\frac{p''}{p} = \frac{1}{L}[(L'' + \beta'^2 L) + (\beta'' L + 2\beta' L')],$$

$$\frac{q''}{q} = \frac{1}{L}[(L'' + \beta'^2 L) - (\beta'' L + 2\beta' L')], \tag{8.110}$$

where the second equation can be obtained from the first by putting $-\beta$ for β. The field equations (**8.102**) are therefore seen to be equivalent to

$$L'' + \beta'^2 L = 0, \quad \text{i.e.,} \quad \beta = \pm \int \left(-\frac{L''}{L} \right)^{1/2} du, \tag{8.111}$$

and the conditions (**8.103**) for flatness *additionally* require

$$\beta'' L + 2\beta' L' = 0. \tag{8.112}$$

If p and q are positive and constant before the passage of the wave (as we may assume without loss of generality), L and β are constant also and $L > 0$. Thus, before L attains a zero, we have $L'' \leq 0$, from (**8.111**). If β *stays* constant, there is no wave; if it changes at all, L'' must be negative *somewhere*, and that ensures that L attains a zero. But when L is zero, so is p—unless e^{β} has a compensating pole there; in that case, however, q is zero. So we cannot avoid the vanishing of at least one of these coefficients, and with it, a singularity in the metric (**8.101**). The history of gravitational radiation has been bedeviled by this singularity, which at one time even caused Einstein to doubt the very existence of gravity waves. In 1956 I. Robinson showed it to be a mere coordinate singularity. We shall presently demonstrate this for a special case.

First note another advantage of the functions L and β. Via Equation (8.111)(ii), these allow us to generate any number of solutions, simply by choosing any L and then finding β by quadrature. [If we use Equation (8.102), we must solve a differential equation.] For the postwave zone it is easiest to revert to p and q by calculating them from L and β just inside the wave zone and continuing them linearly. It should be noted that we need L'' continuous also, since this is linked to the continuity of β' via (8.111).

Consider now the very special choice

$$L = L_0, \quad \beta = \beta_0 \ (u \le -a^2); \quad L = 1 - u, \quad \beta = 0 \ (u \ge 0), \quad (8.113)$$

where L_0 and β_0 are suitable constants, and the two given linear portions of L are to be joined through the wave zone so that L, L', and L'' (<0) are continuous. Note that we have assumed the wave zone to *end* at $u = 0$. The postwave spacetime corresponding to (8.113) has the metric

$$ds^2 = dt^2 - dx^2 - (1 - u)^2(dy^2 + dz^2). \quad (8.114)$$

In this case, *all* test-dust particles originally at rest in a plane $x = x_0$ get "focused" at time $t = 1 + x_0$. [Readers familiar with cosmology will recognize the metric (8.114), with $dx^2 = 0$ for the plane $x = x_0$, as a two-dimensional Robertson–Walker universe with "expansion factor" $R(t) = (1 + x_0 - t)$; cf. Section 9.5.] The following transformation to new co-ordinates X, Y, Z, T,

$$
\begin{aligned}
T &= t - \tfrac{1}{2}(1 - u)(y^2 + z^2) \\
X &= x - \tfrac{1}{2}(1 - u)(y^2 + z^2) \\
Y &= (1 - u)y \\
Z &= (1 - u)z,
\end{aligned}
\quad (8.115)
$$

is regular everywhere except at $u = 1$. It is straightforward to verify that it transforms the metric (8.114) into the Minkowskian form

$$ds^2 = dT^2 - dX^2 - dY^2 - dZ^2, \quad (8.116)$$

which shows that $u = 1$ is a mere coordinate singularity of the metric (8.114). A similar transformation to the Minkowskian form can be found in the general case. It is really the X, Y, Z, T system of coordinates that should be shown in back of the wave in Figure 8.10a.

We can use the X, Y, Z, T system to examine how the various test particles, originally at rest in x, y, z, t space, travel to their respective foci after the wave has passed. Consider a set of such particles on a line $x, z = $ constant in the metric (8.114). To be specific, consider the worldlines

$$x = 0, \quad y = m, \quad z = 0, \quad t = t, \quad (8.117)$$

for various constant values of m, the last equation simply stating that we use t both as coordinate and parameter. The worldines (8.117) translate into

$$X = -\tfrac{1}{2}(1 - t)m^2, \quad Y = (1 - t)m, \quad Z = 0, \quad T = t - (1 - t)m^2. \quad (8.118)$$

We note at once that they are *not* perpendicular to the X axis, but rather have slope $-2/m$. Since they satisfy $x = 0$, the particles all come out of the wave zone at $t = 0$ $(u = 0)$; in the new coordinates, therefore, they come out at $T = -\frac{1}{2}m^2$ and $X = -\frac{1}{2}m^2$, i.e., neither at the same time nor on the same plane. Nevertheless, the back end of the wave at each instant is still a plane, $X = $ constant, since its equation $t = x$ translates into $T = X$. So in X, Y, Z, T space the following picture emerges. A plane wave sweeps over the particles, but its effect is not purely transverse: it also imparts a forward velocity component to the particles, except to those on the X axis. All the particles lying on the parabola $X = -\frac{1}{2}Y^2$, $Z = 0$ (see Figure 8.13) get focused to a single event $(0, 0, 0, 1)$, when the parameter $t = 1$. Their travel time is $1 + \frac{1}{2}m^2$ and they travel a distance $(m^2 + \frac{1}{4}m^4)^{1/2}$; since $(1 + \frac{1}{2}m^2)^2 > m^2 + \frac{1}{4}m^4$, none breaks the relativistic speed limit. [This is of course obvious *a priori* from the timelike nature of (**8.117**).] The square interval along *each* of these paths is unity. If we revolve the parabola about the X axis we get a paraboloid of revolution with the property that *all* particles on it get focused to its vertex at time $T = 1$. A whole series of identical and coaxial paraboloids can be drawn, each corresponding to a locus $t = x = $ constant in X, Y, Z space, and each determining a set of particles which the passing wave focuses. [The general wave (**8.101**) focuses to lines rather than to single points, but the properties of the focusing are essentially the same as those discussed here.]

One last question now remains: why do all the foci lie on the X axis and not on some other line $Y, Z = $ constant? The answer is that *all* particles in a given plane $x = $ constant are totally equivalent. The effect of the wave is to make a homogeneously contracting universe of them. Each, with equal right, can consider itself the "center" of this contraction. Each has an equal right to map this universe in its own (postwave) inertial rest frame. (After all, American and European cartographers each draw *their* respective continents in the middle of the world, and no doubt they would each consider *their* continents

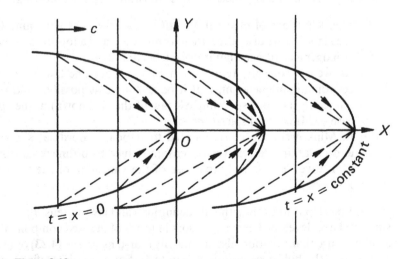

Figure 8.13

at rest if all the world were shrinking.) The inertial system X, Y, Z, T of Equations (8.115) is that associated with the particles on the x axis. We could *just* as well have given preference to the particles on a line $y = y_0, z = z_0$. In that case we would substitute $y - y_0$ and $z - z_0$ for y and z, respectively, in Equations (8.115). An identical picture would emerge, except that the original stay-at-home particles would now be shown as travelers.

Pioneering efforts to detect gravitational waves, apparently coming to us from violent events in the central regions of our galaxy, have been made by Weber at the University of Maryland for more than a decade. Although it appears that his data statistically support the reception of such waves, other researchers in various parts of the world have so far been unable to duplicate his results. One of the current problems is to find a likely theoretical mechanism—black holes have been studied in this context—that could produce gravitational radiation of the enormous power that would be required to show up on Weber's apparatus.

8.9 The Laws of Physics in Curved Spacetime

Our arguments so far have led to the recognition that spacetime is generally curved. In such spacetime, SR—which is a flat-space theory—applies only approximately and locally, just as Euclidean geometry applies only approximately and locally on the surface of a sphere. What, then, are the *exact* laws of (nongravitational) physics in curved spacetime? For example, what laws govern particle collisions or electromagnetic fields in Schwarzschild space? Lest the reader despair at the prospect of yet another major revision, let us quickly state that the transition from flat-space to curved-space laws is essentially trivial. The key ideas for this transition are the following:

(i) The definitions of physical quantities, and the laws governing them, are in the nature of axioms; their formulation and adoption are matters of judgement rather than proof (cf. Section 2.2).

(ii) All we can logically require of a curved-spacetime law is that it be coordinate independent, that it be as simple as possible, and that it reduce to the corresponding SR law (if that is known) in the special case of Minkowski spacetime.

(iii) In Minkowski spacetime referred to standard coordinates, absolute and covariant derivatives of tensors reduce to ordinary and partial derivatives, respectively (because of the vanishing of the Christoffel Γ's).

Our first step is to express the SR definitions and laws in tensor form. And since we have developed general tensors in terms of indexed components, we shall here use that notation also for the SR four-tensors [cf. (4.23) or (4.24)], rather than the bold-face notation as in (4.8). For example, Definition (4.8) of the four-velocity of a particle can be written (if $c = 1$, as we shall here assume) in the form

$$U^\mu = \frac{dx^\mu}{ds}. \qquad (8.119)$$

But dx^μ/ds behaves as a tensor under *all* coordinate transformations. We can therefore widen the definition of U^μ from four-tensor to general tensor in M_4, by simply adopting Definition (8.119) for arbitrary coordinate systems in M_4. The extension of this same definition to curved spacetimes is now a rather natural step. For the same reason, the definition of four-momentum (5.6), written in component form,

$$P^\mu = m_0 U^\mu = m_0 \frac{dx^\mu}{ds}, \qquad (8.120)$$

is adopted in arbitrary spacetime. Next, consider the law of momentum conservation, (5.9),

$$P_1^\mu + P_2^\mu = P_3^\mu + P_4^\mu. \qquad (8.121)$$

This too can be taken over directly to arbitrary spacetimes, with the newly widened definition of P^μ.

A slight complication occurs with the definition of four-force, (5.30),

$$F^\mu = \frac{d}{ds} P^\mu = \frac{d}{ds}\left(m_0 \frac{dx^\mu}{ds}\right). \qquad (8.122)$$

The second and third members are *not* general tensors [cf. (8.12)]. Consider, instead,

$$F^\mu = \frac{D}{ds} P^\mu = \frac{D}{ds}\left(m_0 \frac{dx^\mu}{ds}\right) \qquad (8.123)$$

[cf. (8.15)]. Here the second and third members *are* general tensors, and, moreover, (8.123) reduces to (8.122) in standard coordinates in M_4. So we accept (8.123) as the definition of force in arbitrary spacetimes. Suppose in SR the four-force were derivable from a scalar potential φ according to the law

$$F_\mu = -k\varphi_\mu, \qquad (8.124)$$

in the notation of (8.6). (Note that we have used the *covariant* form of F^μ, $g_{\mu\nu}F^\nu$, since the right member is covariant.) This can be taken over directly to arbitrary spacetimes. However, if the SR vacuum field equations for φ are

$$g^{\mu\nu}\varphi_{,\mu\nu} = 0 \qquad (8.125)$$

(where "$,\mu\nu$" stands for $\partial^2/\partial x^\mu \partial x^\nu$), these can *not* be taken over directly. The general tensor reducing to $\varphi_{,\mu\nu}$ in standard coordinates in M_4 is $\varphi_{;\mu\nu}$ (shorthand for $\varphi_{;\mu;\nu} = \varphi_{\mu;\nu}$). The equation we would therefore adopt in arbitrary spacetime is

$$g^{\mu\nu}\varphi_{;\mu\nu} = 0. \qquad (8.126)$$

The standard procedure for generalizing SR definitions and laws should now be clear:

(iv) SR definitions and laws are taken over directly to arbitrary spacetimes if they involve only expressions directly interpretable as general tensors. Ordinary and partial derivatives of tensors of rank ≥ 1 (which are not so interpretable) must be replaced by absolute and covariant derivatives, respectively.

The generalized definitions and laws will then be coordinate independent, and reduce to the SR forms in M_4 referred to standard coordinates, thereby satisfying our chief requirements. To some extent, the above procedure can be *justified* by the equivalence principle. In its original and rather vague formulation, the EP asserts that the laws of (nongravitational) physics are identical in all LIF's, identical, in fact, to the laws discovered in SR. But LIF'S are not small regions of flat spacetime. Rather, as we have seen in Section 8.1, they are special local coordinate systems in arbitrarily curved spacetime, with the property that at their origin—a single event—the coordinate lines are orthogonal and the Γ's vanish. *A priori*, the limitations on the EP are unclear. However, in light of the above procedure (iv) for generating curved-space laws, we can reformulate the EP much more precisely:

(v) At each LIF-origin, the general definitions and laws of (nongravitational) physics shall reduce to their SR forms—if possible.

This requirement is satisfied by (**8.119**) and all our other examples thus far. But it cannot be maintained when *second* (and higher) absolute or covariant derivatives of tensors of rank ≥ 1 enter the general laws. For these derivatives do *not* (in general) reduce to the corresponding second (or higher) ordinary or partial derivatives at the LIF-origin. Hence the EP seems to have significance only for laws not involving second and higher derivatives of tensors of rank ≥ 1. This is evidently related to the fact that a LIF deviates from being an inertial frame in *second* order of approximation.

When it applies [as formulated in (v) above], the EP acts as a *principle of minimum coupling*, namely of gravity—*qua* curvature—to the rest of physics. For it prohibits the introduction of "unnecessary" curvature terms into the general laws. For example, the field equation (**8.125**) can be generalized not only to (**8.126**), but equally well, say, to

$$g^{\mu\nu}\varphi_{;\mu\nu} = R\varphi, \qquad (8.127)$$

where R is the "curvature invariant" $g^{\mu\nu}R_{\mu\nu}$. This, too, reduces to (**8.125**) in Minkowski space. According to the EP, however, we would reject (**8.127**); for while the Γ's vanish at a LIF-origin, the curvature tensor does not, and so (**8.127**) does *not* reduce to (**8.125**) at a LIF-origin, whereas (**8.126**) does.

The procedure (iv) unfortunately becomes ambiguous when second and higher derivatives of tensors of rank ≥ 1 are involved. Suppose we wish to generalize an SR expression of the form $A^\mu{}_{,\sigma\tau}$, which is, of course, equivalent to $A^\mu{}_{,\tau\sigma}$. But the corresponding generalizations, $A^\mu{}_{;\sigma\tau}$ and $A^\mu{}_{;\tau\sigma}$, are *not* equivalent: they differ by a curvature term [cf. (8.25)]. In all such cases of ambiguity, other criteria must decide our choice. In fact, a problem of this nature arises in the generalization of Maxwell's equations.

One law of particular interest in the sequel is that which governs the behavior of continuously distributed matter. For dust, we define the energy tensor

$$T^{\mu\nu} = \rho_0 U^\mu U^\nu, \tag{8.128}$$

as in (5.49), but now generally. And the generalized form of the law (5.51) is

$$T^{\mu\nu}{}_{;\nu} = T^{\mu\nu}{}_{,\nu} + \Gamma^\mu_{\tau\nu} T^{\tau\nu} + \Gamma^\nu_{\tau\nu} T^{\mu\tau} = 0. \tag{8.129}$$

There is a correspondingly more complicated definition of $T^{\mu\nu}$ for media with internal stresses, but (8.129) still applies, as well as $T^{\mu\nu} = T^{\nu\mu}$.

Note, incidentally, that our procedure (iv) leads from the SR law of free motion, $(d/ds)(dx^\mu/ds) = 0$, to the GR law of geodesics, $(D/ds)(dx^\mu/ds) = 0$. But, of course, like all such transitions, this constitutes no *proof* of the curved-space law: its logical status is that of a suggestion.

By the rules of this section, all the *forces* of SR (electromagnetic, elastic, impact, etc.) have their counterparts in GR, i.e., they are now players on a new stage. Of all the classical forces, only gravity has no counterpart in GR: instead of being one of the players, it has become "part of the stage" (in E. T. Whittaker's phrase). Gravity has become "geometrized." For many years it was hoped that it might be possible also to absorb electromagnetism into the geometry ("unified field theories") by somehow enlarging the geometry, but after the discovery of yet other basic forces (the nuclear forces) those attempts lost much of their attraction. However, recent discoveries of the particle physicists have now given a new impetus to this quest.

Finally, it is of interest to recall that Einstein's original aim in constructing GR was to produce a theory not tied to the Lorentz transformations, but rather valid under *all* coordinate transformations ("principle of covariance.") He thought thereby automatically to satisfy Mach's demand for the exclusion of absolute space, i.e., of any preexisting geometry unaffected by matter. There, however, he was mistaken, as was soon pointed out by Kretschmann, and acknowledged by Einstein (1918). For even if all spacetime were Minkowskian, and SR strictly valid, the SR laws could be written —by the methods of the present section—in a form valid in all coordinate systems. Yet this would in no way diminish the physical preeminence of the inertial frames, and thus of an absolute standard of nonacceleration. In other words, a spacetime may have more structure than the coordinate systems used to describe it. A recent attempt to formulate the criterion that eluded Einstein was made by Anderson.[9] But GR does not stand or fall by it. The postulate of the Riemannian nature of spacetime is today regarded as primary. The need for arbitrary coordinates and tensor laws (Einstein's principle of covariance) then becomes a corollary.

---

[9] J. L. Anderson, *Principles of Relativity Physics*, New York, Academic Press, 1967.

8.10 The Field Equations in the Presence of Matter

The Einstein vacuum field equations $R_{\mu\nu} = 0$ are, as we have seen in Section 8.2, the GR analog of the Laplace equation $\Sigma\varphi_{ii} = 0$, which governs Newton's potential in empty space. If we wish to discuss the gravitational field in the presence of matter—for example, in a dust cloud, or inside the earth, or, indeed, in the universe (where for a first crude oversimplification the actual contents are replaced by continuous dust)—then we need a GR equivalent not of Laplace's equation but of Poisson's equation

$$\Sigma\varphi_{ii} = 4\pi G\rho. \tag{8.130}$$

The GR field equations will have to relate the matter with the *geometry* (the analog of φ). Since in the vacuum case the equations should reduce to $R_{\mu\nu} = 0$ (this we have already accepted), an obvious candidate for the left-hand side of the GR field equations is $R_{\mu\nu}$. And an equally obvious candidate for the right-hand side is the energy tensor $T^{\mu\nu}$, of which, as we have seen, the density ρ is one component (in LIF coordinates). However, we cannot equate covariant with contravariant tensors, and so we first convert $R_{\mu\nu}$ to its contravariant form

$$R^{\mu\nu} = g^{\mu\rho}g^{\nu\sigma}R_{\rho\sigma}. \tag{8.131}$$

The most obvious field equations, then, would be

$$R^{\mu\nu} = -\kappa T^{\mu\nu}, \tag{8.132}$$

where κ is some suitable "coupling" constant, and the minus sign is inserted for later convenience.

An immediate implication of these equations would be $T^{\mu\nu} = 0$ in flat spacetime, which would seem to vitiate our work with continuous media in SR: there could be *no* continuous media in SR. But this was to be expected: the presence of matter curves the spacetime. On the other hand, we have seen in Section 8.3 how very small a curvature results locally from even an enormous mass like the earth, or the sun. Hence the κ in Equations (8.132) would be exceedingly small, and any curvature resulting from the introduction of "ordinary" matter distributions into flat-space laboratories would be totally insignificant.

If we accept (8.129), however, (8.132) implies

$$R^{\mu\nu}{}_{;\nu} = 0. \tag{8.133}$$

On the other hand, it is known in Riemannian geometry (from *Bianchi's* identity) that

$$(R^{\mu\nu} - \tfrac{1}{2}g^{\mu\nu}R)_{;\nu} = 0, \quad (R = g^{\mu\nu}R_{\mu\nu}), \tag{8.134}$$

and thus (8.133) implies

$$R_{;\nu} = R_{,\nu} = 0, \tag{8.135}$$

since the covariant derivative is linear and satisfies Leibniz's rule, and since the covariant derivative of all g's vanishes and that of a scalar is the same as its partial derivative. But (8.132) also implies $R = -\kappa T$ ($T = g_{\mu\nu} T^{\mu\nu}$), and so (8.135) would imply $T_{,\nu} = 0$, which is an unwarranted restriction on $T^{\mu\nu}$. For example, in the dust case, $T = c^2 \rho_0$, and this need not be constant. Hence Equations (8.132) are not acceptable.

However, as reference to (8.134) shows, the equations

$$R^{\mu\nu} - \tfrac{1}{2} g^{\mu\nu} R = -\kappa T^{\mu\nu} \tag{8.136}$$

would meet our present objection: they satisfy $T^{\mu\nu}{}_{;\nu} = 0$ *automatically*. They are, in fact, the equations that Einstein proposed as his general field equations in 1915. (Ironically, the mathematician Hilbert, having only recently become interested in Einstein's work, arrived at them independently, and presented them to the Royal Academy of Sciences in Göttingen just *five days* before Einstein presented his—the fruit of many years of intensive search—to the Prussian Academy in Berlin.[10])

Multiplying (8.136) by $g_{\mu\nu}$, we find (recall that $g_{\mu\nu} g^{\nu\sigma} = \delta^\sigma_\mu$)

$$R - 2R = -\kappa T, \quad \text{i.e., } R = \kappa T, \tag{8.137}$$

and so (8.136) can also be written in the alternative form

$$R^{\mu\nu} = -\kappa(T^{\mu\nu} - \tfrac{1}{2} g^{\mu\nu} T). \tag{8.138}$$

We note: (i) In vacuum $T^{\mu\nu} = 0$, and then these field equations reduce to $R^{\mu\nu} = 0$, which is equivalent to $R_{\mu\nu} = 0$. (ii) The four differential identities (8.134), satisfied by both sides of the field equations, are the ones that ensure that we do not get the g's uniquely but have four degrees of freedom left to apply arbitrary transformations of the coordinates (cf. the end of Section 8.2). (iii) As Ehlers has pointed out, the "correction" (8.132) → (8.136) is quite analogous to the usual introduction of the displacement current into Maxwell's equations: there it is required by charge conservation, here by energy and momentum conservation (i.e., by $T^{\mu\nu}{}_{;\nu} = 0$). (iv) The field equations are nonlinear in the $g_{\mu\nu}$ and their derivatives, which means that solutions cannot be added. For example, the field of a sphere is not the sum of the fields of two hemispheres. It is this nonlinearity which allows the gravitational field to act as its own source *without being represented among the source terms* $T^{\mu\nu}$: nonlinearity gives the "field of the field" as the difference between the field of the whole and the sum of the fields of the parts. Indeed, the energy of the gravitational field *could* not be represented as a source term, since its location is ambiguous: the field at any event can be "transformed away" by the EP. (But *all* other manifestations of energy—e.g., any Maxwell fields present—must be included in $T^{\mu\nu}$.) Nonlinearity also permits the field equations to imply the interaction of the sources, i.e., interactive laws of motion [see the paragraph including (8.147) below].

It can be shown that the field equations (8.136) are the most general that

[10] See J. Mehra, *Einstein, Hilbert, and the Theory of Gravitation*, Reidel Pub. Co., 1974, especially p. 25.

satisfy the following desiderata: (i) to have tensorial character; (ii) to involve no higher derivatives of the g's than the second, and these and $T^{\mu\nu}$ only linearly; (iii) to satisfy $T^{\mu\nu}_{;\nu} = 0$ identically; and (iv) to permit flat spacetime as a particular solution in the absence of matter. Eventually Einstein dropped the last requirement, and then a slightly more general set of equations becomes possible, namely

$$R^{\mu\nu} - \tfrac{1}{2}Rg^{\mu\nu} + \Lambda g^{\mu\nu} = -\kappa T^{\mu\nu}, \tag{8.139}$$

where Λ is a universal constant. This could be positive, negative, or zero, and, like R, it has the dimensions of a space curvature, namely (length)$^{-2}$. It has come to be called the "cosmological" constant, because only in cosmology does it play a significant role; there, however, it may be forced on us by the observations.[11] Current estimates suggest $|\Lambda| < 10^{-54}$ cm^{-2} (see Section 9.11), and we shall show that this makes Λ quite negligible in all noncosmological situations.

Analogously to (8.137), we find from (8.139) that

$$R = \kappa T + 4\Lambda, \tag{8.140}$$

which allows us to rewrite (8.139) in the alternative form

$$R^{\mu\nu} = \Lambda g^{\mu\nu} - \kappa(T^{\mu\nu} - \tfrac{1}{2}g^{\mu\nu}T). \tag{8.141}$$

Clearly these equations do not permit globally *flat* spacetime in the absence of matter, for then they reduce to

$$R^{\mu\nu} = \Lambda g^{\mu\nu}, \tag{8.142}$$

which implies curvature. Any space satisfying a relation of the type (8.142) is called an *Einstein space*; it turns out that every space of constant curvature is an Einstein space, but not vice versa, unless its dimension is three.

The numerical value of κ in Einstein's field equations must be obtained experimentally, just like that of Newton's constant. In practice, however, one benefits from already knowing Newton's constant, and one simply compares Einstein's with Newton's theory in a convenient limiting case where the two converge to each other and where both involve their constants. Such a case is that of a vanishingly weak quasistatic distribution of dust. It turns out that the metric (7.25), with $d\sigma^2$ *flat*, does not satisfy Einstein's field equations even in a limiting way, when φ is a Newtonian potential. A metric which *does* have this property is the following:

$$ds^2 = \left(1 + \frac{2\varphi}{c^2}\right)c^2dt^2 - \left(1 - \frac{2\varphi}{c^2}\right)(dx^2 + dy^2 + dz^2). \tag{8.143}$$

Let us set $(x^1, x^2, x^3, x^4) = (x, y, z, ct)$. (When making approximations it often pays to have all coordinates of the same dimensions and to retain c.) In calculating the Ricci tensor components of (8.143) (by reference to

---

[11] The effect of a *positive* Λ, as we shall see, is to counteract gravity. Einstein at one time needed this because he wanted to construct a *static* universe. Today we may need it to permit an acceleratingly expanding one, if that is what the observations should indicate.

Appendix I), we neglect products of any two of φ/c^2, φ_i/c^2, φ_{ij}/c^2, which are all assumed to be small compared to unity. [We use the indicial notation for derivatives introduced in (8.26) et seq.] Thus we find

$$R^{\mu\mu} = -c^2\Sigma\varphi_{ii}, \quad R^{\mu\nu} = 0 \ (\mu \neq \nu). \tag{8.144}$$

If the matter is quasistatic dust of density ρ, with $T^{\mu\nu}$ given by (8.128), all $T^{\mu\nu}$ vanish except $T^{44} = c^2\rho$; so $T = c^2\rho$, and hence

$$T^{\mu\mu} - \tfrac{1}{2}g^{\mu\mu}T = \tfrac{1}{2}c^2\rho, \quad T^{\mu\nu} = 0 \ (\mu \neq \nu). \tag{8.145}$$

We have already seen (cf. end of Section 7.6) that the φ in (8.143) must be identified with Newton's potential. It will therefore satisfy Poisson's equation (8.130). Consequently (8.144) and (8.145) will satisfy the field equations (8.138) if

$$\kappa = \frac{8\pi G}{c^4} = 2.073 \times 10^{-48} \ \text{sec}^2\text{cm}^{-1}\text{gm}^{-1}. \tag{8.146}$$

This is the accepted value of *Einstein's constant of gravitation*.

We are now in a position to justify the omission of the Λ term in all but cosmological applications. [See also the remark after (8.154) below.] In the solar system, for example, each component of the Riemann tensor $R^\mu_{\nu\rho\sigma}$, being a certain combination of the curvatures at a given point, is of the order of m/r^3 [see before (8.44)], where $m = 1.47$ km is the mass of the sun, in units that make $c = G = 1$, and r is distance from the sun. Up to the orbit of Pluto this exceeds 0.6×10^{-39} cm^{-2}, and is responsible for the dynamics of the system. The effect of the Λ term—roughly speaking—is to cause a "pre-existing" curvature of spacetime [via (8.142)] of the order of Λ, i.e., $< 10^{-54}$ cm^{-2} according to cosmological estimates. Hence its effect in the solar system, and in similarly "strong" fields, is totally insignificant. Another intuitive way of regarding Λ is as a "preexisting" (negative) density—via (8.141). As such, it corresponds to at most 10^{-27} gm/cm^3—less than one hydrogen atom per liter of space. Again, its insignificance in noncosmological applications is apparent. In cosmology, however, where the average curvature is entirely due to the average density of the universe ($\sim 10^{-30}$ gm/cm^3) plus the Λ term, these two "source terms" on the right of (8.141) can become comparable.

The approximate metric (8.143), with φ the Newtonian potential, can be established under much more general conditions than we have assumed. The matter need not be dust, nor need it be static. In the vacuum regions of the metric it is merely necessary that φ and its derivatives be small (for the satisfaction of the field equations). If we wish to apply (8.143) in the source regions also, it is necessary that the predominant component of $T^{\mu\nu}$ be $T^{44} = c^2\rho$. So the sources must be weak, move slowly, and have small internal stress [cf. after (8.183)]. Under these conditions, and for predicting the motions of *slow* test particles, Newton's theory can be regarded as a first approximation to Einstein's theory. On close examination it is even seen to be a very *good* theoretical approximation. Consequently all the classical observations of celestial mechanics, which are in such excellent agreement with Newtonian gravitational theory, can now be adduced as support for Einstein's theory too.

In fact, (8.143) goes beyond the Newtonian approximation, in that it allows us also to predict (with good approximation to full GR theory) the motions of *fast* test particles and even of light. Roughly speaking, for light in vacuum it gives twice the deflection that Newton's theory predicts for a particle of speed c. (If one wishes to use GR for predicting light paths *inside* media, either these media must be sufficiently tenuous not to affect the speed of light or essentially Maxwellian methods must be used in place of the null-geodesic hypothesis.)

If we go one step further, and allow the sources to move at arbitrary speeds, we get, as a first approximation to GR, not Newton's theory, but an extension of Newton's theory with distinctly Maxwellian features. Newton's theory constitutes the "Coulomb" part of that wider theory (see Section 8.12). Since interesting *exact* solutions of the GR field equations are difficult to get except in situations of high symmetry, much work has been done instead with this "linear approximation" (although it is not always clear how good an approximation to the full theory one gets in this way). For example, it was used to obtain verifications of the various "induction" effects predicted on the basis of Mach's principle (cf. Section 1.16). And it was also with this theory that gravitational waves were first investigated.

A most interesting property of Einstein's field equations is that they, by themselves, imply the geodesic law of motion, which had originally been introduced as a separate axiom. Now in theories with *linear* field equations, the interaction between sources *must* be specified by separate axioms (as by Lorentz's force law in Maxwell's theory). For consider two solutions of some set of linear field equations, each with a source moving in some arbitrary way. The superposition of these fields and motions will also satisfy the field equations. Hence they imply no interaction whatsoever between the sources. But in nonlinear theories superposition is impossible: each source "feels" the other. In GR this results in the geodesic law. The general proof is long and difficult; suffice it to say that it has been given (in a series of papers by Einstein and collaborators beginning in 1927), though perhaps still not with sufficient rigor to satisfy the mathematicians. In the special case of *dust* particles, however, the proof is quite easy. This is not really surprising. For the field equations were constructed to imply $T^{\mu\nu}{}_{;\nu} = 0$ automatically, and for dust in *flat* spacetime we already know that this equation implies geodesic motion (see Section 5.13). In general spacetime we have, from (8.128) and (8.129),

$$(\rho_0 U^\mu U^\nu)_{;\nu} = 0,$$

or, in expanded form,

$$U^\mu(\rho_0 U^\nu)_{;\nu} + \rho_0 U^\mu{}_{;\nu} U^\nu = 0. \tag{8.147}$$

Multiplying this by $g_{\mu\tau} U^\tau$, we have

$$c^2(\rho_0 U^\nu)_{;\nu} + \rho_0 g_{\mu\tau} U^\tau U^\mu{}_{;\nu} U^\nu = 0. \tag{8.148}$$

But $U^\mu_{;v} U^v = A^\mu$, the four-acceleration (as can be seen in LIF coordinates), and so the second term in **(8.148)** vanishes, being $\rho_0 \, U \cdot A$. Hence the first term vanishes also. When that is substituted into **(8.147)**, we get

$$U^\mu_{;v} U^v = 0, \quad \text{i.e.,} \quad A^\mu = 0, \quad \text{i.e.,} \quad \frac{D}{ds}\left(\frac{dx^\mu}{ds}\right) = 0,$$

and this is precisely the equation of a geodesic. If a dust particle in a dust cloud (which may be a very *small* dust cloud) follows a geodesic, it seems very likely, of course, that *any* free test particle follows a geodesic.

Finally a word about "solving" the field equations. In GR the situation is very different from that in any other field theory. For example, in Newtonian theory we start with a well-defined coordinate system, relative to which we can specify the density distribution ρ, and then we simply *solve* the relevant Poisson equation for the potential φ. It is much the same in Maxwell's theory. But in GR we do *not* start with a well-defined coordinate system: on the contrary, that is what we are trying to find. Hence we cannot simply "plug in" the given $T^{\mu v}$ on the right-hand side of the field equations, since there is no way of knowing the components of $T^{\mu v}$ except with reference to a coordinate system. Ideally, one could make a catalog by inventing arbitrary $g_{\mu v}$ and listing the corresponding $T^{\mu v}$ from **(8.136)**; in any given situation one would then try to find the relevant pair $(g_{\mu v}, T^{\mu v})$. In practice, one often works from both ends: for example, the symmetry of the physical situation may suggest a certain pattern of g's involving unknown functions; the T's can then be expressed also in terms of these functions, which are finally determined by the field equations when the T's and g's are substituted into them.

8.11 From Modified Schwarzschild to de Sitter Space

In Section 8.3 we found the most general spherically symmetric static metric subject to Einstein's original vacuum field equations. (It will be recalled that according to Birkhoff's theorem the assumption of staticness can be omitted: it is a *consequence* of spherical symmetry.) We can now easily adapt that argument to the case of Einstein's "modified" vacuum field equations **(8.142)**. Again we shall assume staticness, although there exists an analogous Birkhoff theorem which says that the result is again independent of this assumption. Thus we begin with a metric of form **(8.36)** and the relevant $R_{\mu v}$ as given by **(8.37)**–**(8.41)**. Now, however, we require $R_{\mu v} = \Lambda g_{\mu v}$ instead of $R_{\mu v} = 0$. Still, as before, we find $A' = -B'$ and $A = -B$. The equation $R_{22} = \Lambda g_{22}$ then yields

$$e^A(1 + rA') = 1 - \Lambda r^2, \tag{8.149}$$

or, setting $e^A = \alpha$,

$$\alpha + r\alpha' = (r\alpha)' = 1 - \Lambda r^2.$$

Hence

$$\alpha = 1 - \frac{2m}{r} - \tfrac{1}{3}\Lambda r^2, \tag{8.150}$$

where $-2m$ is again a constant of integration. It is easily verified that this solution indeed satisfies *all* the equations $R_{\mu\nu} = \Lambda g_{\mu\nu}$. Thus we have found the following essentially unique metric satisfying our conditions:

$$ds^2 = \left(1 - \frac{2m}{r} - \tfrac{1}{3}\Lambda r^2\right)dt^2 - \left(1 - \frac{2m}{r} - \tfrac{1}{3}\Lambda r^2\right)^{-1}dr^2$$
$$- r^2(d\theta^2 + \sin^2\theta d\phi^2). \tag{8.151}$$

Comparison with the metric (7.24) shows that slow orbits in the space (8.151) approximately correspond to Newtonian orbits under a central potential

$$\varphi = -\frac{m}{r} - \tfrac{1}{6}\Lambda r^2, \tag{8.152}$$

in units in which $c = G = 1$. Its first term is recognized as the Newtonian effect of a mass m, and thus again we must identify the constant m with a spherically symmetric mass centered at the origin. The Λ term corresponds to a repulsive central force of magnitude $\tfrac{1}{3}\Lambda r$, which is quite independent of the central mass. Its effect on the orbits can be found by retracing the calculations that led to the previous orbit equations (8.52) and (8.53). The former reappears unchanged, but instead of the latter we now get

$$\frac{d^2u}{d\phi^2} + u = \frac{m}{h^2} + 3mu^2 - \frac{\Lambda}{3h^2u^3}. \tag{8.153}$$

The main effect of the extra Λ term in this equation can be shown to be an *additional* advance of the perihelion by an amount

$$\Delta = \frac{\pi\Lambda h^6}{m^4} = \frac{\pi\Lambda a^3(1 - e^2)^3}{m}. \tag{8.154}$$

In the case of Mercury, for example, this would be one second of arc per century if Λ were $\sim 5 \times 10^{-42}$ cm^{-2}. Since this would be detectable, Λ *cannot* be that big.

This conclusion is one useful by-product of obtaining the metric (8.151). Another is the space we get on letting $m \to 0$:

$$ds^2 = (1 - \tfrac{1}{3}\Lambda r^2)dt^2 - (1 - \tfrac{1}{3}\Lambda r^2)^{-1}dr^2 - r^2(d\theta^2 + \sin^2\theta d\phi^2). \tag{8.155}$$

It was discovered by *de Sitter* in 1917 (with $\Lambda > 0$) in connection with cosmology, where it still plays an important role. Just like Schwarzschild's metric, de Sitter's metric has a singularity if $\Lambda > 0$, at $r = (3/\Lambda)^{1/2}$. But just like Schwarzschild's singularity, this singularity is of a purely coordinate nature—as we shall see. Apart from the usual possibility of transforming coordinates, this is the *unique* static solution of Einstein's modified vacuum field equations **(8.142)** which is spherically symmetric about a given point and regular at that point: any nonzero m in **(8.151)** makes the origin an intrinsic singularity. According to Birkhoff's theorem quoted above, there is no other solution even if we relinquish staticness. De Sitter space **(8.155)** is therefore the successor of Minkowski space *if* the equations $R_{\mu\nu} = \Lambda g_{\mu\nu}$ govern an empty world. Of course, it would be a serious flaw of these equations if the resulting spacetime could be spherically symmetric about one point only. But, in fact, the metric **(8.155)** represents a pseudosphere of curvature $-\frac{1}{3}\Lambda$, and can be expressed in identical form with *any* event (r, θ, ϕ, t) as the new origin.

To see this, suppose first that Λ is positive, say

$$\frac{3}{\Lambda} = a^2 \quad (a > 0). \tag{8.156}$$

Now consider *five*-dimensional Minkowski space M_5, with coordinates X, Y, Z, W, T and metric

$$ds^2 = dT^2 - (dX^2 + dY^2 + dZ^2 + dW^2), \tag{8.157}$$

and, in it, the pseudosphere S_4^- with equation

$$X^2 + Y^2 + Z^2 + W^2 - T^2 = a^2. \tag{8.158}$$

(This turns out to be a space of constant curvature $K = -1/a^2$.) The space **(8.155)** can be mapped isometrically into S_4^-, by the equations

$$\begin{aligned} X &= r \sin\theta \cos\phi \\ Y &= r \sin\theta \sin\phi \\ Z &= r \cos\theta, \end{aligned} \tag{8.159}$$

and, according as $r \lessgtr a$, respectively,

$$\begin{aligned} W &= a\left(1 - \frac{r^2}{a^2}\right)^{1/2} \cosh\frac{t}{a} \\ T &= a\left(1 - \frac{r^2}{a^2}\right)^{1/2} \sinh\frac{t}{a} \end{aligned} \tag{8.160}$$

or

$$\begin{aligned} W &= a\left(\frac{r^2}{a^2} - 1\right)^{1/2} \sinh\frac{t}{a} \\ T &= a\left(\frac{r^2}{a^2} - 1\right)^{1/2} \cosh\frac{t}{a}, \end{aligned} \tag{8.161}$$

where in all cases we choose the *positive* root. These equations ensure that $X^2 + Y^2 + Z^2 + W^2 - T^2 = a^2$, no matter what r, θ, ϕ, t may be; i.e., *every* event (r, θ, ϕ, t) maps into S_4^-. That the mapping is isometric is easily verified by substituting from **(8.159)**–**(8.161)** into **(8.157)**, whereupon **(8.155)** results. However, we have mapped the space **(8.155)** into only one *half* of S_4^-; for the mapping always satisfies $W + T \geq 0$. We need another copy of

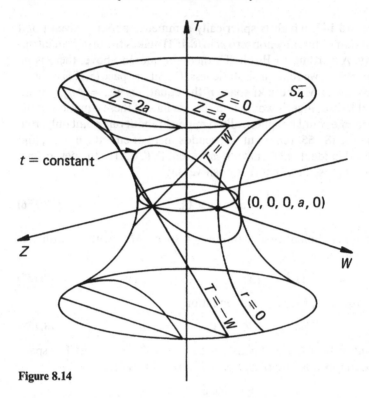

Figure 8.14

(8.155), mapped into the other half of S_4^- [by choosing the *negative* roots in **(8.160)** and **(8.161)**], in order to cover S_4^- completely. Note that the singularity at $r = (3/\Lambda)^{1/2}$ has been "transformed away": the metric **(8.157)** is regular everywhere.

Figure 8.14 shows a diagram of S_4^- with the X and Y dimensions suppressed. Its equation is then $Z^2 + W^2 - T^2 = a^2$, and this is a hyperboloid of revolution when mapped in Euclidean three-space Z, W, T, as in our diagram. It represents a *typical* spatial line, $\theta = 0$, through the spatial origin of **(8.155)**; however, it represents that *entire* line, $-\infty \le r \le +\infty$. Since $Z = r$, the "front" half of S_4^- ($Z > 0$) corresponds to positive values of r. The hyperbolas cut on S_4^- by planes $Z =$ constant correspond to lines of constant r, while the ellipses and hyperbolas cut on S_4^- by planes $T/W =$ constant correspond to lines of constant t. This diagram has many pleasant features. For example, the (straight-line) generators of the hyperboloid (like $T = W$) correspond to light paths. Timelike sections of the hyperboloid by planes through the origin correspond to geodesic worldlines. Timelike sections by arbitrary planes correspond to worldlines with constant proper acceleration.[12] The annotated surface of the hyperboloid of Figure 8.14 is a kind of Kruskal diagram for the de Sitter metric, being related to that metric much as Kruskal's diagram Figure 8.7 is related to the Schwarzschild metric. We

---

[12] Cf. W. Rindler, *Phys. Rev. 120*, 1041 (1960).

leave it to the interested reader to look at the many aspects of this analogy in detail. However, we shall remark on just one of them, the de Sitter singularity at $r = a = (3/\Lambda)^{1/2}$. It was once as thoroughly misunderstood as Schwarzschild's. Even Einstein and Weyl at first regarded it as real and indicative of the presence of mass—hence its early name "mass horizon." But clarification came much sooner than in the Schwarzschild case. Certainly by 1920 Eddington understood it perfectly.[13] In Figure 8.14 it corresponds to the light paths $T = \pm W$. In full de Sitter space it is a spherical light front "standing still" relative to the observer at the origin $r = 0$ (but changing its direction of propagation relative to him half-way through eternity, at $T = 0$). However, whereas in Schwarzschild space the horizon is a *unique* light front "standing still" relative to *all* observers at constant r, θ, ϕ outside of it, in de Sitter space each freely falling observer has his *own* horizon, with himself as center.

We still have to prove our assertion that each event in **(8.155)** is equivalent to every other. In order to do this, we show that S_4^-, as a subspace of M_5, can be moved over itself so as to bring any given one of its points, $(X_0, Y_0, Z_0, W_0, T_0)$, into the position of $(0, 0, 0, a, 0)$. We accomplish this in the following three steps:

$$(X_0, Y_0, Z_0, W_0, T_0) \to (0, 0, -, W_0, T_0) \to (0, 0, 0, -, T_0) \to (0, 0, 0, a, 0).$$

The first can evidently be done by a three-rotation in X, Y, Z, and the second by a two-rotation in Z and W; since any point on S_4^- has a "spacelike" position vector, by **(8.158)**, the last step can be achieved by a suitable Lorentz transformation in W and T. Each of these three transformations preserves the metric **(8.157)** and Equation **(8.158)**. If the correspondence **(8.159)**–**(8.161)** is maintained throughout, the above motions therefore carry an arbitrary event in the metric **(8.155)** into the origin event $r = t = 0$ [corresponding to $(0, 0, 0, a, 0)$ on S_4^-]. The metric **(8.155)** will then be valid around this new point, and our assertion is proved.

We shall meet de Sitter space again, and its horizon, in our discussion of cosmology.

The space **(8.155)** with $\Lambda < 0$ is often referred to as *anti–de Sitter space*. Its cosmological importance is somewhat less than that of de Sitter space, but it is not without theoretical interest. It, too, is a pseudosphere (but with *positive* curvature) and possesses symmetry about each of its points. We can treat it quite similarly to the case $\Lambda > 0$. Suppose now

$$\frac{3}{\Lambda} = -a^2 \quad (a > 0). \tag{8.162}$$

In this case we must consider a five-dimensional *pseudo*-Minkowski space $\tilde{M}_5$ with metric

$$ds^2 = dT^2 + dW^2 - (dX^2 + dY^2 + dZ^2), \tag{8.163}$$

---

[13] See A. S. Eddington, *Space, Time, and Gravitation*, Chapter X, Cambridge University Press, 1920 (and New York, Harper Torchbooks, 1959). Also Eddington, *loc. cit.* (*Mathematical Theory*) p. 166.

and, in it, the pseudosphere S_4^+ with equation

$$X^2 + Y^2 + Z^2 - W^2 - T^2 = -a^2. \tag{8.164}$$

(This turns out to be a space of constant curvature $K = 1/a^2$.) The following transformation,

$$W = a\left(1 + \frac{r^2}{a^2}\right)^{1/2} \cos\frac{t}{a}$$

$$T = a\left(1 + \frac{r^2}{a^2}\right)^{1/2} \sin\frac{t}{a}, \tag{8.165}$$

together with (8.159), maps the space (8.155) [with (8.162)] isometrically onto the *whole* of S_4^+, as one can easily verify.

Like S_4^-, so also S_4^+ can be represented—under suppression of the X and Y dimensions—by a hyperboloid in Euclidean three-space. Its equation is $W^2 + T^2 - Z^2 = a^2$. This looks identical in outline to Figure 8.14, but with the T and Z axes interchanged. A strange property of the mapping (8.165) is that r, θ, ϕ = constant goes into a ("horizontal") *circle* on the hyperboloid, with t/a the angle about the axis of symmetry. Thus, by sitting still at a point r, θ, ϕ = constant, and letting time go by, we would retrace our life history after a time $t = 2\pi a$ has elapsed! This is evidently an undesirable feature of the *mapping*, since no such closing up of worldlines is implicit in the metric (8.155). But it is easily avoided by modifying the mapping topologically. Instead of regarding the locus $W^2 + T^2 - Z^2 = a^2$ as an ordinary hyperboloid, we can regard it as a "hyperboloidal scroll" wrapped round itself; when we have gone round any circle Z = constant ($t : t_0 \to t_0 + 2\pi a$) we simply are on the next layer of the scroll rather than back where we started; and thus we run into no difficulties of causality.

8.12   The Linear Approximation to GR

We end this chapter with a brief discussion of a subject that is very important in many practical applications of GR, from gravitational waves to the physics of black holes: the linear approximation. This approximation to GR is usually much simpler to apply than GR itself, though it must be applied with care: it sometimes gives results which in no way approximate to those of the full theory. Manipulatively, this section is a little harder than the others and should perhaps be omitted at a first reading.

Suppose we have a *weak* gravitational field, i.e., a spacetime which differs little from flat Minkowski space. We can then set

$$g_{\mu\nu} = \eta_{\mu\nu} + h_{\mu\nu}, \tag{8.166}$$

where $\eta_{\mu\nu}$ is the usual Minkowski metric

$$\eta_{\mu\nu} = \text{diag}(-1, -1, -1, c^2), \quad \text{with inverse } \eta^{\mu\nu} = \text{diag}(-1, -1, -1, c^{-2}),$$
$$\tag{8.167}$$

and the h's are regarded as so small that products of them can be neglected. We also assume that the first two derivatives of the h's are of the same order of smallness as the h's themselves. (Recall that the first derivatives of the g's are a measure of the inertial-plus-gravitational force on particles at rest in the coordinate system and that the second derivatives are a measure of the spacetime curvature, i.e., of the tidal forces.)

The Christoffel symbols (8.13) are now given by

$$2\Gamma^\mu_{\nu\sigma} = h^\mu{}_{\nu,\sigma} + h^\mu{}_{\sigma,\nu} - h_{\nu\sigma},{}^\mu. \tag{8.168}$$

In raising and lowering indices on quantities containing h's, only the *principal* parts of $g_{\mu\nu}$, $g^{\mu\nu}$, namely $\eta_{\mu\nu}$, $\eta^{\mu\nu}$, contribute.

From the definition (8.20) we then easily find—neglecting products of Γ's, since they only contain products of h's—that

$$2R_{\mu\nu} = 2R^\sigma_{\mu\nu\sigma} = \Box h_{\mu\nu} + h_{,\mu\nu} - h^\sigma{}_{\mu,\nu\sigma} - h^\sigma{}_{\nu,\mu\sigma}, \tag{8.169}$$

where

$$\Box \equiv {}_{,\mu\nu}\eta^{\mu\nu} \equiv \frac{1}{c^2}\frac{\partial^2}{\partial t^2} - \frac{\partial^2}{\partial x^2} - \frac{\partial^2}{\partial y^2} - \frac{\partial^2}{\partial z^2}. \tag{8.170}$$

With a little manipulation, this can be brought into the form

$$2R_{\mu\nu} = \Box h_{\mu\nu} - \gamma^\sigma{}_{\mu,\nu\sigma} - \gamma^\sigma{}_{\nu,\mu\sigma}, \tag{8.171}$$

where

$$\gamma_{\mu\nu} = h_{\mu\nu} - \tfrac{1}{2}\eta_{\mu\nu}h, \quad h = \eta^{\mu\nu}h_{\mu\nu}. \tag{8.172}$$

A suitable coordinate transformation will now rid us of the γ terms in (8.171). Consider the transformation

$$\tilde{x}^\mu = x^\mu + \xi^\mu(x), \tag{8.173}$$

where the ξ's and their derivatives are of the same order of smallness as the h's. The g's transform as tensors [cf. (8.9)] and so, in an obvious notation

$$\begin{aligned} g_{\mu\nu} &= \tilde{g}_{\rho\sigma}(\partial\tilde{x}^\rho/\partial x^\mu)(\partial\tilde{x}^\sigma/\partial x^\nu) \\ &= \tilde{g}_{\rho\sigma}(\delta^\rho_\mu + \xi^\rho{}_{,\mu})(\delta^\sigma_\nu + \xi^\sigma{}_{,\nu}), \end{aligned}$$

from which we find that

$$\tilde{h}_{\mu\nu} = h_{\mu\nu} - \xi_{\mu,\nu} - \xi_{\nu,\mu}. \tag{8.174}$$

In the theory of differential equations it is known that an equation of the form

$$\Box \xi^\mu = \gamma^{\mu\sigma}{}_{,\sigma} \tag{8.175}$$

can be solved for ξ^μ. Let this determine the ξ^μ in (8.173). A short computation, using (8.174) and (8.175), then shows that

$$\tilde{\gamma}^{\mu\nu}{}_{,\nu} = 0. \tag{8.176}$$

This, in turn, simplifies the Ricci tensor (8.171) in the new coordinates to $\tilde{R}_{\mu\nu} = \frac{1}{2}\Box\tilde{h}_{\mu\nu}$. From now on we omit the tilde in our new coordinate system, and therefore write

$$R_{\mu\nu} = \tfrac{1}{2}\Box h_{\mu\nu}. \tag{8.177}$$

For the curvature invariant R we then find

$$R = \eta^{\mu\nu}R_{\mu\nu} = \tfrac{1}{2}\Box h, \tag{8.178}$$

and for the Einstein tensor,

$$R_{\mu\nu} - \tfrac{1}{2}\eta_{\mu\nu}R = \tfrac{1}{2}\Box(h_{\mu\nu} - \tfrac{1}{2}\eta_{\mu\nu}h) = \tfrac{1}{2}\Box\gamma_{\mu\nu}. \tag{8.179}$$

Consequently Einstein's field equations (8.136) or (8.138) take the respective (and equivalent) forms

$$\Box\gamma_{\mu\nu} = -2\kappa T_{\mu\nu}, \quad \Box h_{\mu\nu} = -2\kappa(T_{\mu\nu} - \tfrac{1}{2}\eta_{\mu\nu}T). \tag{8.180}$$

Equation (8.180)(ii) shows that $h_{\mu\nu}$ and $T_{\mu\nu}$ are of the same order of smallness, so that products of T's and Γ's can be ignored and $T^{\mu\nu}{}_{;\nu} = T^{\mu\nu}{}_{,\nu}$, in this approximation [cf. (8.129)]. But $T^{\mu\nu}{}_{,\nu} = 0$, and so *any* $\gamma_{\mu\nu}$ satisfying (8.180) will automatically satisfy the "coordinate condition" (8.176) and therefore Equation (8.179). Equations (8.180) are *linear* field equations, hence the name "linear approximation" for this theory. In this approximation, at least, both $h_{\mu\nu}$ and $\gamma_{\mu\nu}$ satisfy the wave equation with speed c in empty regions of space-time. Disturbances of the field will therefore be propagated through vacuum at the speed of light. But among these one must distinguish genuine gravitational waves, i.e., waves of curvature, from mere "coordinate waves."

If the sources move negligibly slowly, and have negligible stress, then, from (8.128),

$$T_{\mu\nu} = \mathrm{diag}(0, 0, 0, c^4\rho), \quad T_{\mu\nu} - \tfrac{1}{2}\eta_{\mu\nu}T = \mathrm{diag}\,\tfrac{1}{2}c^2\rho(1, 1, 1, c^2), \tag{8.181}$$

whence, by (8.180)(ii),

$$h_{11} = h_{22} = h_{33} = \frac{h_{44}}{c^2}. \tag{8.182}$$

As a result of the slow motion of the sources, the field will change slowly too, so that $\Box h_{\mu\nu} \approx -\nabla^2 h_{\mu\nu}$ and (8.180)(ii) reduces to

$$\nabla^2 \tfrac{1}{2}h_{44} = 4\pi G\rho. \tag{8.183}$$

We have now recovered the metric (8.143) with $\varphi = \tfrac{1}{2}h_{44}$ the Newtonian potential.

In the general case, Equations (8.180) can be integrated by standard methods. For example, the first yields as the physically relevant solution,

$$\gamma_{\mu\nu} = -\frac{4G}{c^4}\iiint\frac{[T_{\mu\nu}]dV}{r}, \tag{8.184}$$

where [ ] denotes the value "retarded" by the light travel time to the origin of r.

As an example, consider a system of sources in stationary motion (e.g., a rotating mass shell). All γ's will then be time-independent. If we neglect stresses and products of source velocities (which is not really quite legitimate[14]), the energy tensor (8.128) becomes

$$T_{\mu\nu} = \begin{pmatrix} \mathbf{0}_3 & -c^2\mathbf{v} \\ -c^2\mathbf{v} & c^4\rho \end{pmatrix}, \tag{8.185}$$

where $\mathbf{0}_3$ stands for the 3×3 zero matrix, and so, from (8.184),

$$\gamma_{ij} = 0, \quad (i, j = 1, 2, 3). \tag{8.186}$$

For slowly moving test particles, $ds = cdt$. If we denote differentiation with respect to t by dots, the first three geodesic equations of motion become [cf. (8.15)]

$$\ddot{x}^i = -\Gamma^i_{\mu\nu}\dot{x}^\mu\dot{x}^\nu \tag{8.187}$$

$$= -(\gamma^i{}_{\mu,\nu} - \tfrac{1}{2}\gamma_{\mu\nu,}{}^i - \tfrac{1}{4}\eta^i{}_\mu\gamma_{,\nu} - \tfrac{1}{4}\eta^i{}_\nu\gamma_{,\mu} + \tfrac{1}{4}\eta_{\mu\nu}\gamma_{,}{}^i)\dot{x}^\mu\dot{x}^\nu, \tag{8.188}$$

where we have substituted into (8.187) from (8.168) and (8.172) and used $\gamma = \eta^{\mu\nu}\gamma_{\mu\nu} = -h$. Moreover, $\gamma = c^{-2}\gamma_{44}$. Now if we let $\dot{x}^\mu = (u^i, 1)$ and neglect products of the u's, Equation (8.188) reduces to

$$\ddot{x}^i = -\gamma^i{}_{4,j}u^j + \gamma_{j4,}{}^iu^j + \tfrac{1}{4}\gamma_{44,}{}^i.$$

This can be written vectorially in the form

$$\ddot{\mathbf{r}} = \mathbf{grad}\,\varphi - \frac{1}{c}(\mathbf{u} \times \mathbf{curl}\,\mathbf{a}) = -\left[\mathbf{e} + \frac{1}{c}(\mathbf{u} \times \mathbf{h})\right], \tag{8.189}$$

where [cf. (8.184), (8.185)]

$$\varphi = -\tfrac{1}{4}\gamma_{44} = G \iiint \frac{[\rho]dV}{r}, \quad \mathbf{a} = -\frac{c}{4}\gamma^i{}_4 = \frac{1}{c}G \iiint \frac{[\rho\mathbf{u}]dV}{r}, \tag{8.190}$$

and

$$\mathbf{e} = -\mathbf{grad}\,\varphi, \quad \mathbf{h} = \mathbf{curl}\,4\mathbf{a}. \tag{8.191}$$

The formal similarity with Maxwell's theory is striking. The only differences are: the minus sign in (8.189) (because the force is attractive); the factor G in (8.190) (due to the choice of units); and the novel factor 4 in (8.191)(ii).

H. Thirring, in 1918, considered the gravitational field inside a rotating spherical shell of mass m, radius R, and angular velocity $\boldsymbol{\omega}$. Now it is known in Maxwell's theory that the electric field $\mathbf{e}$ inside such a shell of *charge q* is zero, while the magnetic field $\mathbf{h}$ is given by

$$\mathbf{h} = \frac{2}{3}\frac{q\boldsymbol{\omega}}{cR}. \tag{8.192}$$

---

[14] See L. Bass and F. A. E. Pirani, *Philos. Mag.* 46, 850 (1955).

Thus the force on a particle of unit charge moving with velocity $\mathbf{u}$ is given by

$$\mathbf{f} = \frac{1}{c}\mathbf{u} \times \mathbf{h} = \frac{2}{3}\frac{q}{c^2 R}\mathbf{u} \times \boldsymbol{\omega}. \tag{8.193}$$

It is now clear, from the analogy of Formulae (8.189)–(8.191) with their Maxwellian counterparts, that in Thirring's problem the (Coriolis-type) acceleration is given by

$$\ddot{\mathbf{r}} = -\frac{8}{3}\frac{Gm}{c^2 R}\mathbf{u} \times \boldsymbol{\omega}, \tag{8.194}$$

and this, indeed, was Thirring's finding.

Similarly, we can utilize the known result that the magnetic field outside a rotating ball of charge is given by

$$\mathbf{h} = \frac{3}{5}\frac{R^2 q}{c}\left(\boldsymbol{\omega}\cdot\mathbf{r}\,\frac{\mathbf{r}}{r^5} - \frac{1}{3}\frac{\boldsymbol{\omega}}{r^3}\right), \tag{8.195}$$

where R is the radius, q the charge, $\boldsymbol{\omega}$ the angular velocity, and $\mathbf{r}$ the position vector of the field point relative to the center of the ball. By our translation process, we see that the "gravomagnetic" field outside a similar ball of mass m will be

$$\mathbf{h} = \frac{12}{5}\frac{R^2 Gm}{c}\left(\boldsymbol{\omega}\cdot\mathbf{r}\,\frac{\mathbf{r}}{r^5} - \frac{1}{3}\frac{\boldsymbol{\omega}}{r^3}\right), \tag{8.196}$$

and the Coriolis acceleration on a test particle, $\ddot{\mathbf{r}} = -(1/c)(\mathbf{u} \times \mathbf{h})$. This result was first given by Thirring and Lense in 1918.

CHAPTER 9

Cosmology

9.1 The Basic Facts

Modern scientific man has largely lost his sense of awe of the universe. He is confident that, given sufficient intelligence, perseverance, time, and money, he can understand all there is beyond the stars. He believes that he sees here on earth and in its vicinity a fair exhibition of nature's laws and objects, and that nothing new looms "up there" that cannot be explained, predicted, or extrapolated from knowledge gained "down here." He believes he is now surveying a fair sample of the universe, if not in proportion to its size—which may be infinite—yet in proportion to its large-scale features. Little progress could be made in cosmology *without* this presumptuous attitude. And nature herself seems to encourage it, as we shall see, with certain numerical coincidences that could hardly be accidental.

Accordingly, cosmologists construct theoretical "models" that they believe represent the universe as a whole, concentrating on its largest-scale features, and they compare these models with the universe as observed by astronomers. Modern theoretical cosmology found its greatest inspiration in Einstein's general relativity, which provided it with a consistent dynamics and with such exciting possibilities as closed and yet unbounded universes. Even more significantly, GR had already yielded expanding models (de Sitter, Friedmann) by the time astronomers found that these were needed. Yet, it happened in cosmology as it so often happens in mathematics: when old and difficult problems are solved at long last by intricate methods, much simpler solutions become quickly apparent. In prerelativistic days it was thought that Newtonian theory was inadequate for dealing with the dynamics of the universe as a whole, mainly because of apparently un-avoidable infinities in the potential. But a decade after relativistic cosmology had been developed, it was suddenly realized that Newtonian theory (with a few now almost "obvious" modifications) could, after all, be used to obtain many model universes essentially similar to those of GR, and in any case a sufficient variety to survive comparison with the actual universe for a long time to come.

However, before we discuss cosmological theories, we must look at some of the facts. Evidently the first concern of cosmologists must be with the spatial distribution of stars and galaxies. For Copernicus, as for Ptolemy, all the stars were still fixed to a "crystalline" sphere, though now centered on the sun rather than on the earth. But as early as 1576, Thomas Digges boldly replaced that sphere by an infinity of stars extending uniformly through all space, the dimmer ones being farther away. The same extension was also made by Giordano Bruno, and mystically foreshadowed a century earlier by Nicholas of Cusa. But, whereas to Digges the sun was still king of the heavens, one who "raigneth and geeveth lawes of motion to ye rest," Bruno recognized it for what it is: just a star among many. The infinite view was later supported by Newton, who believed that a finite universe would "fall down into the middle of the whole space, and there compose one great spherical mass.[1] But if the matter was evenly disposed throughout an infinite space ... some of it would convene into one mass and some into another .... And thus might the sun and the fixed stars be formed" (1692). The really revolutionary content of this passage is the idea of an *evolving* universe, and of gravity as the mechanism causing condensation.

Newton also realized that the stars are immensely far apart. Assuming that they are essentially like the sun, and knowing that light from luminous objects falls off inversely as the square of their distance, he compared the apparent luminosity of the stars with portions of the sun showing through pinholes and so estimated that the stars must be at least 100,000 times farther away from us than the sun. In fact, the correct figure for the nearest star is about 270,000.

The next revolutionary idea was born sometime around 1750 and has been variously ascribed to Swedenborg, Lambert, Wright, and Kant. They all wrote on the subject of "island universes," recognizing the finiteness of our galaxy and conjecturing the existence of similar stellar systems far out in space. Various "nebulae" seen by the astronomers were candidates for this new role. Kant well understood the shape of our own galaxy ("stars gathered together in a common plane"—as indicated by the Milky Way) and so explained the observed elliptical appearance of some of the nebulae as discs seen obliquely. In 1783 Messier catalogued 103 such nebulae, and William Herschel with his powerful 48-inch reflector telescope located no fewer than

---

[1] This, of course, is not quite true. Given absolute space, a finite universe could avoid collapse by rotation. Ironically, it is the *infinite* universe that is not consistent with Newton's theory of gravitation, as we shall see in Section 9.2.

2500 before his death in 1822. He became the great observational supporter of the multi-island universe theory, though, ironically, many of his arguments turned out to be quite false. Still, he came to foresee the important division of nebulae into two main classes—galactic and extragalactic. Herschel's theory had its ups and downs in favor, but essentially its verification had to wait for the slow development of observational capacity to its huge demands. The waiting period culminated in a historic wrangle, continued at one astronomers' conference after another from 1917 to 1924—until it suddenly ended on January 1, 1925: Hubble, with the help of the new (1917) 100-inch telescope at Mount Wilson, had resolved star images in three of the nebulae, and, as some of these were Cepheids, he was able to establish beyond all doubt their extragalactic distances. Only one main feature of the universe as we know it today was still missing: its expansion. From 1912 onwards, Slipher had observed the spectra of some of the brighter spiral nebulae and found many of them redshifted, which presumably meant that these nebulae were receding. But distance criteria were still lacking. Hubble now applied his "brightest star" measure of distance and, together with Humason, extended the red shift studies to ever fainter nebulae. Finally, in 1929, he was able to announce his famous law: all galaxies recede from us (apart from small random motions) at velocities proportional to their distance from us. The modern era of cosmology had begun.

For another 20 years or so the big 100- and 200-inch telescopes were chiefly responsible for enriching our knowledge of the universe. But radar developments during the war led to radio astronomy after the war; atomic clocks and electronic computers led to new methods of evaluating data; balloon, rocket, and satellite experiments escaped the obscuring atmosphere of the earth; photoelectric scanning circumvented the limitations of photographic emulsions; and x-ray and neutrino astronomy are just emerging. Computers, moreover, made possible previously unthinkable theoretical investigations into stellar and galactic evolution.

However, we must forego an account of the further detailed growth of modern knowledge, and content ourselves with simply listing the main astronomical findings relevant to our purpose.[2] Stars, to being with, are now known to be tremendous thermonuclear reactors, going through reasonably well understood life cycles—whose exact course depends mainly on their mass. Their size can be appreciated by considering that the earth *with* the moon's orbit would comfortably fit into most stars, including the sun. About 7000 of them are visible to the naked eye; about 10^{11} are contained in a typical galaxy. [Most of us lack mental images for numbers of that size. Here is one possibility: consider a row of books a mile long; the number of *letters* in all those books is about 10^{11}. Or consider a cubical room, 15 × 15 × 15 foot; the number of pinheads needed to fill it is about 10^{11}.] Stars within a galaxy are very sparsely distributed, being separated by distances of the order of ten light years. In a scale model in which stars are represented

---

[2] For details see, for example, D. W. Sciama, *Modern Cosmology*, Cambridge University Press, 1971, or A. Unsöld, *The New Cosmos*, Springer-Verlag, 1977, 2nd ed.

by pinheads, these would be 50 km apart, and the solar system (out to the orbit of Pluto) would be a 10-meter circle centered on a pinhead sun. A typical galaxy has a radius of 3×10^4 light years, is 3×10^6 light years from its nearest neighbor, and rotates differentially with a typical period of 100 million years. Like a dime, it has a width only about a tenth of its radius, and dimes spaced about a meter apart make a good model of the galactic distribution. About 10^{11} galaxies are within range of the 200-inch Mount Palomar telescope. *On the dime scale*, the farthest of these are 2.5 km away from us, but the farthest known quasars would be more than 10 km away.

There is one more complication: galaxies, too, are not distributed uniformly throughout space but instead tend to cluster. Single galaxies are exceptional; most belong to clusters of from 2 to 1000 to even 10,000, apparently bound by gravity. There are some indications that the clusters themselves may cluster and so form "superclusters." This is perhaps a convenient point to mention the so-called "hierarchical" model universe of Charlier, which, though it has found little favor, is based on an interesting idea. Already in 1761 Lambert had loosely speculated that various solar systems might revolve about a common center, that such supersystems in turn might combine and revolve about another center, "and where shall we stop?" Analogously to this scheme, Charlier between 1908 and 1922 developed a cosmological theory in which galaxies form clusters, clusters form superclusters, and so on *ad infinitum*. By arranging the dimensions suitably, it is possible in this way to construct a universe with zero average density. [To take an overly simple example, suppose a cluster of order n contains p^n stars of mass m in a volume kq^n; its average density ρ is $(m/k)(p/q)^n$, and so $\rho \to 0$ as $n \to \infty$, if $p < q$.] Thus one can avoid the infinities inherent in the Newtonian treatment of *homogeneous* universes, whose average density, of course, is finite. Charlier universes are *not* homogeneous: no volume V is large enough to be typical, since there will always be clusters larger than V and thus not represented in V. But today these models are little more than a curiosity.

We now turn to a discussion of the apparent isotropy and homogeneity of the universe, its expansion, density, and age.

The galaxies appear to be distributed more or less isotropically around us, and also to recede from us equally in all directions. Allowing for the obscuring matter within our own Milky Way, no direction in the sky seems preferred over any other. Recently our faith in the isotropy of the universe, from our vantage point, has been reinforced by observation of the so-called "background" microwave radiation of temperature $\sim 3°K$—an apparent left-over from a hot explosive origin of the universe. This happens to be incident on earth with a very high degree of isotropy. (See end of Section 9.4.) Such radiation was predicted by Alpher and Herman in 1948 on the basis of Gamow's theory of the "big bang." Its eventual—and accidental—discovery by Penzias and Wilson in 1965 now provides one of the strongest indications that a big bang, in fact, took place.

It is much harder to test the universe for homogeneity, i.e., to check whether distant regions of it are similar to our own. The basic problem is that we see those regions as they were billions of years ago, but we don't know exactly *how* many billions of years ago, and in any case we don't know exactly what *our* region was like then. For we have only an imperfect understanding of the evolution of galaxies, we have no reliable distance indicators on that scale, no *a priori* knowledge of the space geometry, or of the past rate of expansion of the universe. Thus almost all observations of the really distant galaxies could be compatible with homogeneity. However, in our more immediate vicinity, homogeneity seems to prevail among regions of about 10^8 light years in diameter.

For the rate of expansion of the universe, Hubble in 1929 gave the figure of 540 (km/sec)/megaparsec. [The parsec (pc) is a distance unit popular with astronomers and is equivalent to 3.087×10^{18} cm, or 3.26 light years; a megaparsec (Mpc) equals 10^6 parsec.] This figure, *Hubble's constant*, has undergone several drastic revisions, mainly downward, and mainly caused by refinements of the various steps leading to a determination of cosmic distances. The best present estimates are close to 50 (km/sec)/Mpc, which corresponds to an increase of distances by 1 % in 2×10^8 years. It must be stressed that Hubble's "constant" may well vary in time—though not, of course, as quickly as the above estimates! An analogy with a rubber sheet that is isotropically expanded should make this clear: If at any instant two points on the sheet one inch apart have relative velocity H, then points x inches apart will have relative velocity xH, and H is Hubble's constant. But there is no reason why point-pairs one inch apart should always have the same relative velocity; the expansion rate may well vary, and H could even change from positive to negative values (contraction). [Constant H would imply exponential expansion: $x = x_0 \exp(Ht)$.] Two further remarks on Hubble's expansion can be made at once. First, the expansion is not necessarily due to a Λ term in Einstein's modified field equation (**8.139**): a big-bang creation of the universe *must* be followed by a period of expansion, no matter what field equations are assumed. And secondly, gravitationally bound galactic clusters and smaller structures (galaxies, planetary systems, etc.) do not change their dimensions as part of the cosmic expansion. We have already seen that the Schwarzschild metric (and thus its planetary orbits) would be unaffected by the existence of an expanding isotropic mass distribution around it. The general situation is theoretically similar. [Also if *everything* expanded equally—atoms, rulers, observers, stars, galaxies— *nothing* would expand.]

Another datum dependent on local cosmic distance estimates is the average density of the universe. The average density due to its galaxies is now judged to be about 2×10^{-31} gm/cm^3, though values as high as 6×10^{-31} cannot be ruled out. But this, of course, is only a lower bound on the present average density of *all* matter; *that* may well be higher, owing to the presence of undetected intergalactic matter, neutrinos, gravitational waves, etc. But there are other reasons for believing the cosmic density not to be much greater than 6×10^{-31} gm/cm^3. These come from various astrophysical arguments, such as those based on the presently observed relative abundances of helium and deuterium. As we shall see later, the cosmic density is a very critical datum, which may determine whether the universe is positively or negatively curved, and also whether it is infinitely expanding or doomed to recollapse.

An equally critical datum is the age of the universe—that is, if it *has* an age: it could, of course, be infinitely old. The age of the earth since the formation of its crust can be quite accurately determined from considerations of radioactive decay. It turns out to be $(4.5 \pm 0.3) \times 10^9$ years. Evidently the universe cannot be younger than this. Theories of stellar evolution applied to "globular clusters" (of stars, orbiting our galaxy) suggest that their age is of the order of 14×10^9 years. A quite independent estimate for the age of the universe can be made from the relative abundances of certain heavy elements, which are products of radioactive decay. These estimates (based on models of evolution of our galaxy) indicate an age between 6 and 20×10^9 years. Yet another estimate can be made from Hubble's constant: if the universe had been expanding *linearly* at its present rate, it would have begun about 20×10^9 years ago. But (provided there is no Λ repulsion) we would expect its expansion rate to have slowed, so that, in fact, its age would be *less* than 20×10^9 years.

In the remainder of this book we must content ourselves with a study of what may be called postgalactic cosmology—the kinematics and dynamics of the universe of galaxies. Pregalactic cosmology, or "cosmogony"—the study of the formation of the elements, of the stars, and of galaxies, out of a primordial mixture of elementary particles in thermodynamic equilibrium ("hot big bang")—depends on nuclear and atomic physics and is well beyond our present scope. It has, of course, great philosophical interest since, in a sense, it takes over where Darwin left off. If Darwin and modern biology can explain the rise of man from a lifeless earth, cosmogony can explain the rise of earths, suns, and galaxies from amorphous matter, given only the immutable laws of nature and an energizing big bang.

Once this cosmic evolution is granted, it is hard to escape another conclusion of profound philosophical interest: that life must exist throughout the universe. Given the vast number of stars *known* to exist ($\sim 10^{22}$, equal to the number of pinheads needed to fill a 20-km cube), and that planetary systems seem to be an occasional concomitant of stellar condensation, it seems highly probable—even if only one planet in a million, or in a million million, can support life—that such life has, in fact, arisen on billions of planets.[3] This is the culmination of the Copernican revolution: we are in *no* way central to the universe. It may also explain the size of the universe—if one can grant that it has a purpose: possibly no lesser size could guarantee the ultimate survival and full flowering of at least *some* civilizations.

---

[3] Admittedly, a minority of biologists still believe that the probability of a spontaneous origin of life is so minute that the earth *may* be the only life-bearing planet in the universe.

9.2   Apparent Difficulties of Prerelativistic Cosmology

In retrospect, it is hard to understand why nineteenth-century astronomers were so bent on the idea that the universe must be infinite and static. A finite shower of stars (or galaxies) shot out in all directions from some primordial explosion, its expansion finally halted and reversed by gravity, and ending in another holocaust, is certainly well within Newtonian possibilities. And so is a finite universe that rotates forever and thereby avoids gravitational collapse. Was the *appearance* of an infinite static universe so compelling that it must not be questioned? Again, how could generations brought up on the idea of energy conservation—after Joule and Helmholtz— expect the stars to burn forever, without, evidently, receiving comparable energies from their surroundings? But here we must remember that the source of stellar energy was still totally unknown, $E = mc^2$ was undreamed of, and thus it was perhaps still legitimate to imagine an infinite reservoir of energy hidden in a finite mass (like that which results from the indefinitely close approach of two point-particles under Newtonian gravitation).

Whatever the reason, an infinite static universe it had to be. And this led to difficulties. The first was with Newton's theory. Consider an infinite, homogeneous and initially static distribution of mass (e.g., stars) throughout all space; what happens? To judge from the quotation given in Section 9.1, Newton apparently thought that this was an equilibrium configuration. Believing in absolute space, he might well have appealed to symmetry and argued that the resultant force on each star must vanish, whence no motions could occur. [*Without* absolute space, a general homogeneous contraction under gravity of the infinite distribution would be possible: no star would move in a preferred manner *relative to the rest*.]

But there is a snag to Newton's static universe. Suppose we remove a finite sphere of matter from it. What will be the field inside the cavity? No answer can be found by integrating the potential: the integral diverges to infinity. If the field inside were zero, then, when we reintroduce the matter, it would collapse under its own gravity. It seems that the infinite universe outside the cavity must provide a centrifugal force inside. Yet when we integrate the *field* due to an infinite number of spherical mass shells around the cavity, we get zero. The root of the trouble lies in the fact that Poisson's equation allows no solution $\varphi = $ constant.

This led Neumann in 1896 to suggest that the Newtonian potential of a point mass be replaced by

$$\varphi = -\frac{mG}{r} e^{-r\sqrt{\lambda}}, \quad (\lambda = \text{constant} \approx 0). \tag{9.1}$$

(Note that this is formally identical to Yukawa's mesonic potential put forward in 1935.) Instead of Poisson's equation[4] $\nabla^2\varphi = 4\pi G\rho$ we then have

$$\nabla^2\varphi - \lambda\varphi = 4\pi G\rho, \tag{9.2}$$

which possesses the obvious constant solution

$$\varphi = -\frac{4\pi G\rho}{\lambda} \qquad (9.3)$$

in a homogeneous universe. [This results also on integrating (9.1) throughout space for a continuous distribution of matter.] And, indeed, the finite self-force of the matter inside our hypothetical spherical cavity is now precisely balanced by a finite centrifugal force $\sim (4/3)\pi G\rho r$ due to the entire matter outside.

It is interesting to observe the striking *formal* analogy between Einstein's modification (8.139) of his original field equations (8.136) and Neumann's modification (9.2) of Poisson's equation—both made for the same purpose, namely to allow a static universe. Surprisingly enough, in first approximation (8.139) does *not* reduce to (9.2) but rather to another modification of Poisson's equation, namely

$$\nabla^2\varphi + c^2\Lambda = 4\pi G\rho, \qquad (9.4)$$

as can be shown by methods similar to those of Section 8.10. This *also* admits a constant solution in the presence of homogeneous matter, namely $\varphi = 0$, *provided* $c^2\Lambda = 4\pi G\rho$—a relation which obtains exactly in Einstein's static universe.

The most general modification of Poisson's equation, *in the spirit* of Einstein's modification of *his* field equations, would be subject to only three conditions: (i) preservation of the scalar character, (ii) preservation of linearity, (iii) occurrence of no derivatives of the potential higher than the second. And this leads uniquely to the form

$$\nabla^2\varphi + A\varphi + B = 4\pi G\rho, \qquad (9.5)$$

of which *both* (9.2) and (9.4) are special cases. [Within the Newtonian formalism, the coefficients A and B *could* be functions of time—for example, they could be linked to the density of an expanding universe. In Einstein's theory, on the other hand, Λ has no freedom but to be an absolute constant of nature.] If A or B or both are nonzero, Equation (9.5) admits *nonconstant* singularity-free solutions for *empty* space ($\rho = 0$): Assuming isotropy and regularity, let us substitute a power series in r into (9.5) with $\rho = 0$, and solve for the coefficients. The result is

$$\varphi = \varphi_0 - \tfrac{1}{6}(B + A\varphi_0)[r^2 + AO(r^4)]. \qquad (9.6)$$

This would be the solution to use in the spherical cavity that we have discussed. It would counterbalance the self-force of the matter that has been removed.

---

[4] From now on we write $\nabla^2\varphi$ for our previous $\Sigma\varphi_{ii}$ to stress the coordinate-independent meaning of the operation. We recall that in the case of spherical symmetry $[\varphi = \varphi(r)]$, $\nabla^2\varphi = \varphi'' + (2/r)\varphi'$.

In the case of Equation (**9.2**) ($A = -\lambda$, $B = 0$) it can be regarded as due to the outer masses. But in the case of Equation (**9.4**) ($A = 0$, $B = c^2\Lambda$) Λ acts as a "preexisting" space density $-c^2\Lambda/4\pi G$: it is as though space itself repelled matter. Such is the effect of Einstein's Λ, in Newtonian language.

Tampering with Newton's inverse square law (as embodied in Poisson's equation) is only one way out of the difficulty that an infinite universe poses within the classical framework. Another is to abolish absolute space and allow the universe to contract or expand. Though this makes the modification of Poisson's equation superfluous, one can do *both* and so get the most general theory. Something similar happened in GR: Einstein introduced the Λ term to make possible a static universe; when that was no longer relevant, the Λ term was kept anyway—though not by Einstein!

Another famous objection to the prerelativistic static infinite homogeneous universe, quite independent of gravitational theory, is what has come to be known as *Olbers' paradox*.[5] This states that *if* the stars shine with unvarying and equal light, and space is Euclidean, the whole sky in such a universe must appear uniformly as bright as the sun. For consider any narrow cone of rays entering the eye. That cone must come from the surface of *some* star. But since its cross-sectional area varies as r^2, whereas the apparent luminosity of a unit area of star surface decreases as $1/r^2$, the light gathered into this cone is independent of where it originates; hence it is the same as if it originated on the sun, and the paradox is established. Today the most immediately questionable assumption is the eternal burning of the stars. A more popular explanation (Chéseaux, 1744) was absorption by some intersteller medium, but, as Bondi has pointed out, this would soon radiate as much as it absorbs. Nor can a finite and closed geometry help: our cone of rays may have to be traced back several times "round the universe" before it strikes a star, and it will not spread as r^2; but its spread and the stellar luminosity still vary inversely, and the argument is unimpaired. However, a universe with finite past can avoid Olbers' problem; as we see farther out, we see farther back in time, and beyond a certain distance we may see no more: the universe was not yet. Also a universe with sufficient expansion can avoid Olbers' problem, even if—as in the "steady state" theory—it has an infinite past and the average luminosity per unit volume remains constant; for light from distant sources can be so redshifted (and thus, by Planck's relation $E = h\nu$, so deenergized) that the sum remains finite, and even negligible.

9.3 Cosmological Relativity: The Cosmological Principle

Though little value is placed these days on the Euclidicity of a model universe, and none on its staticness, practically all modern cosmologies contain the

---

[5] For its interesting historical antecedents, see S. L. Jaki, "Olbers', Halley's, or Whose Paradox?" *Am. J. Phys. 35*, 200 (1967).

assumption that the universe is homogeneous. This eliminates such "sensible" models as island universes (in which the boundary galaxies are not typical) and hierarchical or Charlier universes (in which no volume is large enough to be typical). Homogeneity is a simplifying hypothesis of great power. Whereas nonhomogeneous model universes involve us in global questions, the beauty of homogeneous models is that they can be studied mainly locally: any part is representative of the whole.

Most modern cosmologies also contain the assumption of isotropy. This can be rather more easily dispensed with than homogeneity, without leading to inordinate difficulties. For example, it is not much harder to discuss a universe that has different expansion rates along the x, y, and z directions than one with isotropic expansion. Still, the most general homogeneous but nonisotropic model is complicated. Since present observations do not seem to require this complication, isotropy is usually assumed. [It is only in the discussion of the very early universe—shortly after the big bang—that nonisotropic models may be needed.] Isotropy from *every* vantage point implies homogeneity (but not vice versa). This can be seen quite easily; for suppose that at a given instant the universe had different properties in the neighborhoods of two points, A and B. Then at a point C, equidistant from A and B, this would show up as a lack of isotropy.

The assumption of large-scale homogeneity, often (as in this book) together with the assumption of large-scale isotropy, is called the *cosmological principle* (CP). As far as present-day observations go, there are certainly no compelling indications to abandon this principle. On the contrary, the fairly good isotropy observed from our position in the universe provides us with an argument *for* the CP: By mere considerations of probability, *we* are hardly likely to be in a special position in the universe; hence we suspect isotropy everywhere. And this, as we have seen, implies homogeneity. It should be noted that the CP is not a "principle" of physics in the same sense as, for example, energy conservation is. Its status is merely that of a *working hypothesis*.

There exists also a *perfect cosmological principle* (PCP), on which the *steady state* cosmology of Bondi and Gold (1948) was based. It asserts that, in addition to being spatially homogeneous and isotropic, the universe is also temporally homogeneous, i.e., it presents the same average aspect at all times. It has no beginning or end—a very attractive feature, philosophically. As the universe expands, sufficient new matter must be created to fill the gaps. This constitutes a deliberate violation of energy conservation, but not by "much": the spontaneous creation of about one hydrogen atom per 60 km^3 of space per year is all that is needed. The steady state theory has the further advantage of leading to a unique model, which, as such, is highly vulnerable to empirical disproof (cf. the end of Section 2.2). It had many adherents and enjoyed great popularity for almost two decades. But the observational evidence against it (radio source counts, the distribution of quasars, the $3°K$ radiation, etc.) has been mounting steadily to the point where few will still defend it.

The CP and the PCP (and all similar symmetry assumptions) may be regarded as defining a "relativity of the universe," i.e., a group of transformations under which the large-scale universe transforms into itself. For this reason alone we may consider *any* cosmology based on such principles as a relativity theory (cf. Section 1.1), and it is in this sense that the term "cosmological relativity" appears in the title of this book and of this section.

The actual universe is manifestly irregular in detail, consisting of vast empty stretches punctuated by concentrations of mass of many different shapes and sizes. To deal with the kinematics and dynamics of such a conglomerate mathematically, we must idealize it. Apart from small proper motions, each galaxy seems to follow a "natural" motion associated with its location in the universe. The smoothed-out pattern of these natural motions is the *substratum*, a concept used to discuss the *kinematics* of the universe. We can think of it as a space-filling set of particles in motion—the *fundamental particles*—each of which is a potential center of mass of a galaxy, or of a cluster of galaxies. When in the sequel we loosely speak of galaxies in a model, we shall really mean the fundamental particles. Each fundamental particle is imagined to carry a *fundamental observer*. It is the substratum that is assumed to be strictly homogeneous and isotropic.

So far we have used the notion of homogeneity purely intuitively. In an expanding universe, however, spatial homogeneity is not quite as simple a concept as it is in static situations. Intuitively we think of it as meaning that all sufficiently large spatial "samples" of the universe are equivalent. But when? Our neighborhood *now* probably differs even from *itself*, let alone from other regions, one hundred million years ago. Thus the comparison involves time, and relatively has taught us to be wary of time. The following definition, due to A. G. Walker, avoids this difficulty: Homogeneity means that the totality of observations that any fundamental observer can make on the universe is identical with the totality of observations that any other fundamental observer can make on the universe. In other words, if throughout all time we here, as well as observers on all other galaxies, could log all observations—e.g., the density and directional distribution of galaxies, their rate of expansion, etc., together with the times at which the observations are made (as measured, say, by standard cesium clocks), then homogeneity would be equivalent to the coincidence of all these logs (up to a possible translation in time, of course).

A most important corollary of such homogeneity is the existence of *cosmic time*, i.e., of an absolute universe-wide sequence of moments. In fact, a homogeneous universe itself, *if it is evolving*, acts as the relevant synchronization agent at each point. For we need merely reset the time origins of the various cesium clocks introduced in the preceding paragraph to make the logs *identical*. Then their readings define τ, the cosmic time; clearly τ is the *proper* time at each galaxy. If the universe is static, or in steady state, the clock setting to cosmic time can be achieved by two-way signaling experiments similar to those of Section 7.6. We shall in all cases assume this has been done.

9.4   Milne's Model

Breaking the historical sequence, we shall now describe a most ingenious and simple model universe, invented by E. A. Milne in 1932, which nicely illustrates many of the features shared by the more complicated models. And, though it will not be immediately apparent, Milne's model in fact satisfies the CP.

Against a background of empty Minkowski space M_4, and totally neglecting gravity, Milne considered an infinite number of test particles (no mass, no volume) shot out (for reasons unknown), in all directions and with all possible speeds, at a unique creation event $\mathscr{C}$. Let us look at this situation in some particular inertial frame $S(x, y, z, t)$, and suppose $\mathscr{C}$ occurred at its origin O at $t = 0$. All the particles, being free, will move uniformly and radially away from O, with all possible speeds short of c. Hence the picture in S will be that of a ball of dust whose unattained boundary expands at the speed of light. At each instant $t = $ constant in S, Hubble's velocity–distance proportionality is accurately satisfied relative to O: a particle at distance r has velocity r/t. Still, at first sight, this seems an unlikely candidate for a modern model universe, since (i) it appears to have a unique center, and (ii) it appears to be an "island" universe. Leaving aside the second objection for the moment, let us dispose of the first: The boundary of the cloud behaves kinematically like a spherical light front emitted at $\mathscr{C}$, and thus each particle, having been present at $\mathscr{C}$, will consider *itself* to be at the center of this front! Moreover, since all particles coincided at $\mathscr{C}$, and since all move uniformly, *each* particle will consider the whole motion pattern to be radially away from itself, and of course uniform. There remains the question whether we can have an isotropic density distribution around each particle.

To study this, let τ denote the proper time elapsed at each particle since creation. Then n_0, the proper particle density at any given particle P, is of the form

$$n_0 = N/\tau^3, \quad (N = \text{constant}), \tag{9.7}$$

because a small sphere around P, containing a fixed number of particles, expands with the constant velocity du of the farthest particles, and thus has radius $du\tau$ and volume $(4\pi/3)du^3\tau^3$. Evidently, for maximum symmetry, we must choose N the same at all particles.

If particle P is at distance r from the origin in S, then for events *at* P we have

$$\tau = \frac{t}{\gamma(u)}, \quad u = \frac{r}{t}. \tag{9.8}$$

Therefore, since a unit proper volume at P is decreased to $1/\gamma(u)$ in S, the particle density at P *relative to* S is given by

$$n = \frac{\gamma(u)N}{\tau^3} = \frac{\gamma^4(u)N}{t^3} = \frac{Nt}{(t^2 - r^2/c^2)^2}. \tag{9.9}$$

It is clear that, conversely, a density *defined* by (**9.9**) relative to the origin particle O reduces to N/τ^3 at each particle, and thus to (**9.9**) relative to any *other* origin particle. This is therefore the density distribution we must require to hold around *any* particle in SR coordinates. Observe how this density approaches infinity at the "edge" $r = ct$. Note also that τ is the cosmic time in Milne's model, as defined at the end of the last section: the universe determines it via (**9.7**).

Milne's model is now seen to satisfy the CP—i.e., homogeneity and isotropy. In prerelativistic kinematics the conflict of "island" universes with the CP arises because galaxies at (or even near) the edge are evidently not typical. Milne's model demonstrates how in relativistic kinematics this conflict can be avoided: there are no galaxies *at* the edge, and there are no galaxies *near* the edge by their own reckoning.

Though the island nature of Milne's universe does not offend against the CP, it does offend against another criterion that can be required of model universes: *completeness*. Clearly in Milne's universe there can be events in the world at which no galaxy is present (namely, outside the "edge" $r = ct$) but which can nevertheless interact with the galaxies (e.g., be seen by them). World models which, in this sense, "possess more space than substratum" are called *incomplete*. Here, however, Milne's model is in good company: quite a few of the seriously considered cosmological models suffer this blemish.

We now give an alternative description—and with it, an alternative metric—of Milne's model, which is not only instructive in itself but which is in fact standard for *all* models satisfying the CP. Thus it will prepare the way for later comparisons. Cosmic time is now taken as the temporal co-ordinate, and the spatial coordinates are chosen to be *comoving*, which means that the fundamental particles, though in relative motion, have fixed space coordinates—like the lattice points in a permanently labeled but expanding Cartesian lattice. For example, the set (u, θ, ϕ) relative to the inertial frame S could serve as comoving coordinates.

Consider, then, the usual metric of M_4 relative to a given frame S in spherical polar coordinates,

$$ds^2 = c^2 dt^2 - dr^2 - r^2(d\theta^2 + \sin^2\theta \, d\phi^2), \qquad (9.10)$$

and suppose the origin $r = t = 0$ coincides with Milne's creation-event $\mathscr{C}$. Now make the following coordinate transformation to cosmic time τ [see (**9.8**)] and to a new comoving coordinate ρ, which happens to be more convenient[6] than u:

$$\tau = t(1 - u^2/c^2)^{1/2}, \quad c\rho = u(1 - u^2/c^2)^{-1/2}, \quad (u = r/t). \qquad (9.11)$$

A simple computation, aided by noting that

$$r = c\tau \sinh \psi, \quad t = \tau \cosh \psi, \quad (\sinh \psi = \rho), \qquad (9.12)$$

[6] Not to be confused with density. The use of ρ in this context is unfortunate but traditional.

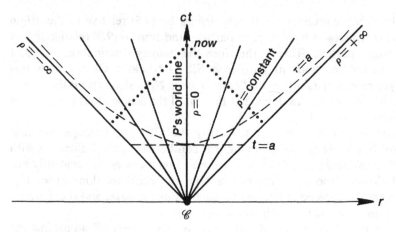

Figure 9.1

then yields the new form of the metric,

$$ds^2 = c^2 d\tau^2 - c^2\tau^2\left\{\frac{d\rho^2}{1 + \rho^2} + \rho^2(d\theta^2 + \sin^2\theta d\phi^2)\right\}. \tag{9.13}$$

Suppressing two spatial dimensions, we can illustrate the relation between the old and the new coordinates by Figure 9.1, in which r and t are the coordinates associated with the fundamental particle P at the origin of S ($r = 0$). The lines $\rho =$ constant *between* $\rho = \pm\infty$ (i.e., $u =$ constant between $\pm c$) represent the worldlines of the fundamental particles, while the lines $\rho = \pm\infty$ represent the unattained edge. Any line $t = a =$ constant represents a map of the universe in P's "private" Euclidean space r, θ, ϕ at an instant of P's "private" time t. All such maps show the universe as a finite ball of radius ct, whose particle density increases without limit towards the edge. On the other hand, any hyperbola $\tau = a =$ constant (i.e., $c^2t^2 - r^2 = c^2a^2$) represents what Milne called a "public" map (or "public space"), i.e., a section of spacetime at an instant of "public" time τ. It coincides with the private map $t = a/P$ over a neighborhood of P of any fundamental particle (see Figure 9.1). In fact, it can be regarded as a composite of local private maps of the universe, all made at the same cosmic time τ. It shows all galaxies on the same footing. It shows the universe as *infinitely* extended, with *uniform* particle density N/τ^3 [cf. (9.7)].

We now show that each public map in Milne's model is a 3-space of constant negative curvature $K = -1/c^2\tau^2$.[7] Its metric $d\sigma^2$ is found by setting $\tau =$ constant in (9.13):

$$d\sigma^2 = c^2\tau^2\left\{\frac{d\rho^2}{1 + \rho^2} + \rho^2(d\theta^2 + \sin^2\theta d\phi^2)\right\}. \tag{9.14}$$

We assert that { } represents a space of constant curvature $K = -1$,

[7] Assuming a present age of the universe $\tau \approx 10^{10}$ years, $K \approx -10^{-56}$ cm^{-2}.

whence $d\sigma^2$ represents a space of constant curvature $-1/c^2\tau^2$. [Multiplying a metric by a constant factor A^2, say, increases all distances by a factor A and thus decreases the curvature everywhere by a factor $1/A^2$, as is most easily seen from Equation (7.6).] Clearly { } is isotropic about the origin $\rho = 0$; and all lines θ, $\phi = $ constant are geodesics (cf. beginning of Section 8.3). Radial distance in { } is given by

$$\psi = \int_0^\rho \frac{d\rho}{(1 + \rho^2)^{1/2}} = \sinh^{-1}\rho, \qquad (9.15)$$

and the perpendicular distance between two geodesics passing through the origin at an angle $d\omega$ is given by $\eta = \rho d\omega = \sinh\psi d\omega$. It follows that $\ddot{\eta} = \eta$ [cf. (7.6)] and $K = -1$. Hence the origin of $d\sigma^2$ is an isotropic point with curvature $-1/c^2\tau^2$. But *every* point of $d\sigma^2$ is equivalent to the origin, since *all* galaxies in Milne's universe are equivalent. So our assertion is established.

Thus, while the metric (9.10) merely gives the spacetime *background* of Milne's model, (9.13) gives that and more. It specifies the substratum of the model as a sequence of public spaces (9.14) of constant curvature, "expanding" proportionately to $c\tau$ as cosmic time goes on. In the following section these ideas will be generalized to arbitrary models satisfying the CP.

The simple Milne model serves well to illustrate the kinematics of the 3°K microwave radiation. It is estimated that about 300,000 years after the big bang—long before galaxies started to condense—the universe had cooled to the point where matter and thermal radiation decoupled and went their separate ways. This cosmic instant can be represented by $\tau = a$ in Figure 9.1. That stippled line is then the effective source of the radiation observed today. Evidently, as time goes on, each observer (say P) receives the radiation (dotted lines in Figure 9.1) from ever farther fundamental particles, and thus with ever greater red shift. The observed isotropy of the incoming microwave radiation indicates that its "sources" must be equally strong and equally far in all directions. Of course, in any direction there is only *one* source observed "now." Still, the isotropy of this subset of sources in an otherwise nonisotropic universe would be hard to explain. Note, incidentally, how P "now" can—theoretically—see fundamental particles at all proper ages $\tau > 0$, however small. But, in Milne's model at least, he can *not* see the creation event.

9.5 The Robertson–Walker Metric

We shall need the following result: if $k = 1$, -1, or 0, the metric

$$dl^2 = \frac{d\rho^2}{1 - k\rho^2} + \rho^2(d\theta^2 + \sin^2\theta d\phi^2) \qquad (9.16)$$

represents a 3-space of constant curvature k. For $k = 0$ the result is trivial, and for $k = -1$ it was established in the last section. If $k = 1$ we can first show that the origin is an isotropic point where the curvature is 1, by adapting the argument following (9.14); now $\psi = \sin^{-1}\rho$. Then we can establish the equivalence of all points of the metric (9.16) by showing—in close analogy to the argument following (8.155)—that the transformation

$$X = \rho \sin \theta \cos \phi \qquad Z = \rho \cos \theta$$
$$Y = \rho \sin \theta \sin \phi \qquad W = (1 - k\rho^2)^{1/2}$$

maps the space (9.16) isometrically into the (pseudo-) sphere with equation

$$X^2 + Y^2 + Z^2 + kW^2 = k \tag{9.17}$$

in (pseudo-) Euclidean 4-space with metric

$$dl^2 = dX^2 + dY^2 + dZ^2 + kdW^2. \tag{9.18}$$

The equivalence of points is then established as in Section 8.11. This argument deals with both cases $k = \pm 1$ and so reestablishes the result found in the last section for $k = -1$.

We also note the following alternative form of (9.16):

$$dl^2 = \frac{dr^2 + r^2(d\theta^2 + \sin^2 \theta d\phi^2)}{(1 + \tfrac{1}{4}kr^2)^2}. \tag{9.19}$$

It is obtained by making the transformation

$$\rho = r/(1 + \tfrac{1}{4}kr^2) \tag{9.20}$$

to a new radial coordinate r. If we multiply (9.16) or (9.19) by a constant factor A^2, we get metrics of spaces of constant curvature k/A^2 [cf. after (9.14)].

The geometric significance of the coordinates ρ and r of a point P *in the case* $k = 1$ is illustrated in Figure 9.2, which represents the sphere obtained from (9.16) or (9.19) by setting $\theta = \pi/2$—a typical "geodesic plane" through the origin O. Figure 9.2, as well as the following two lines of equations, should be self-explanatory.

$$\rho = \sin \psi, \quad r = 2 \tan(\psi/2), \quad \sin \psi \equiv \frac{2 \tan(\psi/2)}{1 + \tan^2(\psi/2)}, \tag{9.21}$$

$$dl^2 = d\psi^2 + \rho^2 d\phi^2 = \frac{d\rho^2}{(1 - \rho^2)} + \rho^2 d\phi^2. \tag{9.22}$$

No such simple interpretation of ρ and r exists in the case $k = -1$.

We are now ready to investigate the kinematics of the most general cosmological models satisfying the CP (i.e., homogeneity and isotropy), within the framework of GR. We shall assume that the galactic distribution is sufficiently sparse not to interfere with the null-geodetic propagation of light. On this basis we derive an important standard form of the metric, in terms of cosmic time and comoving spatial coordinates. That metric was found in 1922 by Friedmann—apparently the first to consider seriously, and study systematically, the kinematics and dynamics of an *expanding* universe.

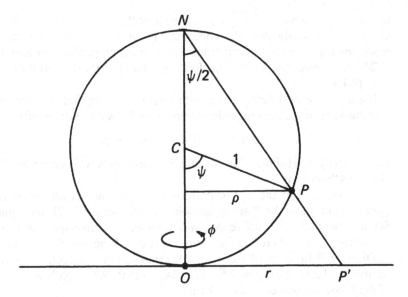

Figure 9.2

We have already seen in Section 9.3 that there will exist a cosmic time (*from now on denoted by* t), which reduces to proper time at each galaxy and which correlates identical observations at different galaxies. If for the moment we denote arbitrary comoving coordinates by x^i ($i = 1, 2, 3$), there can be no cross terms $dtdx^i$ in the metric. For consider a pair of fundamental particles separated, say, by dx^1, with $dx^2 = dx^3 = 0$. Then, as in Section 7.6 (except that now the metric coefficients may depend on t), the presence of a $dtdx^1$ term would imply that signals *emitted* at equal t at the two particles, towards each other, *arrive* at unequal t, which contradicts the CP. Since, furthermore, $dx^i = 0$ (all i) must imply $ds = cdt$ (since t is proper time on each galaxy), the metric must be of the form

$$ds^2 = c^2dt^2 - d\sigma^2, \qquad (9.23)$$

where $d\sigma^2$, the metric of public space, contains the dx^i but not dt, with coefficients possibly dependent on t. Consider next the development in time of a small triangle formed by three neighboring fundamental particles. By isotropy about each vertex, the angles must remain constant, for the total solid angle around each fundamental particle remains 4π. Hence the ratios of the sides must remain constant, since locally we have Euclidean geometry. And this means that time can enter $d\sigma^2$ only through a common factor, say $R^2(t)$. Thus

$$ds^2 = c^2dt^2 - R^2(t)dl^2, \qquad (9.24)$$

where dl^2 is time-independent and comoving and $R(t)$ represents the *expansion function* of the universe, or, more accurately, of public space. We have derived this form of the metric by symmetry arguments only; for consistency with GR we must verify that the comoving point x^1, x^2, x^3 = constant can represent a galaxy, i.e., that it traces out a timelike geodesic in any metric **(9.24)**, or, indeed, **(9.23)**. But this follows directly from the result discussed after **(8.106)**.

If a dot denotes differentiation with respect to t, we find that the instantaneous separation between neighboring galaxies, $d\sigma = Rdl$, satisfies

$$(d\sigma)^{\cdot}/d\sigma = \dot{R}/R, \; = H(t) \quad \text{say}, \tag{9.25}$$

and this is just Hubble's law locally. Consequently we recognize $H(t)$ as Hubble's "constant."

Now, because of the assumed isotropy of the model, all points of the public space must be "isotropic points" (see Section 7.2) and thus, by Schur's theorem, dl^2 must represent a Riemannian three-space of constant curvature. It can therefore be brought into a form which is a constant multiple of **(9.16)** or **(9.19)** (see end of Section 7.3). A constant, however, can be absorbed into the factor $R^2(t)$ in **(9.24)**, so that **(9.24)** finally assumes one of the following two equivalent forms:

$$ds^2 = c^2dt^2 - R^2(t)\left\{\frac{d\rho^2}{1 - k\rho^2} + \rho^2(d\theta^2 + \sin^2\theta d\phi^2)\right\} \tag{9.26}$$

or

$$ds^2 = c^2dt^2 - R^2(t)\left\{\frac{dr^2 + r^2(d\theta^2 + \sin^2\theta d\phi^2)}{(1 + \frac{1}{4}kr^2)^2}\right\}. \tag{9.27}$$

We shall agree that k, the *curvature index*, takes only the values ± 1 or 0. The curvature of public space is now $k/R^2(t)$. The coordinates ρ (or r), θ, and ϕ are, of course, comoving; θ and ϕ are evidently the usual angular coordinates on any sphere ρ (or r) = constant, which permanently consists of the same galaxies. By isotropy, θ and ϕ must be constant along light rays through the origin. (This can be verified from the null-geodetic hypothesis.) Thus θ and ϕ are the familiar visual angular coordinates relative to the origin. And of course, *any* fundamental particle can be taken to be that origin.

It may be noted from Figure 9.2 that, when $k = 1$, the coordinate ρ, as it ranges from 0 to 1, covers only *half* of public space (and causes a coordinate singularity at the "equator"), whereas r, in ranging from 0 to ∞, covers it completely. When $k = -1$, both ρ and r cover public space completely, ranging, respectively, from 0 to ∞ and from 0 to 2.

Milne's model **(9.13)** corresponds to **(9.26)** with $k = -1$ and $R(t) = ct$. It is of course not surprising to find Milne's model again in this context, since it satisfies all the present hypotheses. Like **(9.13)**, **(9.26)** and **(9.27)** specify not merely the underlying spacetime geometry, but also the substratum, i.e., the motion pattern of the galaxies.

Though we have presupposed GR in this section, we have in fact used of that theory so far only (i) that spacetime is Riemannian, (ii) that ds/c measures proper time at each fundamental particle, and (iii) that light propagates along null geodesics. In two independent and important papers, H. P. Robertson and A. G. Walker almost simultaneously discovered (by group-theoretic methods, around 1935) that these properties are actually implicit in the CP, i.e., that the assumed isotropy and homogeneity of the substratum *imply* the existence of a Riemannian metric **(9.27)** with the properties (ii) and (iii). Consequently the metric **(9.27)**—which is now usually called the Robertson–Walker (RW) metric, though in GR it was discovered much earlier by Friedmann—*applies to all models satisfying the CP*, even those outside of GR. In particular, it can be used in the steady state theory, and in "Newtonian" cosmology. Without GR, however, its properties are fewer: Its coefficients are unrestricted by the GR field equations, its timelike geodesics do not necessarily represent the paths of free particles, and $c^{-1} \int ds$ does not necessarily measure proper time along arbitrary paths. But t *does* represent cosmic time, light *does* travel along null geodesics, fundamental particles *do* correspond to fixed r, θ, ϕ, and the spatial part *does* represent a distance element ("radar distance" locally) of public space.

This, then, is as far as we can go with the CP alone. The two *free* elements of the RW metric, k and $R(t)$, can only be determined by additional *assumptions*—if we approach the task theoretically—or possibly by *observation* of the actual universe. Observation without further theory, however, is of very limited scope. The curvature k/R^2 of public space, for example, is somewhat conventional (recall Milne's universe, where each fundamental observer has a *flat* private space) and in any case it is so minute (e.g., $\sim 10^{-56}$ cm^{-2} in a Milne universe today) that fluctuations due to nearby masses would totally swamp it. Nevertheless, as we shall see in Section 9.7, in principle it is obtainable from astronomical observations if these can be pushed "to third order." Such observations also yield the *present* value of Hubble's constant $\dot{R}/R$, and potentially that of the first few of the ratios $\ddot{R}/R$, $\dddot{R}/R$, ... But that is all. Thus, inevitably, we must bring in more theory to make further progress.

One theoretical approach is to appeal to general principles. For example, some cosmologists have felt that Mach's principle requires a closed and finite universe. This, in turn, has sometimes been thought to necessitate the choice $k = +1$, but in fact it does not. Closed and finite if somewhat artificial universes can be constructed also with $k \le 0$ (see Exercise 7.6) by making "topological identifications," which spoil the global isotropy of the model but retain its homogeneity and *local* isotropy. One must bear in mind that a *metric* determines a space only locally. For example, if we are given the metric $dx^2 + dy^2$, it would be very naive to assume that we are necessarily dealing with an infinite plane: we may be dealing with a cylinder, or even with a closed and finite surface, topologically equivalent to a torus, which results on cutting a square from the plane and identifying any two directly opposite points on the edges. Again, the metric of a unit sphere does not necessarily imply that we are dealing with a closed surface of area 4π. We may be dealing with one of area 2π: such a surface results on identifying diametrically opposite points of the sphere; by the time we go half-way round, we are already back where we started. (And, apart from our biased experience, such a surface is, *per se*, no more "unlikely" than a sphere.) These examples illustrate the kind of technique involved in closing the public space of an RW model with $k = -1$ or 0, though in the case $k = -1$ it is not so trivial. In the case $k = 0$, on the other hand, it is quite analogous to the closing of the plane. The opposite process, incidentally, that of "opening up" a space of constant *positive* curvature and making it infinite, is impossible (if the metric is positive-definite).

Another essentially *a priori* assumption is Bondi and Gold's "perfect" cosmological principle: by imposing a further symmetry on the model they made it unique. As we mentioned in Section 9.3, the steady state model not only satisfies the CP but also presents the same aspect at all times. It *must* violate local energy conservation, and thus it *cannot* satisfy GR, which has local energy conservation built into it. Nevertheless, the RW metric (9.27) is applicable. Since $H(t)$ must be constant in this model, (9.25) implies $R = a \exp(Ht)$ for some constant a, which we can absorb into the exponential by a translation of t. And since the curvature $k/R^2(t)$ must also be constant, k must necessarily vanish. Hence the relevant metric is

$$ds^2 = c^2dt^2 - e^{2Ht}\{dr^2 + r^2(d\theta^2 + \sin^2\theta d\phi^2)\}, \tag{9.28}$$

and now the model is fully specified, *kinematically*.

A very different theoretical approach to the narrowing of the RW model is *dynamical*, namely the application of gravitational theory. In GR this amounts to subjecting the metric to the field equations. In a pseudo-Newtonian theory it amounts to applying a Poisson or a pseudo-Poisson equation. We shall discuss these methods in Sections 9.8 and 9.9, and the utilization of observations in Section 9.7. But first we wish to describe a useful "model of the model," which helps us visualize the formalism.

9.6   Rubber Models, Red Shifts, and Horizons

Suppressing one spatial dimension, and choosing the simplest topological spaceforms, we can illustrate the public spaces of the three types of RW metric (corresponding to $k = 1, 0, -1$) by a sphere, an infinite plane, and an infinite surface which is everywhere locally similar to a saddle (see Figure 9.3). Since the public space generally expands or contracts, we shall think of the

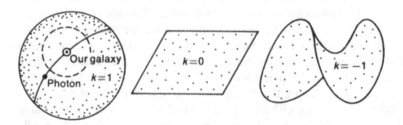

Figure 9.3

sphere as a rubber balloon which can be inflated or deflated at will, and of the other two surfaces as also made of rubber membrane and similarly subject to uniform expansion or contraction. The motion of the balloon is perhaps the easiest to visualize, and we shall mainly talk about it as representative of all three cases. Actual distance on the membrane corresponds to $d\sigma$ in (9.23). The material points of the membrane represent the substratum, and a selected subset of them corresponds to galaxies. These are marked by dots on the membrane, in roughly uniform distribution. Since $\pm 1/R^2(t)$ is the curvature of public space in the cases $k = \pm 1$, $R(t)$ corresponds to the radius of our balloon, or to the radius of either of the two spheres fitting into the saddle (one above, one below). In all cases the distance between galaxies as a function of time is proportional to $R(t)$, and in the plane case this is the *only* interpretation of $R(t)$, since there is then no intrinsic normalization for it.

To model cosmic time, we must assume an *absolute* time throughout the Euclidean three-space E_3 in which the rubber models are embedded, a time which the fundamental clocks keep *without* time dilation, inspite of moving relative to E_3.

It can easily be shown formally that the spatial tracks of null geodesics in the RW metric are ordinary geodesics in the public space; but this is also obvious from symmetry. Hence in the rubber model light propagates along geodesics, e.g., along great circles on the sphere and straight lines in the plane. For a light signal we have $ds^2 = 0$ whence $d\sigma/dt = c$ [see (9.23)]. This means that in the rubber model photons crawl, like ideal bugs, at constant speed c over the membrane, and along geodesics.

All features of the RW metric are now accurately represented by the rubber model. As a first simple application let us establish the cosmological red shift formula

$$1 + z = \frac{R(t_0)}{R(t_1)}, \tag{9.29}$$

where $z = \Delta\lambda/\lambda$, λ being the wavelength of light emitted by a distant galaxy at cosmic time t_1 and received by us at t_0 with wavelength $\lambda + \Delta\lambda$. If two closely successive "bugs" crawl over a nonexpanding track, they arrive as far apart as when they left. But if the track expands—or contracts—proportionally to $R(t)$, then their distances apart at reception and emission will be in the ratio $R(t_0)/R(t_1)$. Replacing the bugs by two successive wavecrests, we get equation (9.29). Note that the cosmological red shift is really an *expansion* effect rather than a *velocity* effect.

The fact that the red shift in the light of all presently observed cosmic objects depends only on the "radius of the universe" at the time when that light was emitted, led Shklovsky in 1967 to suggest an interesting explanation of the then-puzzling predominance of values $z \approx 2$ among quasars; it is merely necessary to assume that the radius of the universe was for a comparatively long time quasistationary at approximately one-third of its present value [see Figure 9.4, where we also introduce the obvious notation $R(t_0) = R_0$, $R(t_1) = R_1$]. Then one quasar could be several times as far away from

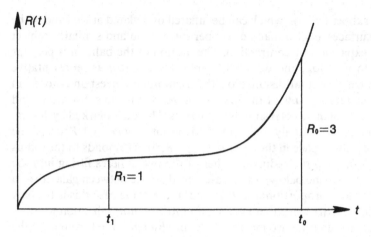

Figure 9.4

us as another, yet as long as the radius of the universe was the same at emission, the red shifts observed will be the same. Although the early clustering of quasar red shifts around $z = 2$ is now recognized as spurious, this example well illustrates that z is not a faithful indicator of cosmic distances.

We can of course derive (**9.29**) directly, though less illuminatingly, from the RW metric (**9.26**). Any light signal satisfies $ds^2 = 0$; hence two successive signals from a galaxy at coordinate ρ_1 to the origin-galaxy will satisfy, respectively,

$$\int_0^{\rho_1} \frac{d\rho}{(1 - k\rho^2)^{1/2}} = \int_{t_1}^{t_0} \frac{c\,dt}{R(t)} = \int_{t_1 + \Delta t_1}^{t_0 + \Delta t_0} \frac{c\,dt}{R(t)}. \tag{9.30}$$

Now, an integral over a short range Δt equals the integrand times Δt, and so (**9.30**) implies

$$\frac{c\Delta t_1}{R(t_1)} = \frac{c\Delta t_0}{R(t_0)}. \tag{9.31}$$

If we identify the two "signals" with successive wavecrests, Formula (**9.29**) results.

As another application of the rubber model we briefly illustrate two of the *horizon* concepts used in cosmology.[8] For definiteness we consider a universe of positive curvature, though the argument applies equally in all three cases. In the first diagram of Figure 9.3 we have marked our own galaxy and a photon on its way to us along a geodesic. It can happen that "the balloon is blown up" at such a rate that this photon never gets to us. As Eddington has put it, light is then like a runner on an expanding track, with the winning post (us) receding forever from him. In such a case there will be two classes of (actual or virtual) photons on every geodesic through us: those that reach us

[8] Further details can be found in W. Rindler, *Mon. Not. R. Astron. Soc. 116*, 662 (1956).

at a finite time and those that do not. They are separated by the aggregate of photons that reach us exactly at $t = \infty$ —shown in the diagram as a dashed circle, but in the full model these photons constitute a spherical light front converging on us. This light front is called our *event horizon*, and its existence and motion depend on the form of $R(t)$. Events occurring beyond this horizon are forever beyond our possible powers of observations (that is, if we remain on our own galaxy). It is sometimes stated that at the horizon galaxies stream away from us at the speed of light, in violation of special relativity. [For the horizon may be (but need not be) "stationary" relative to us, and galaxies clearly must cross *it*—since it crosses them—at the speed of light.] But the SR speed limit applies only to objects in an observer's inertial rest frame; and cosmological observers with horizons do not *have* extended inertial rest frames.

The same diagram can also be used to illustrate the concept of a *particle horizon*. Suppose the very first photons emitted by our own galaxy at a big-bang creation event are still around, and let the dashed circle in the diagram indicate their present position. In the full model this is a spherical light front moving away from us. As it sweeps outward over more and more galaxies, these galaxies see us for the very first time. By symmetry, however, at the cosmic instant when a galaxy sees *us* for the first time, we see *it* for the first time. Hence at any cosmic instant this light front, called our particle horizon, divides all galaxies into two classes relative to us: those already in our view and all others.

It is quite easy to obtain the quantitative formulae from the RW metric. The equation of motion (ρ, t relation) of a photon emitted towards the origin-galaxy ("us") at coordinates ρ_1, t_1 is seen to be, as in **(9.30)**,

$$\int_\rho^{\rho_1} \frac{d\rho}{(1 - k\rho^2)^{1/2}} = \int_{t_1}^t \frac{cdt}{R(t)}. \qquad \textbf{(9.32)}$$

Since both integrands are positive, ρ decreases as t increases. Suppose that, for a fixed pair of values (ρ_1, t_1), ρ tends to a positive limit as $t \to \infty$. Then the photon from (ρ_1, t_1) never reaches the origin, and thus (ρ_1, t_1) is an event beyond the event horizon. The condition for such a horizon to exist[9] is

$$\int_{t_1}^\infty \frac{dt}{R(t)} < \infty, \qquad \textbf{(9.33)}$$

and the coordinate ρ_1 of the horizon at any time t_1 is then given by

$$\psi(\rho_1) \equiv \int_0^{\rho_1} \frac{d\rho}{(1 - k\rho^2)^{1/2}} \equiv \begin{Bmatrix} \sin^{-1} \rho_1 & (k = 1) \\ \rho_1 & (k = 0) \\ \sinh^{-1} \rho_1 & (k = -1) \end{Bmatrix} = \int_{t_1}^\infty \frac{cdt}{R(t)}, \qquad \textbf{(9.34)}$$

where $\psi(\rho_1)$ is defined by the first equation. For the steady state model **(9.28)**, for example, we find that an event horizon exists at

$$\rho_1 = (c/H)e^{-Ht_1}, \qquad \textbf{(9.35)}$$

[9] When $k = 1$, this needs elaboration; see below, after **(9.35)**.

which corresponds to a *constant* ruler distance $\sigma = c/H$ in public space. Milne's model (9.13), on the other hand, has *no* event horizon.

When the model has a *future* big bang [say $R(t_\omega) = 0$, $t_\omega > t_{now}$], the definition of event horizon must be modified: it is then the light front which reaches the observer at $t = t_\omega$. Consequently the infinite upper limits of integration in (9.33) and (9.34) must be replaced by t_ω.

When $k = 1$, ρ is restricted to be less than unity, whereas the time integral in (9.34) may exceed $\pi/2$. A more convenient variable is then ψ itself, the "angular" distance around the universe (cf. Figure 9.2). In this case we may have to go beyond the equator ($\rho = 1$, $\psi = \pi/2$), or beyond the antipode ($\psi = \pi$), or even several times around the universe, before we come to the horizon. Though statements about horizons retain their *formal* correctness also in this case, their physical significance is subject to modification. For example, a photon from the event horizon *does* reach us at $t = \infty$, but it may have passed us several times previously during its trips around the universe.

A particle horizon will exist now ($t = t_1$) at $\rho = \rho_1$ if the light emitted by "us" at creation has "now" reached a finite coordinate ρ_1. For an *outgoing* photon Equation (9.32) applies with a minus sign on one side. If, in that equation, we set $\rho = 0$ and $t = 0$ or $-\infty$ [the latter in cases where the definition of $R(t)$ extends to negatively unbounded values of t] we obtain the coordinate ρ_1—or, more generally, ψ—of the particle horizon,

$$\psi(\rho_1) = \int_0^{t_1} \frac{cdt}{R(t)}, \quad \text{or} \ = \int_{-\infty}^{t_1} \frac{cdt}{R(t)}, \tag{9.36}$$

provided $R(t)$ satisfies the condition for the existence of a particle horizon,

$$\int_0^{t_1} \frac{dt}{R(t)} < \infty, \quad \text{or} \quad \int_{-\infty}^{t_1} \frac{dt}{R(t)} < \infty. \tag{9.37}$$

In models with particle horizon, any two galaxies were causally unconnected in the past, until overtaken by each other's horizon. If there was a big bang, each was shot away from the other faster than light; but since the curvature was infinite, *no* neighborhood was small enough for SR to apply, with its pedantic insistence on $v < c$. All this may seem strange. Yet if there is *no* particle horizon, the model is "incomplete" [cf. after (9.9)]. For where would the creation photons be now? Clearly somewhere in space where there are *no* galaxies. So a big bang entails at least one of these "blemishes."

Some models have an event horizon only (e.g., the steady state model); some have a particle horizon only [e.g., the Einstein–de Sitter model, in which $R(t) \propto t^{2/3}$]; many have both (e.g., all the "inflexional" and "oscillating" nonempty isotropic models of GR); and some have neither (e.g., Milne's model). *All* nonempty isotropic GR big-bang models have a particle horizon.

Some further properties of the event horizon (EH) and particle horizon (PH) may be noted:

(i) Every galaxy but A within A's EH eventually passes out of it. For, if B is such a galaxy, then A's horizon photon in the direction AB is

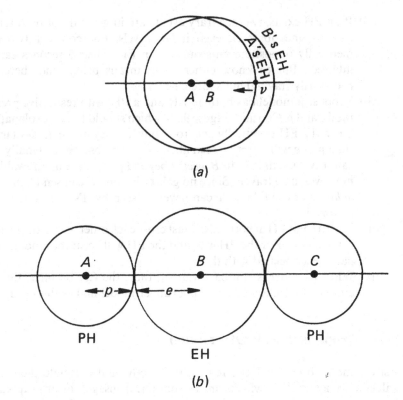

Figure 9.5

within B's EH (see Figure 9.5a), and will therefore reach B at finite cosmic time (or before t_ω). That, of course, is when B passes out of A's EH.

(ii) Every galaxy B within A's EH remains visible forever at A. For the EH itself brings A a last view of B. As B approaches A's EH in models with infinite future, its history, as seen at A, gets infinitely dilated, and its light infinitely red-shifted; for R is infinite at reception [cf. (9.29)]. In collapsing models B's light gets infinitely blue-shifted as B approaches the EH; for R is zero at reception.

(iii) As galaxies are overtaken by A's PH, they come into view at A with infinite red shift in big-bang models, and infinite blue shift in models with unlimited past (e.g., one with $R \propto \cosh t$). For in the former case R was zero, in the latter, infinite, at emission.

(iv) If a model possesses *no* EH, every event at every galaxy is seen on every galaxy. For an invisible event implies the existence of an EH.

(v) If a model possesses *no* PH, every observer—if necessary, by traveling away from his original galaxy—can be present at any one event at any galaxy. For, in principle, his only travel restriction is his forecone (forward light cone) at creation; but that would be a PH if all galaxies were not always within it.

(vi) If an EH exists, two arbitrary events are in general not both knowable to one observer, even if he travels. For consider two diametrically opposite events outside an EH. Their forecones cannot intersect. But to know either event means being (and thereafter necessarily staying) in its forecone.

(vii) Suppose a model has both an EH and a PH, with respective present radii e and p. Consider Figure 9.5b, which should be self-explanatory. Since A's PH eventually gets to B but not beyond, *galaxies* farther than $p + e$ will never be seen at A. Since an observer originally at A can travel maximally to B, *events* beyond $p + 2e$ are unknowable to him, even if he travels. Since no galaxy beyond C can send light to B, *galaxies* beyond $2p + 2e$ can never be seen by A's observer, even if he travels.

(viii) The EH and PH, if both exist, must cross each other within the lifetime of the model. For the PH was, and the EH will be, at the fundamental particle associated with them.

(ix) When a model is run backward in time, the EH becomes the PH and vice versa. Reference to the rubber model makes this clear.

9.7 Comparison with Observation

Astronomers observe galactic red shifts directly, and correlate them with galactic "magnitude," which, in astronomical usage, means apparent luminosity normalized in a certain way. For these and other observations to be compared with theory, it is necessary for theory to predict relations between observables. Several such relations will be derived in the present section.

First we digress briefly to discuss some concepts of "distance" in cosmology. As in Schwarzschild space (cf. Section 8.3), various "reasonable" definitions of distance turn out to be inequivalent. One can, for example, define the *proper distance* between two galaxies P and Q: this would be measured by a chain of fundamental observers on the spatial geodesic between P and Q laying rulers end to end at *one* cosmic instant. (It corresponds to distance on the rubber model.) The proper distance σ from the origin to a galaxy at coordinate ρ, at time t, is then given by

$$\sigma = \int_0^\rho d\sigma = R(t) \int_0^\rho \frac{d\rho}{(1 - k\rho^2)^{1/2}} = R(t)\psi(\rho), \qquad (9.38)$$

[see (9.23), (9.26), (9.34)]. As an example, the event horizon (9.35) in the steady state model is at constant proper distance c/H from the observer. More strikingly, the "spherical boundary" in Milne's universe, which is at distance ct from each observer in his private Minkowskian coordinates, is always at infinite *proper* distance from him.

The practical methods of measuring distance in astronomy are, for example, by parallax, by radar, by apparent size, or by apparent luminosity.

Only the last two will interest us here. The (uncorrected) *distance from apparent size*, S, of an object of known cross-sectional area dA (e.g., a galaxy) that is seen subtending a small solid angle $d\omega$, is *defined* by

$$S = (dA/d\omega)^{1/2}, \qquad (9.39)$$

i.e., by the *naive* formula which deliberately ignores any possible deviation of the space geometry from the geometry of Euclidean space. And, similarly, the (uncorrected) *distance from apparent luminosity*, L, of an object of known intrinsic luminosity B and apparent luminosity b, is *defined* by

$$L = (B/b)^{1/2}, \qquad (9.40)$$

again deliberately ignoring all possible deviations from the classical, static situation; B is energy flux per unit area per unit time at unit distance from the source—here idealized as a point source—and b is the same at the distance of the observer. (Astronomical "magnitude" m is related to b by the equation $m = -2.5 \log_{10} b + $ constant.)

By (9.26), the area of a sphere at coordinate ρ_1 and time t_1 is evidently $4\pi R_1^2 \rho_1^2$, and its solid angle at the origin is 4π. Consequently, by isotropy and proportionality, the solid angle subtended at the origin by the spatial geodesics to an area dA at coordinate ρ_1 at time t_1 is $d\omega = dA/R_1^2\rho_1^2$; and this is the solid angle which dA is seen to subtend when the light emitted at t_1 arrives at the origin. [The balloon model should help to make it clear that the solid angle is determined at the cosmic moment of *emission*: think of a galaxy as a rigid disc whose center is glued to the expanding balloon.] Hence

$$S = R_1\rho_1. \qquad (9.41)$$

Suppose, next, that a source of intrinsic luminosity B is placed at the *origin* and that its light is observed on a sphere centered on the origin, moving with the substratum, and having radial coordinate $\bar{\rho}_1$, at time t_0. If the sphere were stationary, the total flux of energy per unit time through it would be independent of its radius, and thus it would be $4\pi B$, since *near* the source Euclidean geometry holds in the limit. But, because of the motion of the sphere, the energy $E = h\nu$ carried by each photon is diminished by a Doppler factor ("Planck effect") and, moreover, the number of photons arriving per unit time is also diminished by a Doppler factor, since the time between the arrival of "successive" photons varies as the wave length ("number effect"). From (9.29), therefore, the total flux through the sphere is $4\pi B R_1^2/R_0^2$, and that through a unit spherical area is

$$b = \frac{BR_1^2}{\bar{\rho}_1^2 R_0^4}. \qquad (9.42)$$

By symmetry, the radial coordinate ρ_1 of the source (at the origin) relative to the observer (at the sphere) will equal that of the observer relative to the source, and thus we conclude from (9.42), by reference to (9.40) and (9.41), that

$$L = \rho_1 R_0^2/R_1 = S(1 + z)^2. \qquad (9.43)$$

We shall now combine this equation with **(9.29)** in order to relate z with L, or, equivalently, z with b. Introducing, *for the present section only*, the notation

$$R_0 = R, \qquad R_0 - R_1 = \Delta, \qquad (9.44)$$

we can cast **(9.29)** into the form

$$R_1 = R(1 + z)^{-1} = R(1 - z + z^2 - \cdots),$$

provided $|z| < 1$ (which, of course, is *not* true for some recent observations), whence

$$\Delta = R(z - z^2 + \cdots). \qquad (9.45)$$

Next, using **(9.32)** and the notation **(9.34)**, we have, for a signal reaching the origin at t_0,

$$\psi(\rho_1) = \int_{t_1}^{t_0} \frac{c\,dt}{R(t)} = \int_{R_1}^{R_0} \frac{c\,dR}{R(t)\dot{R}(t)}, \qquad (9.46)$$

where a dot here and in the sequel denotes d/dt. Regarding the last integral as a function f of R_1, we can expand it as a Taylor series about R_0 with increment $-\Delta$,

$$\psi(\rho_1) = f(R) - f'(R)\Delta + \tfrac{1}{2}f''(R)\Delta^2 + \cdots,$$

and, remembering that the derivative of an integral with respect to its lower limit is minus the integrand, we find

$$\psi(\rho_1) = \frac{c}{R\dot{R}}\Delta + \frac{c}{2}\frac{R\ddot{R}/\dot{R} + \dot{R}}{R^2\dot{R}^2}\Delta^2 + \cdots, \qquad (9.47)$$

where, consistently with our convention **(9.44)**, R and its derivatives are evaluated at t_0. Since $\rho = 1$ corresponds to the equator relative to the origin in the positive-curvature model, it is clear that $\rho \ll 1$ for most observed galaxies even if $k = 0$ or -1; thus $\psi(\rho_1) = \rho_1 + \tfrac{1}{6}k\rho_1^3 + \cdots \approx \rho_1$ if we retain second powers only. Hence, substituting from **(9.45)** into **(9.47)**, we find

$$\rho_1 = \frac{c}{\dot{R}}z + \frac{c}{2}\frac{R\ddot{R} - \dot{R}^2}{\dot{R}^3}z^2 + \cdots, \qquad (9.48)$$

which in turn can be substituted into the following variant of **(9.43)**:

$$L = R\rho_1(1 + z);$$

this yields

$$\frac{L}{c} = \frac{R}{\dot{R}}z + \frac{1}{2}\frac{R^2\ddot{R} + R\dot{R}^2}{\dot{R}^3}z^2 + \cdots, \qquad (9.49)$$

or finally, on inversion [we substitute $z = a_1(L/c) + a_2(L/c)^2 + \cdots$ into (9.49) and equate coefficients to find the a's],

$$z = \frac{\dot{R}}{R}\frac{L}{c} - \frac{1}{2}\left(\frac{\ddot{R}}{R} + \frac{\dot{R}^2}{R^2}\right)\frac{L^2}{c^2} + \cdots.^{10} \tag{9.50}$$

All "reasonable" distance definitions become equivalent in the limit of small distances; assuming also that for small distances the classical Doppler formula $z = u/c$ applies, we can read off from (9.50), as a check, Hubble's law $u = HL$ [cf. (9.25)] in first approximation. The usual way in which (9.50) is compared with (z, b) observations is to substitute $(B/b)^{1/2}$ for L and to *assume* that B is a certain constant for all galaxies, at all times, at least over the range of the observations.

Substituting from (9.43) into (9.49), we find the following series expansion for S, the distance from apparent size,

$$\frac{S}{c} = \frac{R}{\dot{R}}z + \frac{1}{2}\frac{R^2\ddot{R} - 3R\dot{R}^2}{\dot{R}^3}z^2 + \cdots, \tag{9.51}$$

which, on inversion, yields

$$z = \frac{\dot{R}}{R}\frac{S}{c} - \frac{1}{2}\left(\frac{\ddot{R}}{R} - \frac{3\dot{R}^2}{R^2}\right)\frac{S^2}{c^2} + \cdots. \tag{9.52}$$

When we replace S by $(dA/d\omega)^{1/2}$, and assume that dA is essentially constant (for galaxies, or clusters of galaxies), this formula can be compared with observational $(z, d\omega)$ relations, which, however, are less reliable than the usual (z, b) relations.

Formulae (9.50) or (9.52), which can be continued to any order, could in principle yield through observations the present values of $\dot{R}/R$, $\ddot{R}/R, \ldots$, and of k/R^2 (this appears in the third-order terms). In fact, however, the practical difficulties are so great that only Hubble's constant, $\dot{R}/R$, is known with any certainty; it is not even known for sure whether $\ddot{R}$ is positive or negative. A restriction to bear in mind is that for very distant objects $(z > 1)$ the above series expansions become invalid, and observations must be compared with particular models in nonapproximate (integral) form.

Another empirical relation obtained by the astronomers (both optical and radio) is the number of galaxies, per unit solid angle of sky, whose red shift is less than some given z, or whose luminosity is less than some given b. Such "source counts" evidently probe the galactic distribution in depth, i.e., radially. We shall assume that $R(t)$ is a monotonically increasing function so

---

[10] To gain some idea of the convergence of this (and similar) series, we can calculate the exact values of the terms in (9.50) for some specific models, e.g., (i) $R(t) = \exp(Ht)$, $k = 0$; (ii) $R(t) = t$, $k = 0$. The results for a red shift $z = 0.25$, and to one more term than is shown in (9.50), are as follows:

$$\text{(i) } 0.25 = 0.312 - 0.097 + 0.061 + \cdots,$$
$$\text{(ii) } 0.25 = 0.278 - 0.039 + 0.014 + \cdots.$$

For smaller z, of course, the convergence improves.

that larger z corresponds to smaller t_1 [cf. (9.29)], therefore earlier emission, therefore greater distance. Consider a cone of solid angle ω issuing from the origin of public space, and terminating at radial coordinate ρ_1. As we saw earlier in this section, the area which this cone cuts from a sphere at coordinate ρ at the "present" time t_0 is $\omega R^2\rho^2$, and hence the present proper volume of the cone is given by

$$
V = \omega R^3 \int_0^{\rho_1} \rho^2 \frac{d\rho}{(1 - k\rho^2)^{1/2}}
$$

$$
= \omega R^3 \int_0^{\rho_1} \rho^2 (1 + \tfrac{1}{2}k\rho^2 + \cdots)d\rho.
$$

Multiplying V by the present particle density n_0, which can be estimated locally, we obtain the total number N of galaxies presently in the cone — and this, of course, is the number in it independently of time *if* there is local conservation:

$$
N = n_0 V = n_0 \omega R^3(\tfrac{1}{3}\rho_1^3 + \tfrac{1}{10}k\rho_1^5 + \cdots). \tag{9.53}
$$

The coordinate ρ_1 corresponding to a galaxy now observed with red shift z was obtained in (9.48). Substituting this into (9.53) gives us the required formula for the number N of galaxies seen in a solid angle ω of sky at red shift less than z:

$$
N = n_0 \omega c^3 \left\{ \frac{R^3}{\dot{R}^3}\frac{z^3}{3} + \left(\frac{\ddot{R}R^4}{\dot{R}^5} - \frac{R^3}{\dot{R}^3} \right)\frac{z^4}{2} + \cdots \right\}. \tag{9.54}
$$

The corresponding formula for the steady state theory (where local particle conservation is violated) is easily found to be[11]

$$
N = n_0 \omega c^3 H^{-3} \int_0^z z^2(1 + z)^{-3}dz
$$

$$
= n_0 \omega c^3 H^{-3}(\tfrac{1}{3}z^3 - \tfrac{3}{4}z^4 + \cdots). \tag{9.55}
$$

These formulae can be converted at once into (N, b) relations by substituting from (9.50) and (9.40). Radio-astronomical source counts have long been in apparent conflict with the relation so obtained, from (9.55), for the steady state theory.

---

[11] Let $R(t) = e^{Ht}$, $k = 0$, and $t_0 = 0$. Then, from (9.46) and (9.29), the coordinate ρ, time of emission t, and red shift z are related by

$$
\frac{H}{c}\rho = e^{-Ht} - 1 = z. \tag{I}
$$

The number of galaxies per unit proper volume is constant, say n_0. Hence, the number in a solid angle ω between coordinates ρ and $\rho + d\rho$ at emission time t is $n_0 \omega e^{3Ht}\rho^2 d\rho$. From this and (I), (9.55) follows at once.

9.8 Cosmic Dynamics According to Pseudo-Newtonian Theory

We now turn to the *dynamical* study of the RW model. Since the "lumpiness" of the material contents of the actual universe is not very amenable to mathematical treatment, one generally adopts the theoretical device of grinding these contents into dust and redistributing the dust homogeneously. The assumption is that such a "smoothed out" dust universe, and another in which the same dust is gathered into lumps, will, on the whole, behave identically under their own gravitation. It will be remembered (cf. Section 5.13) that "dust" refers technically to a continuous medium that has density but no internal stress, not even pressure. The proper motion of the galaxies, and the possible intergalactic presence of the undetected neutrinos, magnetic fields, cosmic rays, quanta (e.g., gravitons), hydrogen, etc., are not usually considered to add a significant pressure. (As we shall see, the theoretical effect of pressure is in any case only through the minute mass-equivalent of its energy, which *slows* expansion, rather than through any direct elastic action.) Hence the idealization by dust would seem to be justified, except in the very early stages of a universe that was much denser then.

In order to apply Newtonian gravitational theory to the universe as a whole, some modifications of that theory are necessary. These were begun by Milne and McCrea in 1934 and later made rigorous by Heckmann and Schücking.[12] It is necessary to give up the idea of absolute space, but *not* that of absolute time. Public space and each private space is the usual flat and infinite E_3. However, a fundamental observer's private E_3 is *inertial* only locally, i.e., in his neighborhood. Widely separated fundamental observers may accelerate relative to each other. Newton's inverse square law applies unchanged in its local form, i.e., as Poisson's equation. (The overall consistency of this entire theory can be achieved by modifying the transformation properties of the potential; but since this does not affect our present calculations we shall not go into it.) Light is assumed to propagate rectilinearly in public space, with constant local velocity c relative to the fundamental observers; this is equivalent to classical propagation in an ether that partakes of the expansion of the universe. And for particles, we assume the usual law of motion, $\ddot{\mathbf{r}} = -\mathbf{grad}\ \varphi$.

It is instructive to study the resulting "pseudo-Newtonian" theory *before* taking up general-relativistic cosmology, because it leads to essentially the same equations and to some identical models; and since we understand the Newtonian equations and models more intuitively, they help us understand the GR models and equations, which arise in a much more abstract way. With a little hindsight we can even see why local Newtonian theory *should* give models which, even in the large, coincide with GR models: In the limit

---

[12] For a thought-provoking review, see the article "Cosmology" by E. L. Schücking, p. 218 of *Lectures in Applied Mathematics*, Vol. 8 (Relativity Theory and Astrophysics, I), J. Ehlers, editor, Am. Math. Soc., 1967.

of slow motions and weak fields, GR reduces to Newtonian theory; but locally the cosmic fields *are* weak and the cosmic motions *are* slow, hence locally Newton's and Einstein's cosmic dynamics are equivalent; and finally, in a homogeneous universe local knowledge *is* global knowledge.

Since we assume public space to be flat E_3, the argument involving a triangle of fundamental particles that led up to (9.24) shows that in the Newtonian case the substratum is given by a metric of the form

$$d\sigma^2 = R^2(t)(dx^2 + dy^2 + dz^2),$$

or, transforming to polar coordinates,

$$d\sigma^2 = R^2(t)\{dr^2 + r^2(d\theta^2 + \sin^2 \theta d\phi^2)\},$$

where in both cases the spatial coordinates are comoving, and t is absolute time. Now construct the Riemannian spacetime metric

$$ds^2 = c^2dt^2 - R^2(t)\{dr^2 + r^2(d\theta^2 + \sin^2 \theta d\phi^2)\}. \tag{9.56}$$

It has the right substratum; its null geodesics are precisely the light paths we have postulated; and ds/c measures absolute time at each fundamental particle. We have, in fact, here established the Robertson–Walker theorem (cf. Section 9.5) for the special case of pseudo-Newtonian cosmology.

The starting point for our dynamics is Poisson's equation

$$\nabla^2\varphi = 4\pi G\rho, \tag{9.57}$$

with the density ρ a function of absolute time but not of position. Since we assume spherical symmetry about any fundamental observer P, and since φ cannot locally become infinite in a continuous medium, we seek (as in Section 9.2) a solution of the form $\varphi = a + br + cr^2 + dr^3 + \cdots$, where r is distance from P in P's private Euclidean space. We find that this solution is unique, apart from an arbitrary additive constant, namely

$$\varphi = \tfrac{2}{3}\pi G\rho r^2. \tag{9.58}$$

For radial motion the Newtonian equations of motion reduce to $\ddot{r} = -d\varphi/dr$, and so, in our case,

$$\ddot{r} = -\tfrac{4}{3}\pi G\rho r = -\frac{GM'}{r^2}, \tag{9.59}$$

where M' is the mass contained in a sphere of radius r. Thus a particle is attracted to the center of any sphere through it as though the mass of that sphere were concentrated at its center and none were outside. If r refers to the position of a nearby galaxy Q relative to P, then $M' = $ constant (in time), since galaxies neither enter nor leave this sphere as public space expands. (This is the relevant version of the Newtonian "equation of continuity.") Now the distance r between a pair of galaxies such as P and Q is a multiple of $R(t)$, say $hR(t)$; substituting this into (9.59) yields the following differential equation for $R(t)$:

$$\ddot{R} = -\frac{GM}{R^2}, \tag{9.60}$$

where $M = M'/h^3$ is the mass inside a sphere of radius $R(t)$. (From now on we again write, where convenient, R, $\dot{R}$, $\ddot{R}$, etc., for the *general* values of these functions, *not* specialized by $t = t_0$ as in Section 9.7).

In order to facilitate later comparison with the relativistic models, we can generalize the theory one step further by assuming a modified Poisson equation (9.5). As we have seen in (9.6), this leads to an additional r^2 term in φ (superposition principle!), which, however, is arbitrary unless $A = 0$. We therefore prefer the more restrictive form (9.4), whose only regular spherically symmetric solution is

$$\varphi = \tfrac{2}{3}\pi G\rho r^2 - \tfrac{1}{6}c^2\Lambda r^2, \tag{9.61}$$

apart from an additive constant. If we now proceed as from (9.58) to (9.60), we obtain, instead of (9.60),

$$\ddot{R} = -\tfrac{4}{3}\pi G\rho R + \frac{c^2\Lambda R}{3} = -\frac{GM}{R^2} + \frac{c^2\Lambda R}{3}. \tag{9.62}$$

Multiplying this differential equation by $2\dot{R}$, we immediately find a first integral,

$$\dot{R}^2 = \frac{C}{R} + \frac{\Lambda c^2 R^2}{3} - \tilde{k}c^2, \tag{9.63}$$

where

$$C = 2GM = \tfrac{8}{3}\pi G\rho R^3, \tag{9.64}$$

and $-\tilde{k}c^2$ is simply a constant of integration, written in this form for later convenience. Since R is not intrinsically normalized, we can now impose a normalization on it, at least in the cases where $\tilde{k} \neq 0$: we choose the scale of R so that

$$\tilde{k} = \pm 1 \text{ or } 0. \tag{9.65}$$

In fact, $\tilde{k}$ is an indicator of energy. If $\Lambda = 0$, we see from (9.63) that two galaxies reach infinite separation with finite or zero relative velocity according as $\tilde{k} = -1$ or 0; if $\tilde{k} = +1$, that separation cannot become infinite, and the universe falls back on itself. Thus, since the Λ term must be negligible locally, $\tilde{k} = \pm 1$ distinguishes between galaxies having locally less or more than the "escape" velocity.

We shall presently derive an equation formally identical with (9.63) on the basis of GR; accordingly we postpone the further discussion of the solutions of this equation until then.

9.9 Cosmic Dynamics According to General Relativity

The dynamics of general relativity are expressed in Einstein's field equations, which for cosmological purposes we shall take in their general form (**8.139**), i.e., *with* the so-called cosmological term $\Lambda g^{\mu\nu}$. Various arguments have at times been given *against* the inclusion of this term: (i) that it was only an afterthought of Einstein's (but: better discovered late than never); (ii) that Einstein himself eventually rejected it (but: authority is no substitute for scientific argument); (iii) that with it the well-established theory of SR is not a special case of GR (but: locally the Λ term is totally unobservable); (iv) that it is *ad hoc* (but: from the formal point of view it belongs to the field equations, much as an additive constant belongs to an indefinite integral); (v) that similar modifications could be made to Poisson's equation in Newton's theory and Maxwell's equations in electrodynamics (but: locally the Λ term is ignored, and cosmologically Poisson's and Maxwell's equations may well need similar modification); (vi) that it represents a space expansion *uncaused* by matter, i.e., a field which acts but cannot be acted on (but: in GR, matter and space are intimately related by the field equations, and no mechanical picture is correct); (vii) that one should never envisage a more complicated law until a simpler one proves untenable (but: in cosmology—especially in the RW case—the technical complication is slight, and several recent investigations have suggested that the Λ term indeed may be needed to account for the observations); (viii) (and more technically) that a Λ term in the geometry would destroy the possibility of quantizing gravity [but: Zeldovich has suggested that an energy tensor $(\Lambda/\kappa)g_{\mu\nu}$ may arise naturally out of quantum fluctuations *in vacuo*, so that the Λ term could be regarded as part of the sources rather than part of the geometry.] In this connection we may also note Eddington's mystical argument *for* the Λ term: "An electron could never decide how large it ought to be unless there exsited some length independent of itself for it to compare itself with [namely, $\Lambda^{-1/2}$]."[13]

These general field equations, then, must be satisfied jointly by the metric of spacetime and by the energy tensor—relative to this metric—of the contents of spacetime. In cosmology we are fortunate to be able to restrict the metric considerably by symmetry arguments alone. In fact, as we have seen, the RW metric applicable to all homogeneous and isotropic model universes contains but *two* free elements: the expansion function (or "radius of the universe") $R(t)$, and the curvature index k. The field equations will be seen to impose restrictions on these two elements.

As we explained in the last section, it is usual to treat the cosmic matter as "dust." The energy tensor of dust is given by (**8.128**), i.e., $T^{\mu\nu} = \rho_0 U^\mu U^\nu$, where ρ_0 is the proper density and U^μ is the four-velocity $dx^\mu/d\tau$ of the dust. In the RW model the dust is at rest everywhere with respect to the local fundamental observer, and so we write ρ for ρ_0; U^μ is the velocity of the

---

[13] *Loc. cit.* (*Mathematical Theory*), page 154.

substratum, whence $U^\mu = (0, 0, 0, 1)$ if we take

$$(x^1, x^2, x^3, x^4) = (x, y, z, t), \tag{9.66}$$

and use the form (9.27) of the RW metric in Cartesian coordinates, i.e., with $dx^2 + dy^2 + dz^2$ in the denominator of { }. Consequently the only non-vanishing component of $T^{\mu\nu}$ is $T^{44} = \rho$, which implies

$$T_{\mu\nu} = \text{diag}(0, 0, 0, c^4\rho). \tag{9.67}$$

The main labor in applying the field equations (8.139), which we shall use in their "covariant" form (i.e., with subscripts rather than superscripts), lies in the computation of the (modified) "Einstein tensor" components

$$G_{\mu\nu} = R_{\mu\nu} - \tfrac{1}{2}Rg_{\mu\nu} + \Lambda g_{\mu\nu}.^{14} \tag{9.68}$$

However, we need merely refer to the values listed in Appendix I. For the metric (9.27) (in Cartesian form) we thus find

$$\frac{G_{11}}{g_{11}} = \frac{G_{22}}{g_{22}} = \frac{G_{33}}{g_{33}} = -\frac{2\ddot{R}}{Rc^2} - \frac{\dot{R}^2}{R^2c^2} - \frac{k}{R^2} + \Lambda, \tag{9.69}$$

$$\frac{G_{44}}{g_{44}} = -\frac{3\dot{R}^2}{R^2c^2} - \frac{3k}{R^2} + \Lambda, \tag{9.70}$$

and $G_{\mu\nu} = 0$ when $\mu \neq \nu$. Substituting (9.69), (9.70), and (9.67) into the field equations

$$G_{\mu\nu} = -\frac{8\pi G}{c^4} T_{\mu\nu}, \tag{9.71}$$

we obtain the following two conditions:

$$\frac{2\ddot{R}}{Rc^2} + \frac{\dot{R}^2}{R^2c^2} + \frac{k}{R^2} - \Lambda = 0, \left(= -\frac{8\pi Gp}{c^4} \right), \tag{9.72}$$

$$\frac{\dot{R}^2}{R^2c^2} + \frac{k}{R^2} - \frac{\Lambda}{3} = \frac{8\pi G\rho}{3c^2}. \tag{9.73}$$

In (9.72) we have parenthetically exhibited—without proof—a pressure term on the right-hand side instead of zero, to indicate the only modification that Equations (9.72) and (9.73) would suffer if a pressure p were present.

For later reference, we here record the equation obtained by subtracting (9.73) from (9.72):

$$\frac{\ddot{R}}{R} = \frac{\Lambda c^2}{3} - \frac{4\pi G\rho}{3} \left(-\frac{4\pi Gp}{c^2} \right). \tag{9.74}$$

Multiplying the left-hand side of (9.73) by R^3 and differentiating, we get

---

[14] The R in this formula must not be confused with the expansion factor R. Unfortunately the traditional notations clash here.

$\dot{R}R^2$ times the left-hand side of Equation (9.72), and thus zero. Consequently

$$\tfrac{8}{3}\pi G\rho R^3 = C = \text{constant} \tag{9.75}$$

[cf. Equation (9.64)]. This evidently expresses the constancy of the mass contained in a *small* comoving sphere of radius hR, say. (For large spheres, volumes may be non-Euclidean, frames noninertial, and kinetic energy contributing to mass, all of which complicates the interpretation.) Hence (9.75) is the relativistic equation of continuity, and it should be noted how in GR it is a corollary of the field equations, rather than a separate assumption as in Newton's theory. Substituting (9.75) into (9.73) yields an equation formally identical with the pseudo-Newtonian Equation (9.63):

$$\dot{R}^2 = \frac{C}{R} + \frac{\Lambda c^2 R^2}{3} - kc^2. \tag{9.76}$$

Again, it should be noted how this "equation of motion" for the cosmos results from the field equations of GR, without the need to assume equations of motion as in the pseudo-Newtonian theory [cf. after (9.58)].

Equations (9.72) and (9.73), then, imply (9.75) and (9.76). Conversely, (9.75) and (9.76) imply (9.73), obviously, and also (9.72) *unless* $\dot{R} = 0$ [for the result of differentiating (9.76) is equivalent to $\dot{R}$ times (9.72)]. Equation (9.76) is known as *Friedmann's differential equation*, after its discoverer. Together with (9.75) it essentially represents the GR restrictions on the "dust-filled" RW model. We shall discuss its solutions in the next section. But first we comment on its similarity to the pseudo-Newtonian Equation (9.63).

In GR the curvature index k replaces the Newtonian "energy index" $\tilde{k}$ in Friedmann's equation. In all Newtonian models we assume *flat* public space, i.e., $k = 0$. Hence only those GR models which have $k = 0$ can have exact Newtonian analogs geometrically; if we then choose $\tilde{k} = 0$ to match the dynamics, we have *identical* models. GR models with $k = \pm 1$, though locally similar to their Newtonian counterparts with $\tilde{k} = \pm 1$, i.e., having the same functional form of $R(t)$, have *curved* public space. Our comparison shows that an excess of the local kinetic energy over the escape energy produces negative curvature in GR, whereas a deficit produces positive curvature. Given Hubble's constant, the escape energy is clearly proportional to the cosmic density; thus a high density will lead to a closed universe, a low density to an open one.

Another interesting result of the Newtonian analogy is the *interpretation* of the relativistic effect of a possible pressure. Equation (9.74) without the pressure term is equivalent to (9.62), in which the right-hand side is essentially the *force*, producing the *acceleration* on the left. The two force terms on the right of (9.62) have already been interpreted. The presence of the relativistic pressure term would correspond to a *third* force term

$$-\frac{4\pi G p R}{c^2} \tag{9.77}$$

in (9.62), acting in the same direction as gravity. Thus pressure *slows* the expansion, and moreover, it slows it in two ways. First, because it increases the density ρ (namely, by the mass equivalent of the compressive energy which it generates); this effect is implicitly contained in the ρ term of (9.74). The *explicit* pressure term is hard to interpret intuitively. Working back from (9.62)(i) *with* the pressure term, we find that it results from a modified Poisson equation

$$\nabla^2 \varphi + c^2 \Lambda = 4\pi G\left(\rho + \frac{3p}{c^2}\right). \tag{9.78}$$

But why does pressure *qua* pressure have no expansive effect? The explanation is that a uniform pressure never causes motion; only a pressure *gradient* does—at least in a given inertial frame. Still, why cannot pressure push the inertial frames apart, just as gravity pulls them together? The answer, of course, is that gravity is unique among forces in this property.

Our method of getting the equation of continuity (9.75) in the absence of pressure, yields the equation

$$\frac{d}{dt}\left(\frac{8\pi G\rho R^3}{3c^2}\right) = -\frac{8\pi Gp\dot{R}R^2}{c^4} \tag{9.79}$$

in the presence of pressure. Letting V stand for the volume $\frac{4}{3}\pi(hR)^3$ of a small comoving sphere, we can write this last equation in the form

$$d(c^2\rho V) = -p\,dV = -pS\,d\sigma, \tag{9.80}$$

where S is the surface area and σ the radius of the sphere. The interpretation is again one of "continuity": as the small sphere expands, the pressure inside does work $pS\,d\sigma$ on the matter outside; hence a corresponding amount of energy must be lost inside (recall that $c^2\rho$ stands for the *total* energy density). In pseudo-Newtonian theory, Poisson's equation and the equation of continuity are independent; so we cannot utilize (9.80)—which it would be natural to assume—in order to justify the modified Poisson equation (9.78) other than by analogy with GR.

Before we proceed, it will be well to make a numerical spot check on our equations, just to see whether we are on the right track. If, for simplicity, we assume $k = \Lambda = 0$, we can write Equation (9.73), by reference to (9.25), in the form

$$H^2 = \tfrac{8}{3}\pi G\rho.$$

Taking for H its presently accepted value of 50 (km/sec)/Mpc, we find $\rho = 4 \times 10^{-30}$ gm/cm^3. This is about ten times larger than we would like, but the difference can be made up by various reasonable choices of k, Λ, and R. Thus our dynamics would appear to be viable.

9.10 The Friedmann Models

We shall now discuss the solutions of Friedmann's differential equation (**9.76**), with a view to obtaining and classifying *all* GR "dust" universes that are isotropic and homogeneous. These are the "Friedmann models." It will be convenient to employ units in terms of which $c = 1$, so that Friedmann's equation reads

$$\dot{R}^2 = \frac{C}{R} + \frac{\Lambda R^2}{3} - k = :F(R), \quad (C = \tfrac{8}{3}\pi G \rho R^3). \tag{9.81}$$

The symbol $F(R)$ is simply an abbreviation for the three terms preceding it; the parenthesis is a repeat of Equation (**9.75**), whose sole function is to determine ρ once R is found from the differential equation. We can *formally* write down the solution at once by quadrature,

$$t = \int \frac{dR}{\sqrt{F}}, \tag{9.82}$$

and we could proceed to the full solution by using elliptic functions. In special cases the solution can be obtained in terms of elementary functions, as we shall see. However, in the general case it will be enough for us, and more instructive, to give a qualitative rather than an exact analysis. We preface our discussion with three general remarks:

(i) A Friedmann model is uniquely determined by a choice of the parameters C, Λ, k, an "initial" value $R(t_0)$, and the sign of $\dot{R}(t_0)$. For Equation (**9.81**) then gives $\dot{R}(t_0)$ and thus, in principle, the solution can be iterated uniquely—unless $\dot{R}(t_0) = 0$. In that case, however, we can fall back on Equation (**9.72**) to get $\ddot{R}(t_0)$, and then again the solution can be iterated uniquely.

(ii) Since $R = 0$ is a singularity of the Friedmann equation, no regular solution $R = R(t)$ can pass *through* $R = 0$. Regular solutions are therefore entirely positive or entirely negative. Moreover, the solutions occur in matching pairs $\pm R(t)$: this is because, for physical reasons, we must insist on $\rho \geq 0$, which implies that the sign of C must be chosen to be the same as that of R—but then Equation (**9.81**) is insensitive to the change $R \to -R$, and this proves our assertion. Since only R^2 occurs in the RW metric, we therefore exclude no solutions by insisting, as we shall, that $C \geq 0$ and $R \geq 0$.

(iii) Equation (**9.81**) also enjoys invariance under either of the changes $t \to -t$ or $t \to t + $ constant. The first implies that to every solution $R(t)$ there corresponds a "time-reversed" solution $R(-t)$ (e.g., to every expanding universe there corresponds a collapsing one); and the second implies that each solution $R(t)$ represents a whole set of solutions $R(t + $ constant), which simply expresses the "homogeneity of time," i.e., that the zero-point of time is physically irrelevant.

Bearing these properties in mind, we shall so normalize our solutions that of the pair $R(\pm t)$ we exhibit the expanding one in preference to the collapsing one, and of the set $R(t + \text{constant})$ we exhibit, if possible, that member which has $R = 0$ at $t = 0$.

As a consequence of time-reversibility and (i) above, we note that every *oscillating* Friedmann model (i.e., one having a past *and* future big bang) is time-symmetric. For, given R_{max} and C, Λ, k, the model is unique and time-reversible about R_{max}.

The Static Models

It is well to clear out of the way the static models first, i.e., those which have $\dot{R} \equiv 0$. As we remarked after (9.76), this is the exceptional case in which Friedmann's equation is insufficient and both its parent equations (9.72) and (9.73) must be satisfied separately. These equations permit $\dot{R} \equiv 0$ provided

$$\frac{k}{R^2} = \Lambda = 4\pi G\rho, \quad (\sim 10^{-58} \text{ cm}^{-2}). \tag{9.83}$$

(In parenthesis we give the value of $4\pi G\rho$ corresponding to $\rho = 10^{-31}$ gm/cm^3, which illustrates the typical smallness of Λ.) Equations (9.83) of course imply $\rho = \text{constant}$, and for a physically meaningful solution we need $\rho > 0$, and thus $k = +1$. This gives the so-called *Einstein universe*, the very first GR model to be proposed (by Einstein, in 1917). Note how the spatial part of the model is a 3-sphere as discussed in Section 7.2. To its inventor at that time it had every desirable feature. As a realistic model, however, it had to be abandoned with the discovery of the expansion of the universe.

The only other possibility of satisfying (9.83)—less physical but still acceptable as a limiting case—is $k = \Lambda = \rho = 0$, $R = $ any constant. The transformation $Rr \to r$ in the RW metric then leads to a standard metric for Minkowski space, and the model consequently represents static "test" dust ($\rho = 0$) homogeneously filling an infinite inertial frame.

The Empty Models

Models with zero density, like Milne's or the above, are unrealistic, but they provide instructive and transparent examples of various geometric and kinematic possibilities. We therefore classify them next. Setting $C = 0$ reduces (9.82) to the elementary form

$$t = \int (\tfrac{1}{3}\Lambda R^2 - k)^{-1/2} dR, \tag{9.84}$$

which has the following solutions [apart from (a)—which we obtained above]:

(a) $\Lambda = 0, k = 0,$ $R =$ arbitrary constant
(b) $\Lambda = 0, k = -1, R = t$
(c) $\Lambda > 0, k = 0,$ $R = \exp(t/a)$
(d) $\Lambda > 0, k = 1,$ $R = a \cosh(t/a)$
(e) $\Lambda > 0, k = -1, R = a \sinh(t/a)$
(f) $\Lambda < 0, k = -1, R = a \sin(t/a)$

$$a = |3/\Lambda|^{1/2}$$

Models (a) and (b) we already know: (a) is the empty static model and (b) is Milne's model; these have identical spacetime backgrounds (M_4) but different substrata (i.e., motion patterns). The same is true also of the three models (c), (d), and (e): these all have the de Sitter space S_4^- (see Section 8.11) for their spacetime background.[15] [Such a situation can *only* arise with empty models; nonempty models with different substrata have different spacetimes.] That models (c), (d), and (e) must share de Sitter spacetime is *a priori* clear, since, as we saw in Section 8.11, S_4^- is the *unique* empty spacetime which satisfies Einstein's modified field equations with $\Lambda > 0$ and which is spatially isotropic about every point; and each of the three models in question has these properties. Of course, their substrata differ. Figure 9.6 illustrates this. Its first three diagrams show the hyperboloid of Figure 8.14, which represents de Sitter spacetime with two spatial dimensions suppressed. As we mentioned at the time, geodesic worldlines correspond to timelike plane sections of the hyperboloid through the center. In the notation of Figure 8.14, it can be shown that the substrata of (c), (d), and (e) correspond, respectively, to plane sections containing the line $T + W = 0$, $Z = 0$; the T axis; and the W axis. Lorentz-type transformations of the embedding (flat) spacetime can carry each of these fundamental-particle worldlines into the "central" one in models (c) and (e), which shows the equivalence of all these lines.

Model (c) is the well-known *de Sitter universe*. Kinematically it is identical with the steady state universe (9.28). Thus, for example, it has an event horizon. Though unrealistic in GR because of its emptiness, it constitutes a kind of limit to which *all* indefinitely expanding models with $\Lambda > 0$ must tend. For indefinite expansion ($R \to \infty$) in a general model causes the Λ term ultimately to predominate on the right-hand side of Equation (9.81), which implies $R \sim \exp(\tfrac{1}{3}\Lambda)^{1/2}t$ (a multiplicative constant can be absorbed by a translation of t); the curvature k/R^2 of the model and its density $3C/8\pi G R^3$ become ultimately small, and thus even $k = 0$ and $\rho = 0$ provide a good approximation *locally*. This proves our assertion. For example, *all* indefinitely expanding models with $\Lambda > 0$ consequently possess an event horizon. The horizon associated with the "central" fundamental observer in Figure 9.6 (c) is shown as a pair of stippled lines. In (d) and (e) the event horizon can be similarly represented by a pair of parallel generators. Its significance as

[15] Since de Sitter spacetime allows a static metric [namely, (8.155)] up to the horizon, these models are sometimes confusingly referred to as "static" models.

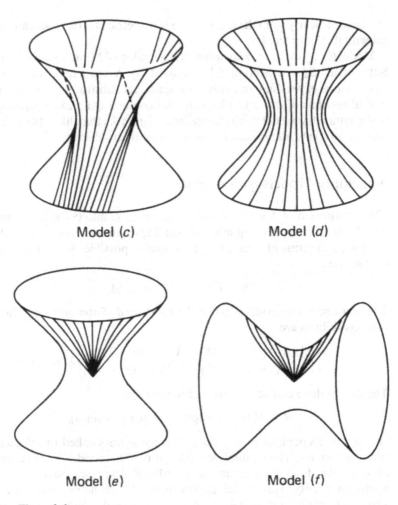

Model (c)   Model (d)

Model (e)   Model (f)

Figure 9.6

the light front which reaches the associated fundamental observer in the infinite future is well brought by the diagram. It is easily seen that the "de Sitter singularity" discussed in Section 8.11 is nothing but the event horizon of the observer at the origin.

Model (d) is the analog of the static model (a) in S_4^- —but it is "static" for only an instant, corresponding to the waist circle of the hyperboloid. Its substratum fills all of S_4^-. Its fundamental particles collapse to a state of minimum approach and then expand back to infinite separation under Λ repulsion.

Model (e) is the analog of Milne's model (b) in S_4^-; within S_4^- it consists of an expanding ball of test dust, bounded by a spherical light front. Its expansion, however, is speeded by the Λ repulsion. The analogy with Milne's model can be seen clearly from the fact that the RW metric of (e) is practically identical with (**9.13**) for small t.

Model (f), by the same reasoning, is the analog of Milne's model in *anti*-de Sitter space S_4^+ (cf. Section 8.11); slowed by Λ *attraction*, this dust ball finally stops expanding and collapses again. Its substratum corresponds to central sections of the (anti-) hyperboloid containing an axis perpendicular to the symmetry axis. It is easy to see from Figure 9.6 that, like Milne's model, (c), (e), and (f) are all *incomplete* ("more space than substratum").

Nonempty Models with $\Lambda = 0$

The substitution of $\Lambda = 0$ and a finite value for C into (**9.81**) leaves us with essentially three types of equation, depending on the choice of k. In each case a solution in terms of elementary functions is possible. When $k = 0$ this is very simple:

$$R = (\tfrac{9}{4}C)^{1/3}t^{2/3}, \quad (k = 0).$$

The corresponding model is called the *Einstein–de Sitter universe*. The other two possibilities are

$$\left.\begin{array}{l} t = C[\sin^{-1}\sqrt{X} - \sqrt{(X - X^2)}], \quad (k = 1) \\ t = C[\sqrt{(X + X^2)} - \sinh^{-1}\sqrt{X}], \quad (k = -1) \end{array}\right\} X = R/C.$$

The first of these can be reexpressed in the form

$$R = \tfrac{1}{2}C(1 - \cos \psi), \quad t = \tfrac{1}{2}C(\psi - \sin \psi),$$

and is thus recognized as a cycloid; it is sometimes called the *Friedmann–Einstein universe*. The qualitative behavior of the second is most easily read off from the differential equation directly: $\dot{R}$ decreases from infinity and approaches unity. This model approximates the Milne universe for large t. The graphs of all three models are shown in Figure 9.7. Note that for small R the C term *always* dominates the right-hand side of (**9.81**), whence *all* nonempty big-bang models share the behavior $R \sim (\tfrac{9}{4}C)^{1/3}t^{2/3}$ near $t = 0$. In

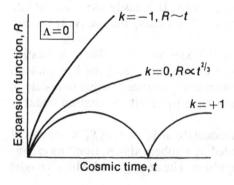

Figure 9.7

particular, therefore, it can be seen that they all have a particle horizon. Moreover, because of the time-symmetry of oscillatory models, and because a particle horizon becomes an event horizon under time reversal, it is seen that all nonempty oscillatory models have *both* a particle and an event horizon.

Nonempty Models with $\Lambda \neq 0$

If we allow arbitrary values of Λ, the variety of possible models increases substantially. We again assume $C \neq 0$. A qualitative solution of Friedmann's differential equation can now be obtained by plotting the level curves

$$f(R, \Lambda) = m = \text{constant} \tag{9.85}$$

of the function [cf. (9.81)]

$$f(R, \Lambda) = \frac{C}{R} + \frac{\Lambda R^2}{3} = \dot{R}^2 + k \tag{9.86}$$

for some fixed value of $C > 0$, in an (R, Λ) diagram, as in Figure 9.8. In other words, we plot Λ against R, where

$$\Lambda = \left(m - \frac{C}{R} \right) \frac{3}{R^2} = \frac{3(mR - C)}{R^3}, \tag{9.87}$$

giving m a succession of values, e.g., $-1, 0, 1/2, 1, 3/2, \ldots$. These curves have one of two characteristic shapes, according as $m \leq 0$ or $m > 0$. In the former case they lie entirely below the R axis, but approach that axis monotonically

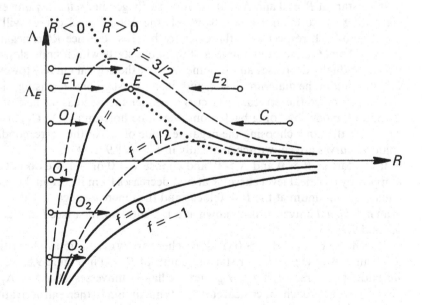

Figure 9.8

from below. In the latter case they cross the R axis (at $R = C/m$), proceed to a maximum, and then approach the R axis asymptotically from above. In all cases $\Lambda \sim -3C/R^3$ near $R = 0$.

The level curves $f = -1, 0, 1$ are of particular importance: they correspond to the locus of $\dot{R}^2 = 0$ in the cases $k = -1, 0, 1$, respectively.

From (9.86), the locus of $\ddot{R} = 0$ (marked in the diagram by a stippled line) is given by

$$2\ddot{R} = \frac{\partial f}{\partial R} = -\frac{C}{R^2} + \frac{2\Lambda R}{3} = 0, \quad \text{i.e.,} \quad \Lambda = \frac{3C}{2R^3}. \tag{9.88}$$

It coincides with the line of maxima of the level curves, since, for them, $d\Lambda/dR = -(\partial f/\partial R)/(\partial f/\partial \Lambda) = 0$. Above this locus, $\ddot{R} > 0$, below it, $\ddot{R} < 0$.

The maximum point of the level curve $f = 1$ (marked E in the diagram) has "coordinates"

$$R = \frac{3C}{2}, \quad \Lambda = \frac{4}{9C^2} = :\Lambda_E, \tag{9.89}$$

i.e., precisely the values corresponding to Einstein's static universe [cf. (9.83) and (9.81)(ii)]—hence the notation Λ_E. It follows that at E *all* derivatives of $R(t)$ vanish.

Now each Friedmann model is characterized by $\Lambda = $ constant. We can therefore obtain solutions by choosing any physically possible starting point ($\dot{R}^2 > 0$!) in Figure 9.8, and proceeding horizontally. Because of Friedmann's differential equation and the definition of f, the values on the level curves tell us $\dot{R}^2$ (once we have chosen k) and thus the slope of the solution curve, up to sign.

If we start at $R = 0$ and $\Lambda < 0$, we necessarily get an *oscillating* universe. According as we choose $k = 1, 0,$ or -1, the critical level curves will be $f = 1, 0,$ or -1, respectively; these cannot be crossed, since $\dot{R}^2$ is negative beyond them. Our solution curve $R(t)$ therefore starts with infinite slope $\dot{R}$, which gradually decreases and becomes zero at the critical curve. However, at that point in the diagram, $\ddot{R} < 0$, and thus $\dot{R}$ must go on decreasing; since we cannot cross the critical level curve, we get the decreasing half of an oscillating model by going back along the same horizontal ($O_1, O_2,$ or O_3) to $R = 0$, this time choosing the negative value of $\dot{R}$. All the corresponding solution curves are shown schematically in Figure 9.9 as O.

If we start at $R = 0$ and $\Lambda > 0$, and choose $k = 0$ or -1, we do not get stopped by a critical level curve; $\dot{R}$ at first decreases from its original infinite value to a minimum at the $\ddot{R} = 0$ locus and then increases again. The result is an *inflexional* universe, also known as a Lemaître universe (see I in Figures 9.8 and 9.9).

The choice $k = 1$ and $\Lambda > 0$ yields a richer variety of solutions. The critical level curve is now $f = 1$. For starting points at $R = 0$ in Figure 9.8, we get inflexional universes (I) if $\Lambda > \Lambda_E$, and oscillatory universes (O) if $\Lambda < \Lambda_E$. If $\Lambda = \Lambda_E$, we approach the critical curve at its maximum E: there, all derivatives of R vanish, and the solution curve flattens out into a straight line. This

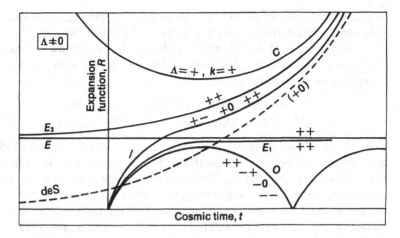

Figure 9.9

corresponds to a big-bang universe which approaches the Einstein static universe asymptotically (E_1 in Figures 9.8 and 9.9). Note that, by choosing $\Lambda > \Lambda_E$ but *close* to Λ_E, we can construct inflexional models with arbitrarily long quasistationary periods, as in Figure 9.4. If we start on the other side of the "hump" of the critical curve (i.e., with large R) and proceed horizontally to the left in Figure 9.8, $\dot{R}$ decreases from large values to zero (and $\ddot{R} > 0$), and we get one half of a *catenary* universe (C); the other half corresponds to going back along the same horizontal with the opposite sign of $\dot{R}$. Like oscillatory universes, catenary universes possess time symmetry about their stationary point, and for the same reason. Observe that for identical C, Λ (and $k = 1$) we can get two different types of model, depending on the initial value of R. Similarly, if for $\Lambda = \Lambda_E$ we approach the critical point E from the right, we get a universe which decreases from infinite extension and approaches the Einstein static universe asymptotically; we prefer to exhibit this model in time-reversed (expanding!) form: E_2 in Figures 9.8 and 9.9. Thus, for $\Lambda = \Lambda_E$ (and $k = 1$) there are *three* different models, including the Einstein static universe.

Figure 9.9 shows the various solution curves schematically; they are labeled according to the signs of Λ and k, in this order. (The dashed curve, contrary to our premise $C \neq 0$, represents the empty de Sitter universe; it is included here to show its role as an "asymptote" to all indefinitely expanding models with $\Lambda \neq 0$.) It can be seen from both Figures 9.8 and 9.9 that the Einstein universe is unstable: a slight perturbation will set it on the course E_1 (in collapsing form), or on the course E_2 (expanding). The physical reason for this instability is clear: a slight contraction causes the density ρ to increase, whereas the "permanent" density $-\Lambda/4\pi G$ stays the same, which ends their balance and results in further contraction. The opposite perturbation for the same reason leads to further expansion. In fact, the expanding universe E_2

is sometimes regarded as the result of a disturbed Einstein universe, and on this interpretation it is called the *Eddington–Lemaître universe*.

As we have seen, all models with $C = 0$ or with $\Lambda = 0$ allow representation by elementary functions. The same is true of models with $k = 0$, and for completeness we give the results:

$$R^3 = (3C/2\Lambda)[\cosh(3\Lambda)^{1/2}t - 1], \quad (\Lambda > 0),$$
$$R^3 = (3C/-2\Lambda)[1 - \cos(-3\Lambda)^{1/2}t], \quad (\Lambda < 0).$$

Finally we make a remark about the initial singularity $R = 0$ of the big-bang models. At one time it was thought to be merely a result of the rather artificial assumption that the fundamental particles move *exactly* radially. If, instead, they moved along slightly skew lines, they could perhaps—as we trace the model back in time—miss each other and rebound. But it has been shown more recently (Penrose, Hawking, Ellis) that, even under very general and reasonable conditions, cosmological models must have one or more singularities in the past. However, the possibility remains that the singularities can be avoided by some or even most of the matter, thus making a kind of rebound possible.

9.11 Once Again, Comparison with Observation

Now that we have subjected the RW metric to dynamical conditions, we still find ourselves left with an embarrassingly large choice of possibilities. But whereas observations have little impact on the unrestricted RW models, they can, in principle, decide between the dynamically possible ones. It is convenient to work with the following three functions of cosmic time:

$$H = \dot{R}/R, \quad q = -\ddot{R}/RH^2 = -R\ddot{R}/\dot{R}^2, \quad \sigma = 4\pi G\rho/3H^2, \quad (9.90)$$

where q and σ are dimensionless, while H has the dimension (time)$^{-1}$. H, of course, is Hubble's "constant," or better: *Hubble's parameter*; q is called the *deceleration parameter*, and σ—which is necessarily positive—is called the *density parameter*. In principle, the present values of these parameters can be determined by observation: H and q from relations such as (9.50), and σ from an estimate of ρ.

In terms of these functions we can now rewrite (i) Equation (9.75), (ii) Equation (9.74), and (iii) the difference of (9.72) and three times (9.73) (we still work in units in which $c = 1$):

(i) $\quad \sigma = C/2H^2R^3;$ $\qquad\qquad C = 2\sigma_0 H_0^2 R_0^3$ $\qquad$ (9.91)

(ii) $\quad \sigma - q = \Lambda/3H^2;$ $\qquad\qquad \Lambda = 3H_0^2(\sigma_0 - q_0)$ $\qquad$ (9.92)

(iii) $\quad 3\sigma - q - 1 = k/H^2R^2;$ $\qquad k = H_0^2 R_0^2(3\sigma_0 - q_0 - 1).$ $\quad$ (9.93)

The second entry in each line represents a solution for one of the constants C, Λ, k; and since these *are* constants, we can evaluate their representative functions at *any* time, in particular at the present time $t = t_0$: that is the significance of the suffix zero. Note that

$$\Lambda = 0 \leftrightarrow \sigma = q, \tag{9.94}$$

and *if* this obtains,

$$k \gtreqqless 0 \leftrightarrow \sigma, q \gtreqqless \tfrac{1}{2}, \quad (\Lambda = 0). \tag{9.95}$$

From the second entries in **(9.91)**–**(9.93)** it appears that if we know H_0, q_0, and σ_0, we can first determine Λ and k/R_0^2 (which yields k *and* R_0 unless $k = 0$, in which case R_0 is arbitrary anyway) and finally C. And, of course, Λ, k, C and R_0 determine a *unique* Friedmann model. But unfortunately this persuasively simple plan does not work out in practice: the uncertainties in the current determinations of q_0 and σ_0, and, to a lesser extent, of H_0, are so great that no direct conclusions are possible from these equations. H. P. Robertson had the idea of coupling these three uncertain data with a fourth: t_0, the age of the universe. Some models can then be eliminated simply because they are too young or too old.

We first remark on a rather specialized "age problem" that arises with high estimates for H_0. Consider a graph of $R(t)$ against t, as in Figure 9.4. The tangent at t_0 intersects the t axis at a point $R_0/\dot{R}_0 = 1/H_0$ units to the left of t_0 (we assume $\dot{R}_0 > 0$). If $\ddot{R}$ has been negative *until* t_0, the curve lies to the right of its tangent, and the model has a present age *less* than its "Hubble age" $1/H_0$. If—as was believed only a few years ago—H_0 were as big as 100 (km/sec)/Mpc, the Hubble age would be 9.7×10^9 years, and this would conflict with the estimated age of globular clusters, 14×10^9 years. Of course, the conflict arises only *if* $\ddot{R}$ was negative until now, which is necessarily the case if one *assumes* $\Lambda = 0$ [cf. Figure 9.7 or Equation **(9.74)**]. But the conflict could be avoided then by allowing positive values of Λ (cf. Figure 9.9), and with today's lower estimates for H_0 it does not arise even if $\Lambda = 0$.

Returning now to the general case, and essentially following Robertson, we substitute from **(9.91)**–**(9.93)** into Friedmann's Equation **(9.81)**, obtaining

$$\dot{y}^2 = H_0^2\{2\sigma_0 y^{-1} + (\sigma_0 - q_0)y^2 + 1 + q_0 - 3\sigma_0\}, \quad y = R/R_0. \tag{9.96}$$

Like the original equation, this can be solved at once by quadrature; and since we are interested here in *present* ages of models with *finite* age, we assume $R(0) = 0$ and perform a definite integration:

$$H_0 t_0 = \int_0^{t_0} H_0\, dt = \int_0^1 \{\ \}^{-1/2} dy = f(\sigma_0, q_0); \tag{9.97}$$

the empty brace denotes the braced expression of **(9.96)**, and $f(\sigma_0, q_0)$ is defined by the last equation. This function can easily be machine evaluated, and thus one can tabulate corresponding values of σ_0, q_0, and $H_0 t_0$.

As far as **(9.96)** goes, σ_0 and q_0 determine $R(t)$ up to independent scale changes in R and t, since it can be written $\dot{R}^2/\dot{R}_0^2 = \{\ \}$. But Equation **(9.93)** determines $\dot{R}_0$ uniquely (unless $k = 0$), and so the scale changes in R and t must be the *same* (unless $k = 0$). Therefore, since σ_0, q_0, and $H_0 t_0$ are related by (9.97), any two of them serve to determine a big-bang Friedmann model up to the above scale change. (For certain σ_0, q_0 there are *two* Friedmann models—one with finite and one with infinite past, like O and C in Figures 9.8 and 9.9. The present analysis omits the latter.) Following Tinsley, we choose $H_0 t_0$ and log σ_0 as "coordinates" for the Robertson diagram,[16] Figure 9.10, on which each point corresponds to a Friedmann model. The limitations on ρ_0 and t_0 then single out a zone in which all acceptable models must occur; and these limitations appear at present to be more firmly established (cf. Section 9.1) than those on q_0. Nevertheless, we have superimposed on the Robertson diagram some level curves of q_0[17] thus adding the fourth "observable" to σ_0, t_0, and H_0.

The diagram shows the important demarkation lines between models with $\Lambda > 0$ and $\Lambda < 0$, and between models with $k > 0$ and $k < 0$; they are easily obtained from (9.92) and (9.93), respectively. Also strictly separated in the diagram are oscillating and nonoscillating models: As is clear from Figure 9.8, the relevant demarkation line, on or above which models do not oscillate, is $\Lambda = 0$ for models with $k \leq 0$ and $\Lambda = \Lambda_E$ for models with $k > 0$. The equation of the former curve is

$$\sigma_0 - q_0 = 0, \tag{9.98}$$

while that of the latter is

$$27(\sigma_0 - q_0)\sigma_0^2 = (3\sigma_0 - q_0 - 1)^3, \tag{9.99}$$

as we find from (9.91)–(9.93) by setting $\Lambda = \Lambda_E = \frac{4}{9}C^{-2}$ and $k = 1$. These two curves, as well as the curve

$$3\sigma_0 - q_0 - 1 = 0, \tag{9.100}$$

corresponding to $k = 0$, all cross each other at the point $\sigma_0 = q_0 = \frac{1}{2}$. It is therefore seen that in the Robertson diagram all points below the line compounded of the locus $\Lambda = 0$ for $\sigma_0 < \frac{1}{2}$ and the locus (9.99) for $\sigma_0 > \frac{1}{2}$ (dotted in the diagram) correspond to oscillating models, whereas all points on or above this line correspond to nonoscillating models. [For each oscillating model there are, of course, *two* cosmic times (symmetric with respect to the time of maximum extension) that correspond to any pair of values σ_0, q_0. The Robertson diagram gives the earlier time, corresponding to the expansive period; for it is based on the *positive* root in (9.97), i.e., on $H_0 > 0$, the one *certain* empirical fact.]

---

[16] Robertson used t_0 and log ρ_0, assuming a definite value for H_0. But Tinsley's coordinates have the advantage that the diagram need not be recalibrated whenever a new value of H_0 is announced. Furthermore, the empirical density restrictions are on σ_0 rather than on ρ_0: The dynamical mass determinations of clusters of galaxies depend on observing relative motions, and these satisfy $v^2 \propto m/r$; but v is directly observable as red shift, whence $m \propto r$. The density involves a further division by (distance)^3, so that $\rho \propto$ (distance)$^{-2} \propto H^2$—since the uncertainty in Hubble's parameter $\dot{R}/R$ is only in the denominator, the numerator being again observable as a red shift.

[17] Making use of computational results given by R. Stabell and S. Refsdal, *Mon. Not. R. Astron. Soc.* 132, 379 (1966).

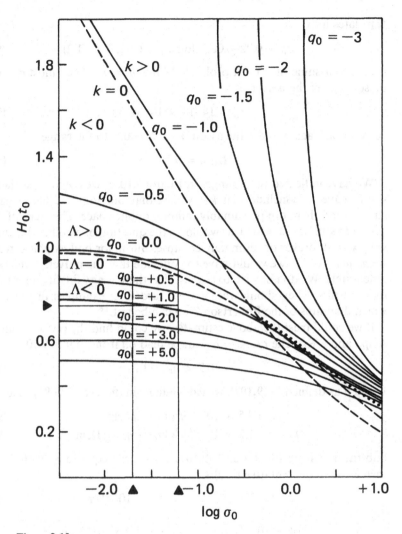

Figure 9.10

There is a convenient numerical identity,

$$\sigma_0 = 2.66 \times 10^{28} \frac{\rho_0}{h^2}, \tag{9.101}$$

where ρ_0 is in gm/cm^3 and

$$h = \frac{H_0}{100} \tag{9.102}$$

is in (km/sec)/Mpc. By use of this, the "most probable" present density range of the universe,

$$\rho_0 = 2\text{–}6 \times 10^{-31} \text{ gm/cm}^3 \tag{9.103}$$

translates into

$$\sigma_0 = 0.02\text{--}0.06, \quad \log \sigma_0 = (-1.7)\text{--}(-1.2), \tag{9.104}$$

if we also assume the "most probable" value $h = 0.5$. The "most probable" present age of the universe,

$$t_0 = 14\text{--}18 \times 10^9 \text{ years} \tag{9.105}$$

corresponds, again under the assumption $h = 0.5$, to the range

$$H_0 t_0 = 0.74\text{--}0.95. \tag{9.106}$$

We have indicated these ranges in Figure 9.10. *If* we could trust them, we would have to conclude that $k < 0$, i.e., that our universe has negatively curved (and therefore presumably infinite) public space. The signs of Λ and q_0 would still be undecided, as would be the question of whether the universe will expand forever or ultimately collapse. But if, for philosophical reasons, we *assume* $\Lambda = 0$, we would have to conclude that the universe will expand indefinitely. We do not wish to suggest that there is any finality about any of these results. They should rather be regarded as examples of the uses to which diagrams of the Robertson kind can be put.[18]

If we could trust the above estimates, we could find the possible ranges of Λ, q_0, and k/R_0^2. The range of q_0 can be read off from Figure 9.10:

$$-0.2 < q_0 < 1.2. \tag{9.107}$$

Using h as defined in (9.102), we calculate from (9.92) and (9.93) that

$$\Lambda = 3.5 \times 10^{-56} h^2 (\sigma_0 - q_0) \text{cm}^{-2}, \tag{9.108}$$
$$k/R_0^2 = 1.2 \times 10^{-56} h^2 (3\sigma_0 - q_0 - 1) \text{cm}^{-2}. \tag{9.109}$$

Substitution from (9.104) and (9.107) into these equations (with $h = 0.5$) then leads to the tentative results,

$$-10^{-56} \text{cm}^{-2} < \Lambda < 0.2 \times 10^{-56} \text{cm}^{-2} \tag{9.110}$$

$$-6.4 \times 10^{-56} \text{cm}^{-2} < \frac{k}{R_0^2} < -1.9 \times 10^{-56} \text{cm}^{-2}. \tag{9.111}$$

9.12 Mach's Principle Reexamined

Now that we have come to the end of our survey of relativity, it may be fitting to return briefly to Mach's principle, and to see whether this principle, which strongly influenced Einstein in his invention of GR, has in fact been vindicated by GR. It will be recalled that, according to Mach, space as such played no

---

[18] To get an idea of the manifold arguments that have recently been used to narrow down the model, and of the "instability" of these arguments, see, for example, J. R. Gott III, J. E. Gunn, D. N. Schramm, and B. M. Tinsley, *Astrophys. J.* **194**, 543 (1974) and also *Sci. Am.* **234**, 62, March 1976; and J. E. Gunn and B. M. Tinsley, *Nature* **257**, 454 (1975).

role in physics and thus did not exist. In GR, on the other hand, space *does* play a role, though now in the guise of four-dimensional spacetime. Spacetime *fully* determines all free motions (i.e., motions under inertia and gravity); the question is whether the matter distribution, in turn, *fully* determines spacetime. For if it did, spacetime could be regarded as a mere mathematical auxiliary, and, as demanded by Mach, only the relative configuration of masses would count.

Before we enter into that question, however, we may note that GR has at least met what many regard as the strongest of the Mach–Einstein objections to *absolute* space, namely that it acts but cannot be acted upon: spacetime both acts *on* mass (as a guiding field) and is acted upon *by* mass (suffering curvature). And it may be that Mach's followers will have to be content with such a nonabsolute space rather than with no space at all.

For, as things stand, spacetime certainly is *not* a mere auxiliary. For example, according to the field equations, flat Minkowski space is consistent with the total absence of matter, and yet it determines all free motions in it. Furthermore, there are *other* nonsingular solutions beside Minkowski space of the unmodified vacuum field equations (e.g., Taub–NUT space[19] and Ozsváth–Schücking space[20]). Hence the *same* matter configuration (namely, *no* matter) can give rise to inequivalent fields. This is possibly connected with the existence of gravitational waves: Minkowski space with a gravitational wave (curvature!) going through it is *not* the same as Minkowski space; and yet it also satisfies the vacuum field equations.

The total absence of matter, of course, is not a realistic situation. But even when matter is present, we have no guarantee of a unique solution. Einstein's field equations are *differential* equations, and thus their solutions necessarily involve some arbitrariness which can only be removed through further information, e.g., boundary conditions. In the case of Schwarzschild's solution, as we have seen, spherical symmetry suffices to ensure uniqueness. In cosmology, too, if we assume the cosmological principle (and a value of Λ), spacetime is uniquely determined by the density and motion pattern at one cosmic instant. It is not clear, however, that in the case of more general (i.e., less symmetrical) mass distributions, different boundary conditions could not lead to different solutions. In other words, it is conceivable that there could be two *different* fields corresponding to the *same* matter content— e.g., the same local field joined smoothly to two different vacuum continuations to infinity.

There even exist explicit "anti-Mach" solutions of the field equations with matter. By this we understand model universes in which the local "compass of inertia" (e.g., a Foucault pendulum) rotates relative to the global mass distribution. The best-known among these are the Gödel universe[21] (im-

[19] See, for example, C. W. Misner, p. 160 of *Lectures in Applied Mathematics*, Vol. 8 (Relativity Theory and Astrophysics, I), J. Ehlers, editor, *Am. Math. Soc.*, 1967.

[20] I. Ozsváth and E. Schücking, in *Recent Developments in General Relativity*, p. 339, New York, Pergamon Press, 1962.

[21] K. Gödel, *Rev. Mod. Phys. 21*, 447 (1949).

portant also in cosmology as an example of a homogeneous though nonisotropic model) and Kerr's metric (representing a single *rotating* body in an otherwise empty universe).

Mach and Einstein regarded Mach's principle as a selection rule among gravitational theories. A more modest interpretation is to regard it as a selection rule among *solutions* of gravitational theories, to eliminate "un-Machian" solutions. Such selection rules have been developed by Raine and others. They can, of course, be regarded as "boundary conditions" necessary to *complete* a theory based on differential equations. Another approach, due to Sciama, is to express inertia by certain integrals which show—albeit implicitly—how it arises from the sources in certain universes. However, both Sciama's and Raine's methods are difficult to apply except in situations with special symmetries.

On balance, it seems hard to avoid the conclusion that GR has realized only *part* of Mach's program. Instead of abolishing space altogether, Einstein merely made it nonabsolute; and, ironically, instead of explaining inertial forces as gravitational, i.e., as matter pulling on inertial "charge" in the spirit of Mach, Einstein explained gravitational forces as inertial, i.e., as "space-guided."

It must be said, in fairness, that today's quantum theoreticians (as well as many others) have little sympathy with Mach's principle. They point out that not only matter is the stuff of physics, but also fields, and that the whole of spacetime is occupied by the fields of the elementary particles. Even in the absence of matter, the fields of the *virtual* particles constitute an all-pervasive background which can in no way be eliminated. In fact, matter is only a small perturbation of it. This background, which possesses Lorentz invariance locally, can be looked upon as a modern ether. Since it possesses no net energy it makes no contribution to curvature, and hence it has no direct effect in general relativity. But it does suggest the *a priori* existence of spacetime, which matter merely modifies and does not create.

Be this as it may, if Einstein really sought to find a framework for Mach's principle in his general relativity, one would probably have to conclude that he, like Columbus, failed in his purpose. And yet, in the one case as in the other, how rich the actual discovery and how forgotten, by comparison, the original purpose!

Appendices

Appendix I
Curvature Tensor Components for the Diagonal Metric

One of the most tedious calculations in GR is the determination of the Christoffel symbols $\Gamma^\mu_{\nu\sigma}$, the Riemann curvature tensor $R_{\mu\nu\rho\sigma}$, the Ricci tensor $R_{\mu\nu}$, the curvature invariant R, and the Einstein tensor $G_{\mu\nu}$, for a given spacetime metric. Various computational shortcuts exist, but rather than start from scratch each time, it is well to have tables for certain standard forms of the metric. In this appendix we shall deal with the general 4-dimensional *diagonal* metric,

$$ds^2 = A(dx^1)^2 + B(dx^2)^2 + C(dx^3)^2 + D(dx^4)^2, \tag{I.1}$$

where A, B, C, D are arbitrary functions of all the coordinates, and as a byproduct—without extra labor—we shall get the various curvature components also for the 2- and 3-dimensional diagonal metrics

$$ds^2 = A(dx^1)^2 + B(dx^2)^2, \quad ds^2 = A(dx^1)^2 + B(dx^2)^2 + C(dx^3)^2. \tag{I.2}$$

Rule 1. The components $\Gamma^\mu_{\nu\sigma}$, $R_{\mu\nu\rho\sigma}$, $R_{\mu\nu}$, and R for the 2- and 3-dimensional metrics (**I.2**) are obtained from the 4-dimensional formulae by setting $D = 1$ (not zero!) in the 3-dimensional case, and $C = D = 1$ in the 2-dimensional case, and treating the remaining coefficients as independent of x^4, or of x^3 and x^4, respectively.

We remind the reader that *all* 2- and 3-dimensional metrics can be "diagonalized," i.e., brought to the respective forms (**I.2**) by a suitable transformation of coordinates, and that every *static* spacetime metric can be so diagonalized [cf. after (**8.36**)], as well as many others.

We shall use the following symbols

$$\alpha = \frac{1}{2A}, \quad \beta = \frac{1}{2B}, \quad \gamma = \frac{1}{2C}, \quad \delta = \frac{1}{2D}, \tag{I.3}$$

and the notation typified by

$$A_\mu = \frac{\partial A}{\partial x^\mu}, \quad B_{12} = \frac{\partial^2 B}{\partial x^1 \partial x^2}, \quad \text{etc.} \tag{I.4}$$

Then, directly from the definition (8.13), we find (for $\mu = 1, 2, 3, 4$)

$$\Gamma^1_{23} = 0, \quad \Gamma^1_{22} = -\alpha B_1, \quad \Gamma^1_{1\mu} = \alpha A_\mu. \tag{I.5}$$

From these three typical Γ's *all* others can be obtained by obvious permutations (e.g., $\Gamma^2_{33} = -\beta C_2$, $\Gamma^4_{44} = \delta D_4$, etc.). We strongly advise the reader to prepare a complete table of components in this way and to keep it handy. The same remark applies throughout this appendix.

From the Γ's we now obtain the curvature tensor $R_{\mu\nu\rho\sigma}$ as defined by (8.20) and the equation preceding (8.21). We find the following typical components:

$$R_{1234} = 0 \tag{I.6}$$

$$2R_{1213} = -A_{23} + \alpha A_2 A_3 + \beta A_2 B_3 + \gamma A_3 C_2 \tag{I.7}$$

$$2R_{1212} = -A_{22} - B_{11} + \alpha(A_1 B_1 + A_2^2) + \beta(A_2 B_2 + B_1^2)$$
$$- \gamma A_3 B_3 - \delta A_4 B_4. \tag{I.8}$$

Again, all other components can be obtained from these by permutation, and, of course, by making use of the symmetries (8.21)(i). Rule 1 applies for extracting the 2- and 3-dimensional formulae.

Next, we calculate the Ricci tensor $R_{\mu\nu}$, as defined by (8.31). Typically, we find

$$
\begin{aligned}
R_{12} = \quad & \gamma C_{12} && + \delta D_{12} \\
& -\gamma^2 C_1 C_2 && - \delta^2 D_1 D_2 \\
& -\alpha\gamma A_2 C_1 && - \alpha\delta A_2 D_1 \\
& -\beta\gamma B_1 C_2 && - \beta\delta B_1 D_2
\end{aligned} \tag{I.9}
$$

$$
\begin{aligned}
R_{11} = \quad & \beta A_{22} && + \gamma A_{33} && + \delta A_{44} \\
& + \beta B_{11} && + \gamma C_{11} && + \delta D_{11} \\
& - \beta^2 B_1^2 && - \gamma^2 C_1^2 && - \delta^2 D_1^2 \\
-\alpha A_1 (0 & + \beta B_1 && + \gamma C_1 && + \delta D_1) \\
-\beta A_2 (\alpha A_2 & + \beta B_2 && - \gamma C_2 && - \delta D_2) \\
-\gamma A_3 (\alpha A_3 & - \beta B_3 && + \gamma C_3 && - \delta D_3) \\
-\delta A_4 (\alpha A_4 & - \beta B_4 && - \gamma C_4 && + \delta D_4)
\end{aligned} \tag{I.10}
$$

All other components can be found by making the obvious permutations on these two. The dashed lines indicate the terms that remain in the 2- and 3-dimensional cases, respectively, according to our Rule 1.

For the curvature invariant $R = g^{\mu\nu}R_{\mu\nu}$ we find

$$\tfrac{1}{4}R = \alpha\beta(A_{22} + B_{11} - \alpha A_2^2 - \beta B_1^2 - \alpha A_1 B_1 - \beta A_2 B_2 + \gamma A_3 B_3 + \delta A_4 B_4)$$
$$+ \alpha\gamma(A_{33} + C_{11} - \alpha A_3^2 - \gamma C_1^2 - \alpha A_1 C_1 + \beta A_2 C_2 - \gamma A_3 C_3 + \delta A_4 C_4)$$
$$+ \beta\gamma(B_{33} + C_{22} - \beta B_3^2 - \gamma C_2^2 + \alpha B_1 C_1 - \beta B_2 C_2 - \gamma B_3 C_3 + \delta B_4 C_4)$$
$$+ \alpha\delta(A_{44} + D_{11} - \alpha A_4^2 - \delta D_1^2 - \alpha A_1 D_1 + \beta A_2 D_2 + \gamma A_3 D_3 - \delta A_4 D_4)$$
$$+ \beta\delta(B_{44} + D_{22} - \beta B_4^2 - \delta D_2^2 + \alpha B_1 D_1 - \beta B_2 D_2 + \gamma B_3 D_3 - \delta B_4 D_4)$$
$$+ \gamma\delta(C_{44} + D_{33} - \gamma C_4^2 - \delta D_3^2 + \alpha C_1 D_1 + \beta C_2 D_2 - \gamma C_3 D_3 - \delta C_4 D_4).$$

$$(I.11)$$

Note: the factor $\tfrac{1}{4}$ on the left remains unchanged even in the 2- and 3-dimensional cases obtained by Rule 1.

Lastly, for the Einstein tensor $G_{\mu\nu} = R_{\mu\nu} - \tfrac{1}{2}Rg_{\mu\nu}$, we find the following typical components:

$$G_{12} = R_{12} \qquad\qquad\qquad (I.12)$$
$$\alpha G_{11} =$$
$$\beta\gamma(-B_{33} - C_{22} + \beta B_3^2 + \gamma C_2^2 - \alpha B_1 C_1 + \beta B_2 C_2 + \gamma B_3 C_3 - \delta B_4 C_4)$$
$$+ \beta\delta(-B_{44} - D_{22} + \beta B_4^2 + \delta D_2^2 - \alpha B_1 D_1 + \beta B_2 D_2 - \gamma B_3 D_3 + \delta B_4 D_4)$$
$$+ \gamma\delta(-C_{44} - D_{33} + \gamma C_4^2 + \delta D_3^2 - \alpha C_1 D_1 - \beta C_2 D_2 + \gamma C_3 D_3 + \delta C_4 D_4).$$

$$(I.13)$$

All other components can be obtained from these by permutation. Note, however, that Rule 1 does *not* apply here.

Appendix II
How to "Invent" Maxwell's Theory

In this appendix we develop Maxwell's theory as a "natural" field theory within SR. As a prerequisite, the reader should have worked through at least part of Chapter 8 to gain some familiarity with basic tensor manipulations. The tensors we shall use here are the four-tensors of SR referred to standard coordinates x, y, z, t, but in the notation of Chapter 8. It should be remembered that partial differentiation is then a covariant operation. Of course, this work can easily be extended to curved spacetime, in the manner outlined in Section 8.9.

We have seen [cf. after (5.31)] that a 3-force **f** which acts on particles independently of their velocity in one inertial frame, transforms into a velocity-dependent 3-force in another inertial frame. Velocity-independence is therefore not a Lorentz invariant condition we can impose on a 3-force, or on a field of 3-force.

We could, however, suppose that there exists a field of 4-force F^μ which acts on any given particle independently of its 4-velocity U^μ. Then, by (5.31), the 3-force would be

$$f = \gamma^{-1}(u)(F^1, F^2, F^3), \tag{II.1}$$

and this would depend, as expected, on the particle's 3-velocity $\mathbf{u}$. However, it would not then be true that $\mathbf{F} \cdot \mathbf{U} = 0$ for arbitrary particle 4-velocities $\mathbf{U}$ (unless $\mathbf{F}$ vanished) and so the field would change the rest mass of the particles on which it acts [cf. (5.33)]—a situation we here reject. (Nevertheless, fields like the above play a role in the theory of mesons.)

The next simplest case—and the one that actually applies in Maxwell's theory—is that of a field of force which is a *linear* function of the velocity of the particles on which it acts. In SR it is natural to make this requirement on the respective 4-vectors F^μ and U^μ:

$$F^\mu = \frac{q}{c} A^\mu{}_\nu U^\nu, \tag{II.2}$$

where the "coefficients" $A^\mu{}_\nu$ in this linear relation must be tensorial to make the equation Lorentz invariant, q is the "charge" associated with the particle on which the force acts, and c is inserted for later convenience. We must stipulate that q—unlike mass, for example—is velocity independent, for to ensure the tensor character of (II.2) q must be a scalar, i.e., the same in all frames as in the rest frame of the particle. We regard $A^\mu{}_\nu$ as the *field tensor* which exists independently of the test charges influenced by it via the force (II.2).

If we demand that the force (II.2) leave unaltered the rest mass of the particles on which it acts, then, by (5.33), we need

$$\mathbf{F} \cdot \mathbf{U} = 0, \quad \text{i.e.,} \quad g_{\sigma\mu} F^\sigma U^\mu = F_\mu U^\mu = A_{\mu\nu} U^\mu U^\nu = 0. \tag{II.3}$$

This implies $(A_{\mu\nu} + A_{\nu\mu})U^\mu U^\nu = 0$, and since it must be true for *all* U^μ,

$$A_{\mu\nu} = -A_{\nu\mu}, \tag{II.4}$$

i.e., the field tensor must be skew-symmetric.

The field affects charges in accordance with Equation (II.2). How do charges, reciprocally, affect the field? The answer is given by the *field equations*. One of the characteristics of a field theory is that the action of the sources can (though it need not) spread through the field at a finite speed, and it is the *field* which eventually acts on test particles, rather than the sources directly by some "action at a distance." The field equations are therefore *differential* equations, telling how sources affect the field in their vicinity, and how one part of the field affects a neighboring part. Newton's theory can be expressed as a field theory too (where, however, effects spread instantaneously through the field), with field equations div $\mathbf{g} = -4\pi G\rho$ [cf. (8.28) and (8.26)].

In our present theory, the analog of div $\mathbf{g}$ is $A^{\mu\nu}{}_{,\nu}$—not $F^\mu{}_{,\mu}$, since the field equations cannot involve quantities depending on the velocities of the *test* charges. On the other hand, we would expect the velocities of the *sources* to affect the field, simply by reciprocity, since the field affects test charges according to *their* velocity. Hence we are led to posit the following field equations

$$A^{\mu\nu}{}_{,\nu} = k\rho_0 U^\mu = kJ^\mu, \tag{II.5}$$

where k is some constant, U^μ the 4-velocity of the source distribution, ρ_0 its *proper charge density* (i.e., the charge per unit comoving volume—a scalar), and where the 4-*current density* J^μ is defined by this equation.

A natural assumption to make is that charge is conserved, i.e., that the difference between incoming and outgoing charge must accrue in any given volume of space. By reasoning quite analogous to that following Equation (5.52), we can then establish an equation which is also quite analogous to (5.52):

$$J^\mu{}_{,\mu} = 0, \tag{II.6}$$

the *equation of continuity* of our theory.

It is a pleasant property of the choice (**II.5**) of the field equations that the equation of continuity (**II.6**) is *implied* by them. For we have

$$kJ^\mu{}_{,\mu} = A^{\mu\nu}{}_{,\nu\mu} = 0, \tag{II.7}$$

because of the skew-symmetry of $A^{\mu\nu}$ and the symmetry of the second derivatives.

But: There are *six* independent components of the field tensor $A_{\mu\nu}$ (down from the 16 of a general second rank four-tensor because of the skew symmetry) and only *four* independent components of the source vector. Additional equations are therefore needed. One way out is to assume that $A_{\mu\nu}$ is determined by a *four-potential* Φ_μ through a relation of the form

$$A_{\mu\nu} = \Phi_{\mu,\nu} - \Phi_{\nu,\mu}. \tag{II.8}$$

Then we can think of Φ_μ as determined by J^μ, via the equation

$$g^{\nu\sigma}(\Phi_{\mu,\nu\sigma} - \Phi_{\nu,\mu\sigma}) = kJ_\mu, \tag{II.9}$$

which results from substituting (**II.8**) into (**II.5**). A necessary consequence of (**II.8**) is the relation

$$A_{\mu\nu,\sigma} + A_{\nu\sigma,\mu} + A_{\sigma\mu,\nu} = 0, \tag{II.10}$$

as can be verified at once. In the theory of differential equations the converse is also well known: (**II.10**) is the *sufficient* condition for a potential Φ_μ satisfying (**II.8**) to exist. [Compare the analogous condition $B_{\mu,\nu} - B_{\nu,\mu} = 0$ for the existence of a *scalar* potential φ such that $B_\mu = \varphi_{,\mu}$.]

Different potentials Φ_μ and $\tilde{\Phi}_\mu$ can give rise to the same field $A_{\mu\nu}$. If they do, the vector $\psi_\mu = \tilde{\Phi}_\mu - \Phi_\mu$ must evidently satisfy

$$\psi_{\mu,\nu} - \psi_{\nu,\mu} = 0, \quad \text{whence} \quad \psi_\mu = \varphi_{,\mu}, \quad \text{i.e.,} \quad \tilde{\Phi}_\mu = \Phi_\mu + \varphi_{,\mu} \tag{II.11}$$

for some scalar φ. So any two four-potentials differ by a gradient. Now this gradient, the "gauge," can be chosen so as to make

$$\bar{\Phi}^{\mu}{}_{,\mu} = 0. \tag{II.12}$$

It is merely necessary to satisfy

$$g^{\mu\nu}\Phi_{\mu,\nu} + g^{\mu\nu}\varphi_{,\mu\nu} = 0, \quad \text{i.e.} \quad \Box\varphi = -g^{\mu\nu}\Phi_{\mu,\nu}, \tag{II.13}$$

where

$$\Box \equiv \frac{1}{c^2}\frac{\partial^2}{\partial t^2} - \frac{\partial^2}{\partial x^2} - \frac{\partial^2}{\partial y^2} - \frac{\partial^2}{\partial z^2}. \tag{II.14}$$

In the theory of differential equations this is known to be solvable for φ [cf. (8.180) and (8.184)]. We can therefore assume without loss of generality that the potential satisfies (II.12), and then the field equations (II.9) simplify to

$$\Box\Phi_{\mu} = kJ_{\mu}. \tag{II.15}$$

Note that in charge-free regions this reduces to the wave equation with speed c for Φ_{μ}. Hence disturbances of the potential, and thus of the field, will propagate at the speed of light.

Maxwell's theory now stands in essence before us. Let us define 3-vectors $\mathbf{e}$ and $\mathbf{h}$ thus

$$A_{\mu\nu} = \begin{pmatrix} 0 & h_3 & -h_2 & ce_1 \\ -h_3 & 0 & h_1 & ce_2 \\ h_2 & -h_1 & 0 & ce_3 \\ -ce_1 & -ce_2 & -ce_3 & 0 \end{pmatrix}. \tag{II.16}$$

Then, substituting from (5.35) and (4.10) into (II.2), we find

$$\mathbf{f} = q\left(\mathbf{e} + \frac{\mathbf{u} \times \mathbf{h}}{c}\right), \tag{II.17}$$

and so (II.2) is equivalent to Lorentz's force law. Defining a charge density ρ and a 3-current vector $\mathbf{j}$ by the equations [cf. before (5.48)]

$$J^{\mu} = \rho_0 U^{\mu} = \rho_0 \gamma(u)(\mathbf{u}, 1) = (\mathbf{j}, \rho), \tag{II.18}$$

we find, if we choose $k = 4\pi/c$, that (II.5) is equivalent to

$$\text{div } \mathbf{e} = 4\pi\rho, \quad \text{curl } \mathbf{h} = \frac{1}{c}\frac{\partial \mathbf{e}}{\partial t} + \frac{4\pi\mathbf{j}}{c}, \tag{II.19}$$

while (II.10) is equivalent to

$$\text{div } \mathbf{h} = 0, \quad \text{curl } \mathbf{e} = -\frac{1}{c}\frac{\partial \mathbf{h}}{\partial t}. \tag{II.20}$$

Equations (**II.19**) and (**II.20**) are, of course, Maxwell's equations in standard form. Finally, defining a scalar potential φ and a 3-vector potential **a** by the equation

$$\Phi_\mu = (-\mathbf{a}, c\varphi), \tag{II.21}$$

we have, from (**II.8**), the familiar expressions

$$\mathbf{e} = -\operatorname{grad} \varphi - \frac{1}{c}\frac{\partial \mathbf{h}}{\partial t}, \qquad \mathbf{h} = \operatorname{curl} \mathbf{a}. \tag{II.22}$$

Exercises

The order of the exercises is roughly that in which the topics appear in the text, rather than that of ascending difficulty. *Note*: Problems in relativity should generally be worked in units in which $c = 1$. The "missing" factor c can either be inserted later throughout, or simply at the answer stage by using it to make the physical dimensions balance. (See the final remarks of Section 2.8.)

Chapter 1

1.1 A river of width l flows at speed v. A certain swimmer's speed relative to the water is $V(>v)$. First he swims from a point P on one bank to a point on the same bank a distance l downstream, and thereupon immediately swims back to P. Prove that the total time taken is $2Vl/(V^2 - v^2)$. Next he swims from P to a point directly opposite on the other bank, and back to P. Prove that now the total time taken is $2l/(V^2 - v^2)^{1/2}$.

1.2 In the light of the preceding exercise, consider the following simplified version of the Michelson–Morley experiment. A laboratory flies at speed v through the supposed ether. Inside there are two identical rods, one placed along the direction of the "ether drift," and the other at right angles to it. *If* light travels at constant speed c relative to the ether, and *if* lengths in the direction of the ether drift shrink by a factor $[1 - (v^2/c^2)]^{1/2}$ (Lorentz theory), prove that light signals sent along either rod and reflected back from the far end take the same total times. [Anyone who remembers his trigonometry and doesn't shrink from a little calculating can now attempt to show that for a given rod placed in *any* direction in this laboratory the to-and-fro light travel times are the same; he must remember that only that component of the rod's length is shortened which is parallel to the ether drift.] In practice, a beam of light was split and sent to and fro along the two equal and orthogonal "arms" of the experiment; on returning to the vertex the beams were recombined and made to interfere. Then the experiment was rotated and observed for changes in the interference pattern, i.e., for directional variations of the light travel times.

1.3 Fill in the details of the following analysis of the Kennedy–Thorndike experiment. It was predicated on the assumption that the sun has a finite velocity through the ether. (Is this a reasonable assumption? Consider that our galaxy rotates and also has a small random motion among the other galaxies.) Consequently the earth's speed v through the ether could be expected to vary from day to day. (Why?) The experiment consists of two orthogonal arms of unequal lengths l_1 and l_2 issuing from a common vertex; light of frequency ν from a source at that vertex is split and sent to and fro along the two arms respectively; on returning to the vertex, the two beams are made to interfere. By length contraction (as in the Michelson–Morley experiment), the times between the splitting and recombination of the signals along the two arms will be $2c^{-1}\gamma l_1$ and $2c^{-1}\gamma l_2$, respectively, where $\gamma = (1 - v^2/c^2)^{-1/2}$. The difference between the number of waves along these paths is $2c^{-1}\gamma\nu(l_1 - l_2)$. This varies with v, since γ does. A null result (i.e., no change in the interference pattern over periods of up to one month) therefore implies $\gamma\nu =$ constant, i.e., time dilation.

1.4 According to the Lorentz theory, length contraction and time dilation assure that the measured two-way speed of light equals c in all directions in a laboratory S even if this moves through the ether frame AS at some speed $v < c$. Show, however, that if the clocks in S are synchronized so that simultaneous events in AS correspond to simultaneous events in S, the *one*-way speed of light in S varies with direction. Then prove that even the one-way speed of light in S can be made c in all directions provided the clock settings in S are arranged so that clock simultaneity on each plane perpendicular to the line of motion agrees with simultaneity in AS, but clocks on planes a distance x apart in S differ in their zero settings by vx/c^2 S-time units from those of AS. Prove this for signals along and across the line of motion only, if the general case is too difficult. The more ambitious reader can now prove that the transformation equations between S and AS are the Lorentz equations as given in Section 2.6.

1.5 It is well known that a straight electric current gives rise to circular magnetic lines of force around it, so that a compass needle suspended above a horizontal current takes up a horizontal position at right angles to the current. Using the relativity principle, deduce that in general a small magnet experiences a torque when it is moved through a static electric field. Would this be easy to prove directly from Maxwell's theory?

1.6 An electric charge moving through a static magnetic field experiences a force orthogonal to both its motion and the field. From the relativity principle deduce that it must therefore be possible to set a stationary electric charge in motion by moving a magnet in its vicinity. Would this be easy to prove directly from Maxwell's theory?

1.7 In a certain Cartesian coordinate system at rest relative to the fixed stars, a bar magnet occupies the z axis from $z = -1$ to $z = 1$; an electrically charged pith ball is placed at the point $(1, 0, 0)$; and this system is surrounded by a spherical mass shell centered on the origin. According to Mach's principle, which way will the pith ball tend to move as the shell is rotated about the z axis? (Take into account the polarization of the magnet, the sign of the charge, and the sense of the rotation.)

1.8 If Mach's principle is correct, the centrifugal field strength f at any point inside a massive shell rotating with angular speed ω (cf. Figure 1.2b) should be proportional to ω^2 and to the perpendicular distance r from the axis of rotation—at least in first approximation. Why? Assume, moreover, that it is proportional to the mass M of the shell and to the reciprocal of its radius R, i.e., $f = k\omega^2 rM/R$, where k is a constant. If the universe is oversimplified to an enormous ball of tenuous matter of density 10^{-31} gm/cm^3 and radius 10^{28} cm, prove that $k \approx 1.6 \times 10^{-28}$ in cgs units. (In several theories of cosmology there exists such an "effective" radius of the universe around *each* observer.) Thirring, in a well-known paper based on general relativity theory, making *no* assumptions as to the density and radius of the universe, obtained a result corresponding to $k = \sim 10^{-29}$. The order-of-magnitude agreement of these values is far more significant than their apparent discrepancy.

1.9 In Newtonian theory, inertial mass (m_I) and active and passive gravitational mass (m_A, m_P) are not only numerically equal in suitable units, but also all positive. Consider now the possibility of negative masses. Could the m_I of *all* particles be negative? Could particles exist which have negative m_I while others have positive m_I? (Consider a head-on collision.) Now assume m_I is positive for all particles. How would a normal particle and an "abnormal" particle (with $m_A = m_P < 0$) move under their mutual gravitation? How would two abnormal particles move under their mutual gravitation? How would an abnormal particle move in the earth's gravitational field? Would the existence of abnormal particles violate either Newton's third law or Galileo's principle? How would a normal particle and one with (i) $m_A > 0$, $m_P < 0$, (ii) $m_A < 0$, $m_P > 0$, move under their mutual gravitation? Would the existence of particles of these latter kinds violate either Newton's third law or Galileo's principle?

1.10 Knowing that radiation issues from an electric charge which accelerates through an inertial frame, would you expect, from the equivalence principle, that a charge held at rest in the earth's gravitational field radiates? Does your answer violate energy conservation? From the EP, would you expect that a charge allowed to fall freely in the earth's gravitational field radiates? Would a prerelativistic physicist have expected either of such charges to radiate? [Note that radiation is evidently not as simple and absolute a phenomenon as one may have thought; cf. A. Kovetz and G. E. Tauber, *Am. J. Phys.* **37**, 382 (1969), and literature cited there.]

1.11 In the elevator experiment discussed in the second paragraph of Section 1.21, prove that the relative velocity of observers A and B is gl/c, where g is the gravitational acceleration and l the height of the cabin. There exist Mössbauer apparatuses capable of measuring with tolerable error optical Doppler shifts corresponding to source velocities of 10^{-4} cm/sec (or 4 mm per hour!). A source of light at rest at the top of a tower is examined by such an apparatus placed at its foot. How high must the tower be to make the "gravitational Doppler effect" measurable? [~ 30 meters.]

Chapter 2

2.1 Assume you know about the conservation of energy, but not of momentum. Use Newtonian relativity (as in Section 2.3) and symmetry considerations to prove that if one billiard ball hits a second stationary one head-on, and no energy is dissipated, the second assumes the velocity of the first while the first comes to a total stop. Could you solve this problem directly, without appeal to relativity?

2.2 A heavy plane slab moves with uniform speed v in the direction of its normal through an inertial frame. A ball is thrown at it with velocity u, from a direction making an angle θ with its normal. Assuming that the slab has essentially infinite mass (no recoil) and that there is no dissipation of energy, use Newtonian relativity to show that the ball will leave the slab in a direction making an angle ϕ with its normal, and with a velocity w, such that

$$\frac{u}{w} = \frac{\sin \phi}{\sin \theta}, \quad \frac{u \cos \theta + 2v}{u \sin \theta} = \cot \phi.$$

2.3 In Newtonian mechanics the mass of each particle is *invariant*, i.e., it has the same measure in all inertial frames. Moreover, in any collision, mass is *conserved*, i.e., the total mass of all the particles going into the collision equals the total mass of all the particles (possibly different ones) coming out of the collision. Establish this law of mass conservation as a *consequence* of mass invariance, momentum conservation, and Newtonian relativity. [*Hint*: Let Σ^* denote a summation which assigns positive signs to terms measured *before* a certain collision and negative signs to terms measured *after* the collision. Then momentum conservation is expressed by $\Sigma^* m\mathbf{u} = 0$. Also, if primed quantities refer to a second inertial frame moving with velocity $\mathbf{v}$ relative to the first, we have $\mathbf{u} = \mathbf{u}' + \mathbf{v}$ for all $\mathbf{u}$.] Prove similarly that if in any collision the kinetic energy $\frac{1}{2}\Sigma mu^2$ is conserved in *all* inertial frames, then mass *and* momentum must also be conserved.

2.4 (i) Draw a reasonably accurate graph of $\gamma(v)$ against v for speeds v between zero and c.
   (ii) Establish the approximation for γ mentioned in Section 2.7(iv). [*Hint*: $1 - v^2/c^2 = (1 - v/c)(1 + v/c)$.]

2.5 Establish the following useful formulae:

$$\gamma v = c(\gamma^2 - 1)^{1/2}, \quad c^2 d\gamma = \gamma^3 v \, dv, \quad d(\gamma v) = \gamma^3 dv.$$

2.6 If two events occur at the same point in some inertial frame S, prove that their temporal sequence is the same in all inertial frames, and that the least time separation is assigned to them in S. Solve this problem by algebra and illustrate it by a Minkowski diagram. [In Figure 2.3, it can be shown that the tangent to any hyperbola $t^2 - x^2 = $ constant at the point of intersection with the t' axis is parallel to the x' axis, and vice versa.]

2.7 Prove that the temporal order of two events is the same in all inertial frames if and only if they can be joined by a signal traveling at or below the speed of light. [See the paragraph containing (2.13).] Illustrate the result by a Minkowski diagram.

2.8 Consider the two events whose coordinates (x, y, z, t) relative to some inertial frame S are $(0, 0, 0, 0)$ and $(2, 0, 0, 1)$ in units in which $c = 1$. Find the speeds of frames in standard configuration with S in which (i) the events are simultaneous, (ii) the second event precedes the first by one unit of time. Is there a frame in which the events occur at the same point? [$\frac{1}{2}c$, $4c/5$.]

2.9 Invent a more realistic arrangement than the garage and pole described in Section 2.12 whereby length contraction could be verified experimentally, if only our instruments were delicate enough.

2.10 A "light clock" consists of two mirrors at opposite ends of a rod with a photon bouncing between them. Without assuming time dilation, prove that this clock will go slow by the expected factor as it travels (i) longitudinally, (ii) transversely, through an inertial frame. [*Hint:* In case (i) use the "mutual" velocities between the photon and the mirrors.]

2.11 In the situation illustrated in Figure 2.4, find the relative velocity between the frames S and S', the distance between neighboring clocks in either frame, and the relative velocity between S' and S''. [$(2\sqrt{2}/3)c$, $\sqrt{2} \times 300{,}000$ km, $\sqrt{2}c/2$.]

2.12 A computer at MIT has been programmed to exhibit on a cathode ray display screen the effect of an *active* Lorentz transformation through an arbitrary velocity $v < c$ on events in the (x, t) plane. Orthogonal x and t axes are permanently displayed with units chosen so that $c = 1$; the operator marks various points, representing events in a first frame S, with his lightpen and then "presses" the "velocity button" until the desired value of v—which increases continuously from zero—is reached; as he does so, the events move to their new positions, determined by their new x and t coordinates in the frame S' having velocity v relative to S. Prove: (i) each individual point traces out a hyperbola, (ii) points on the bisectors b_1 and b_2 of the angles between the axes move along these bisectors, (iii) any three collinear points remain collinear, (iv) two parallel lines of points remain parallel, (v) a line of points perpendicular to b_1 (other than b_2) moves transversely along b_1. Finally suggest how this facility can be used to demonstrate length contraction and time dilation.

2.13 S and S' are in standard configuration. In S' a straight rod parallel to the x' axis moves in the y' direction with velocity u. Show that in S the rod is inclined to the x axis at an angle $-\tan^{-1}(\gamma uv/c^2)$. Show also that a rod moving arbitrarily while remaining parallel to the y' axis in S' remains parallel to the y axis in S. [There are many ways of doing the first part; one is to treat the distances of various points of the rod from the x axes as clocks.]

2.14 S and S' are in standard configuration. In S a slightly slanting guillotine blade in the (x, y) plane falls in the y direction past a block level with the x axis, in such a way that the intersection point of blade and block travels at a speed in excess of c. In some S', as we have seen, this intersection point travels in the *opposite* direction along the block. Explain, from the point of view of S', how this is possible.

2.15 Suppose the rod discussed at the end of Section 2.7 is at rest on the x' axis, between $x' = 0$ and $x' = l'$, for all $t' < 0$, and that the acceleration begins at $t' = 0$. Indicate on an x, y diagram a sequence of snapshots of this rod in S, beginning from its being straight and on the x axis.

2.16 In a frame S' a straight rod in the x', y' plane rotates anticlockwise with uniform angular velocity ω about its center, which is fixed at the origin. It lies along the x' axis when $t' = 0$. Find the exact shape of the rod in the usual second frame S

at the instant $t = 0$, and draw a diagram to illustrate this shape. Also show that when the rod is orthogonal to the x' axis in S' it appears straight in S.

2.17 In Section 2.14 we discussed two twins, A and B, sending each other signals at equal time intervals by their own reckoning, B staying at rest in an inertial frame while A travels away, turns round, and comes back. Draw a Minkowski diagram, using B's coordinates, which shows the worldlines of the twins and their signals. Also draw a set of A's simultaneities, both on his outward and return trips. Note that a finite portion in the middle of B's worldline has no temporal existence for A if he turns round instantaneously.

2.18 In a given inertial frame, two particles are shot out simultaneously from a given point, with equal speeds v and in orthogonal directions. What is the speed of each particle relative to the other? $[v[2 - (v^2/c^2)]^{1/2}.]$

2.19 For collinear velocities, and in the notation of Section 2.15, prove that $u = v + u'$ implies $u' = u + (-v)$. Prove that, if two particles move collinearly with respective velocities u and v in a given frame, their relative speed is $|u + (-v)|$.

2.20 Two particles, 100 m apart, move collinearly, the front one with velocity 999 km/sec, the back one with 1000 km/sec, relative to a frame S. How long will it take, in S-time, for the particles to collide?

2.21 A rod of proper length 8 cm moves longitudinally at speed $.8c$ in an inertial frame S. It is passed by a particle moving at speed $.8c$ through S in the opposite direction. In S, what time does the particle need to pass the rod?

2.22 The *rapidity* ϕ, of a particle moving with velocity u, is defined by $\phi = \tanh^{-1}(u/c)$. Prove that *collinear* rapidities are additive, i.e., if A has rapidity ϕ relative to B, and B has rapidity ψ relative to C, then A has rapidity $\phi + \psi$ relative to C.

2.23 How many successive velocity increments of $\frac{1}{2}c$ from the instantaneous rest frame are needed to produce a resultant velocity of (i) $.99c$, (ii) $.999c$? [5, 7. *Hint*: $\tanh 0.55 = 0.5$, $\tanh 2.65 = 0.99$, $\tanh 3.8 = 0.999$.]

2.24 If $\phi = \tanh^{-1}(u/c)$, and $e^{2\phi} = z$, prove that n consecutive velocity increments u from the instantaneous rest frame produce a velocity $c(z^n - 1)/(z^n + 1)$.

2.25 A particle moving in an inertial frame S has acceleration α in its own rest frame, perpendicular to its velocity $\mathbf{v}$ in S. What is its acceleration in S? Is it perpendicular to $\mathbf{v}$?

2.26 A certain piece of elastic breaks when it is stretched to twice its unstretched length. At time $t = 0$, all points of it are accelerated longitudinally with constant proper acceleration α, from rest in the straight unstretched state. Prove that the elastic breaks at $t = \sqrt{3}c/\alpha$.

2.27 Consider the rectilinear motion with constant proper acceleration α described in Equations (2.32)–(2.34). Let τ be the proper time elapsed on the moving particle, with $\tau = 0$ when $t = 0$. Then $d\tau/dt = [1 - (u^2/c^2)]^{1/2}$. Now establish the following formulae:

$$\alpha t/c = \sinh(\alpha\tau/c), \qquad \alpha x/c^2 = \cosh(\alpha\tau/c),$$
$$u/c = \tanh(\alpha\tau/c), \qquad \gamma(u) = \cosh(\alpha\tau/c).$$

Note that the third of these can be written as $c\phi = \alpha\tau$, where ϕ is the rapidity.

2.28 Given that g, the acceleration of gravity at the earth's surface, is ~ 980 cm/sec^2, and that a year has $\sim 3.2 \times 10^7$ seconds, verify that, in units of years and light years, $g \approx 1$. A rocket moves from rest in an inertial frame S with constant proper acceleration g (thus giving maximum comfort to its passengers). Find its Lorentz factor relative to S when its own clock indicates times $\tau = 1$ day, 1 year, 10 years. Find also the corresponding distances and times traveled in S. If the rocket accelerates for 10 years of its own time, then decelerates for 10 years, and then repeats the whole maneuver in the reverse direction, what is the total time elapsed in S during the rocket's absence? [$\gamma = 1.0000036$, 1.5431, 11013; $x = 0.0000036$, 0.5431, 11012 light years; $t = 0.0027$, 1.1752, 11013 years; $t = 44052$ years. To obtain some of these answers you will have to consult tables of $\sinh x$ and $\cosh x$. At small values of their arguments a Taylor expansion suffices.]

2.29 Two inertial frames S and S' are in standard configuration while a third, S'', moves with velocity v' along the y' axis, its axes parallel to those of S'. If the line of relative motion of S and S'' makes angles θ and θ'' with the x and x'' axes, respectively, prove that $\tan \theta = v'/v\gamma(v)$, $\tan \theta'' = v'\gamma(v')/v$. [Hint: use (2.26).] The inclination $\delta\theta$ of S'' relative to S is defined as $\theta'' - \theta$. If $v, v' \ll c$, prove that $\delta\theta \approx vv'/2c^2$. If a particle describes a circular path at uniform speed $v \ll c$ in a given frame S, and consecutive instantaneous rest frames, say S' and S'', always have zero relative inclination, prove that after a complete revolution the instantaneous rest frame is tilted through an angle $\pi v^2/c^2$ in the sense opposite to that of the motion. ["*Thomas precession.*"]

Chapter 3

3.1 In prerelativistic days, a simple telescope on an earth supposedly flying through the ether was aimed at a star and then filled with water: the star was still found centered on the cross-hairs. Show qualitatively that this could happen, on the ether theory, only if the ether were partially dragged along by the water. [Airy's experiment, 1871.]

3.2 Describe a situation in which the classical and the relativistic formulae predict Doppler shifts in opposite directions, i.e., $D > 1$ and $D < 1$. [*Hint*: cf. (3.3).]

3.3 A source of light moves with speed $c/2$ along the y axis, and an observer moves with speed $c/2$ along the x axis. The source, when it emitted the signal, was as far from the origin as the observer is when he receives it. What Doppler effect does he see? [$(2\sqrt{2} + 1)/(2\sqrt{2} - 1)$.]

3.4 A circular platform of radius r rotates with angular speed ω, its center fixed in an inertial frame. Calculate the Doppler shift in the light from a source at the center to an observer fixed on the rim, by *both* methods mentioned in the paragraph following (3.5), and compare with (3.6).

3.5 On the hyperbolically moving rod discussed in Section 2.16, a light signal is sent from a source at rest on the rod at $X = X_1$ to an observer at rest on the rod at $X = X_2$. Prove that the Doppler shift λ_2/λ_1 in the light is given by X_2/X_1. [*Hint*: Refer to Figure 2.5, and transform to a frame in which the observer is at rest at reception.]

3.6 From (3.5) and (3.8) derive the following interesting relation between Doppler shift and aberration: $\lambda/\lambda' = \sin\alpha/\sin\alpha'$.

3.7 Let Δt and $\Delta t'$ be the time separations in the usual two frames S and S' between two events occurring at a freely moving photon. If the photon has frequencies v and v' in these frames, prove that $v/v' = \Delta t/\Delta t'$. Would you have guessed the ratio the other way round? [*Hint*: Use the result of the preceding exercise.]

3.8 By interchanging primed and unprimed symbols and replacing v by $-v$ in the Doppler formula (3.5), obtain the aberration formula $\gamma^2[1 + (v/c)\cos\alpha] \times [1 - (v/c)\cos\alpha'] = 1$. Derive the same result from (3.8). Why is this a valid technique for obtaining new formulae?

3.9 A circular disc, whose axis coincides with the x axis of a frame S, is fixed at a negative value of x. Describe precisely how the disc appears to various observers traveling along the x axis through the origin at speeds varying from $-c$ to c. Consider especially large positive speeds. Is the situation radically different in prerelativistic theory? The results are perhaps less surprising when we regard the observer as fixed and the disc as moving.

3.10 Two momentarily coincident observers travel towards a small and distant object. To one observer that object looks twice as large as to the other. Find their relative velocity. [$3c/5$.]

3.11 A firework exploding from rest scatters sparks uniformly in all directions. But a firework exploding at high speed will evidently scatter most of its sparks in the forward direction. It is much the same with a source of light that radiates isotropically in its rest frame. Prove that when it moves at speed v, one-half of the total number of photons will be radiated into a forward cone whose semi-angle is given by $\cos\theta = v/c$. [This angle may be quite small; for obvious reasons the phenomenon is called the "headlight effect."]

3.12 A ray of light is reflected from a plane mirror that moves in the direction of its normal with velocity v. Prove that the angles of incidence and reflection, θ and ϕ, are related by $(\tan\tfrac{1}{2}\theta)/(\tan\tfrac{1}{2}\phi) = (c + v)/(c - v)$, and the wavelengths before and after reflection, λ_1 and λ_2, by

$$\frac{\lambda_1}{\lambda_2} = \frac{\sin\theta}{\sin\phi} = \frac{c\cos\theta + v}{c\cos\theta - v} = \frac{c + v\cos\theta}{c - v\cos\theta}.$$

3.13 A particle moves uniformly in a frame S with velocity **u** making an angle α with the positive x axis. If α' is the corresponding angle in the usual second frame S', prove

$$\tan\alpha' = \frac{\sin\alpha}{\gamma(v)[\cos\alpha - (v/u)]} = \frac{\tan\alpha}{\gamma(v)[1 - (v/u)\sec\alpha]},$$

and compare this with (3.7) and (3.8).

3.14 In a frame S, consider the equation $x\cos\alpha + y\sin\alpha = -ct$. For fixed α it represents a plane propagating in the direction of its normal with speed c, that direction being parallel to the (x, y) plane and making an angle α with the negative x axis. We can evidently regard this plane as a light front. Now transform x, y, and t directly to the usual frame S', and from the resulting equation deduce at once the following aberration formula for the wave normal: $\tan\alpha' = \sin\alpha/\gamma[\cos\alpha + (v/c)]$, which is equivalent to (3.7) and (3.8). Thus in SR the aberration of light fronts is the same as the aberration of rays, in vacuum.

3.15 Make a rough sketch, using Figure 3.1 and projections, to demonstrate how some inertial observers could see the outline of a uniformly moving wheel (or disc) boomerang-shaped.

3.16 A cube with its edges parallel to the coordinate axes moves with Lorentz factor 3 along the x axis of an inertial frame. A "supersnapshot" of this cube is made in a plane z = constant by means of light rays parallel to the z axis. Make an *exact* scale drawing of this supersnapshot.

Chapter 4

4.1 A four-vector has components (V_1, V_2, V_3, V_4) in an inertial frame S. Write down its components (i) in a frame which coincides with S except that the directions of the x and z axes are reversed; (ii) in a frame which coincides with S except for a 45° rotation of the (x, y) plane about the origin followed by a translation in the z direction by 3 units; (iii) in a frame which moves in standard configuration with S.

4.2 Use the fact that $U = \gamma(u, 1)$ transforms as a four-vector to rederive the transformation equations (2.25) and (2.28). Recall that our earlier derivation of (2.28), which is essentially equivalent to (2.27), was quite tedious.

4.3 "In three dimensions a particle has a three-velocity u *relative* to some observer; in four dimensions a particle has a four-velocity U, period." Comment. [Distinguish between a vector and its components.]

4.4 An inertial observer O has four-velocity U_0 and a particle P has (variable) four-acceleration A. If $U_0 \cdot A \equiv 0$, what can you conclude about the speed of P in O's rest frame? [*Hint*: Look at the datum in O's rest frame.]

4.5 A particle moves along the circle $x^2 + y^2 - r^2 = 0 = z$ at constant speed u in an inertial frame S. Find the components of its four-acceleration when it crosses the negative y axis. Find the corresponding components in the rest frame of the particle (with axes parallel to those of S) and also the three-acceleration in the rest frame. [In all cases the second component is $\gamma^2 u^2/r$, and all others vanish.]

4.6 A particle moves rectilinearly with constant proper acceleration α. If U and A are its four-velocity and four-acceleration and $c = 1$, prove that $(d/d\tau)A = \alpha^2 U$, where τ is the proper time of the particle. [*Hint*: Exercise 2.27.] Prove, conversely, that this equation, *without* the information that α is the proper acceleration or constant, implies both these facts. [*Hint*: Differentiate the equation $A \cdot U = 0$ and show that $\alpha^2 = -A \cdot A$.] And finally show, by integration, that the equation implies rectilinear motion in a suitable inertial frame, and thus, in fact, hyperbolic motion.

4.7 Using the simplifications discussed at the end of Section 4.5, prove that any four-vector orthogonal to a timelike or null vector (other than the null vector itself in the latter case) must be spacelike, but that two spacelike vectors can be orthogonal to each other.

4.8 Prove that the sign of the fourth component of a timelike or null four-vector is invariant under *general* LT's. According as this sign is negative or positive, such vectors are said to be past- or future-pointing. Illustrate this with a three-dimensional Minkowski diagram.

4.9 Prove that the sum of two timelike four-vectors which are isochronous (i.e., both pointing into the future or both into the past) is a timelike vector isochronous with them. Is this result still true if one or both vectors are null? Illustrate with a three-dimensional Minkowski diagram.

4.10 The aggregate of events considered by any inertial observer to be simultaneous at his time $t = t_0$ is said to be the observer's instantaneous 3-space $t = t_0$. Show that the join of any two events in such a space is perpendicular to the observer's world-line, and that, conversely, any two events whose join is perpendicular to the observer's worldline are considered simultaneous by him.

4.11 Let four events be specified in spacetime, none of which lies inside or on the (fore or aft) light cone of another. Prove that there is an inertial frame in which these four events are simultaneous, and that, in general, this frame is unique.

4.12 For any two future-pointing timelike vectors $\mathbf{V}_1$ and $\mathbf{V}_2$, prove that $\mathbf{V}_1 \cdot \mathbf{V}_2 = V_1 V_2 \cosh \phi_{12}$, where ϕ_{12}, the "hyperbolic angle" between $\mathbf{V}_1$ and $\mathbf{V}_2$, equals the relative rapidity of particles having $\mathbf{V}_1$ and $\mathbf{V}_2$ as worldlines. [*Hint*: (**4.21**), (**4.27**).] Moreover, prove that ϕ is additive, i.e., for any three *coplanar* vectors $\mathbf{V}_1, \mathbf{V}_2, \mathbf{V}_3$ (corresponding to three collinearly moving particles, in that order), $\phi_{13} = \phi_{12} + \phi_{23}$. For two spacelike vectors $\mathbf{W}_1$ and $\mathbf{W}_2$ we can write $\mathbf{W}_1 \cdot \mathbf{W}_2 = W_1 W_2 \cos \theta_{12}$; what is the physical meaning of θ_{12}?

4.13 Derive the wave aberration formula (**4.34**) by using the invariance of $\mathbf{L} \cdot \mathbf{U}$, where $\mathbf{U}$ is the four-velocity of the frame S'.

4.14 "Even if every local physical experiment were fully deterministic, and an observer had at his disposal the most perfect data gathering instruments and the most perfect records left by devoted predecessors, he could never predict his future with certainty." Justify. [*Hint*: Consider the observer's light cone.]

Chapter 5

5.1 How fast must a particle move before its *kinetic* energy equals its *rest* energy? [$0.866\,c$.]

5.2 How fast must a 1-kg cannon ball move to have the same kinetic energy as a cosmic-ray proton moving with γ factor 10^{11}? [~ 5 m/sec.]

5.3 The mass of a hydrogen atom is 1.00814 amu, that of a neutron is 1.00898 amu, and that of a helium atom (two hydrogen atoms and two neutrons) is 4.00388 amu. Find the binding energy as a fraction of the total energy of a helium atom.

5.4 Radiation energy from the sun is received on earth at the rate of 1.94 calories per minute per cm^2. Given the distance of the sun (150,000,000 km), and that one calorie $= 4.18 \times 10^7$ ergs, find the total mass lost by the sun per second, and also the force exerted by solar radiation on a black disc of the same diameter as the earth (12,800 km) at the location of the earth. [4.3×10^9 kg, 5.8×10^{13} dyne. *Hint*: Force equals momentum absorbed per unit time.]

5.5 A particle with four-momentum $\mathbf{P}$ is observed by an observer with four-velocity $\mathbf{U}_0$. Prove that the energy of the particle relative to the observer is $\mathbf{U}_0 \cdot \mathbf{P}$. [*Hint*: Look at components in a suitable frame.]

5.6 A particle of rest mass M (the "bullet") moving at velocity u_1 collides elastically and head-on with a stationary particle of rest mass m (the "target"). After collision the bullet moves with velocity u_2. Prove that in Newtonian mechanics $u_2/u_1 = (M - m)/(M + m)$, so that, for two given particles, the ratio of the initial to the final energy of the bullet is independent of its initial speed. In relativity, prove that $\gamma_2 = (\gamma_1 + k)/(k\gamma_1 + 1)$, where $k = 2mM/(m^2 + M^2)$ and $\gamma_1 = \gamma(u_1)$, $\gamma_2 = \gamma(u_2)$. Note that $\gamma_2 \to 1/k$ monotonically as $\gamma_1 \to \infty$, so that there is now an absolute limit to the energy which the bullet can retain; thus at high incident speeds the relative transfer of energy from the bullet to the target is almost total, even if their rest masses differ greatly. [*Hint*: Write down the equations of conservation of energy and momentum, eliminate—by squaring—the v and γ of the target, express u_1 and u_2 in terms of γ_1 and γ_2, and square once more to obtain a quadratic in γ_1 and γ_2. Note that $\gamma_1 - \gamma_2$ should be expected as a factor of this quadratic—why?]

5.7 The position vector of the center of mass of a system of particles in any inertial frame is defined by $\mathbf{r}_{CM} = \Sigma m\mathbf{r}/\Sigma m$. If the particles suffer only collision forces, prove that $\dot{\mathbf{r}}_{CM} = \mathbf{u}_{CM}$ ($\dot{} \equiv d/dt$); i.e., the center of mass moves with the velocity of the CM frame. [*Hint*: Σm, $\Sigma m\dot{\mathbf{r}}$ are constant; $\Sigma \dot{m}\mathbf{r}$ is zero between collisions, and *at* any collision we can factor out the $\mathbf{r}$ of the participating particles: $\mathbf{r}\Sigma \dot{m} = 0$.]

5.8 A particle of instantaneous rest mass m is at rest and being heated in a frame S', so as to increase its energy at the rate of q units per second. From the definition (5.29) find the relativistic force on the particle in the usual second frame S. What is that force if, after being heated, the particle *cools* at a rate q?

5.9 Prove that the four-force $\mathbf{F}$ and four-velocity $\mathbf{U}$ pertaining to a particle of variable rest mass m_0 are related by $\mathbf{F} \cdot \mathbf{U} = c^2 dm_0/d\tau$, where τ is the particle's proper time. Compare with (5.33). [*Hint*: Look at $\mathbf{F} \cdot \mathbf{U}$ in a suitable frame.]

5.10 A particle of constant proper mass moves *rectilinearly* in some inertial frame. Show that the product of its proper mass and its instantaneous proper acceleration equals the magnitude of the relativistic three-force acting on the particle in that frame. [*Hint*: (2.30), (5.29).] Show also that this is not necessarily true when the motion is not rectilinear.

5.11 From the result of the preceding exercise deduce that the relativistic force on a particle in rectilinear motion is invariant among inertial frames moving along the same line as the particle. Then rederive this result from the transformation properties of the four-force $\mathbf{F}$.

5.12 An otherwise free particle is being pulled by a weightless string. Prove that, whereas the three-force $\mathbf{f}$ on the particle is generally *not* collinear with the string, the three-acceleration $\mathbf{a}$ always is. [*Hint*: In the rest frame of the particle the acceleration *must* be along the string.] Compare with Exercise 6.10 below.

5.13 (i) Prove that, in relativistic as in Newtonian mechanics, the time rate of change of the angular momentum $\mathbf{h} = \mathbf{r} \times \mathbf{p}$ of a particle with linear momentum $\mathbf{p}$ at vector distance $\mathbf{r}$ from an arbitrary point P is equal to the couple $\mathbf{r} \times \mathbf{f}$ of the applied force about P.
(ii) For motion under a rest-mass-preserving inverse square force $\mathbf{f} = -k\mathbf{r}/r^3$ (k = constant), derive the energy equation $\gamma c^2 - k/r =$ constant. [*Hint*: (5.34).]

5.14 (*The relativistic harmonic oscillator.*) A particle of constant proper mass m_0, moving along the x axis of an inertial frame, is attracted to the origin by a relativistic force $-m_0 k^2 x$. If the amplitude of the resulting oscillation is a, prove that the period T is $(4/c)\int_0^a \gamma(\gamma^2 - 1)^{-1/2}dx$, where $\gamma = 1 + \frac{1}{2}k^2 c^{-2}(a^2 - x^2)$. [*Hint*: Supply the details of the following calculation: $\gamma^3 du/dt = -k^2 x$, where $\gamma = [1 - (u^2/c^2)]^{-1/2}$; $u\gamma^3 du = -k^2 x dx$, $\gamma = $ constant $-\frac{1}{2}k^2 x^2 = $ answer given; $T = 4\int_0^a (1/u)dx = $ answer given.]

5.15 By comparing (4.35) with the result of Exercise 3.13 verify that the aberration of the wave normal of a wave traveling at speed c^2/u is precisely the same as that of the track of a particle traveling at speed u.

5.16 Prove that the outcome of a collision between two distinct particles (of finite *or* zero rest mass) cannot be a single photon, and that no particle of finite rest mass can disintegrate into a single photon. [*Hint*: Conservation of momentum.]

5.17 If a photon with four-momentum **P** is observed by two observers having respective four-velocities $\mathbf{U}_0$ and $\mathbf{U}_1$, prove that the observed frequencies are in the ratio $\mathbf{U}_0 \cdot \mathbf{P}/\mathbf{U}_1 \cdot \mathbf{P}$. Hence rederive Formula (3.5).

5.18 In an inertial frame S, two photons of frequencies v_1 and v_2 travel along the x axis in opposite directions. Find the velocity of the CM frame of these photons. [$v/c = (v_1 - v_2)/(v_1 + v_2)$.]

5.19 Suppose a part, m, of the earth's total rest mass M is transmuted into photons and radiated directly ahead in the earth's orbital direction, so that by the reaction the remaining mass comes to a complete halt (and thereupon falls into the sun). If the earth's orbital speed is u (~ 18.5 mile/sec) show that $(M - m)/M = (c - u)^{1/2}/(c + u)^{1/2}$, whence $m/M \approx u/c \approx 1/10,000$. Show also that if mass is ejected in the form of matter rather than light, a greater proportion of rest mass must be ejected. [*Hint*: Let the *total* energy and momentum of the ejected photons be E and E/c and appeal to the conservation of energy and momentum.]

5.20 If one neutron and one pi-meson are to result from the collision of a photon with a stationary proton, find the threshold frequency of the photon in terms of the rest mass n of a proton or neutron (here assumed equal) and that, m, of a pi-meson. [$v = c^2(m^2 + 2mn)/2hn$.]

5.21 A particle of rest mass m decays from rest into a particle of rest mass m' and a photon. Find the separate energies of these end products. [*Answer*: $c^2(m^2 \pm m'^2)/2m$. *Hint*: Use a four-vector argument.]

Chapter 6

6.1 Write down the equations inverse to (6.11) and (6.12), i.e., those giving the unprimed components of the field in terms of the primed.

6.2 If at a certain event an electromagnetic field satisfies the relations $\mathbf{e} \cdot \mathbf{h} = 0$, $e^2 < h^2$, prove that there exists a frame in which $\mathbf{e} = 0$. Then prove that infinitely many such frames exist, all in collinear relative motion. [*Hint*: Choose the spatial axes suitably and then use the transformation equations (6.11).]

6.3 If at a certain event $\mathbf{e} \cdot \mathbf{h} \neq 0$, prove that there exists a frame in which **e** and **h** are parallel. Then prove that infinitely many such frames exist, all in collinear relative motion. [*Hint*: One special frame moves in the direction $\mathbf{e} \times \mathbf{h}$ relative to the general frame.]

6.4 Prove that the vacuum Maxwell equation

$$\frac{\partial e_1}{\partial x} + \frac{\partial e_2}{\partial y} + \frac{\partial e_3}{\partial z} = 0 \tag{1}$$

(i.e., div **e** = 0) can hold in *all* inertial frames only if also the Maxwell equations

$$\frac{\partial h_3}{\partial y} - \frac{\partial h_2}{\partial z} = \frac{\partial e_1}{\partial t} \tag{2}$$

$$\frac{\partial h_1}{\partial z} - \frac{\partial h_3}{\partial x} = \frac{\partial e_2}{\partial t} \tag{3}$$

$$\frac{\partial h_2}{\partial x} - \frac{\partial h_1}{\partial y} = \frac{\partial e_3}{\partial t} \tag{4}$$

(i.e., **curl h** $= \partial e/\partial t$) hold in all frames. (The units are chosen so as to make $c = 1$.)
[*Hint:* Use

$$\frac{\partial}{\partial x} = \gamma\left(\frac{\partial}{\partial x'} - v\frac{\partial}{\partial t'}\right), \quad \frac{\partial}{\partial y} = \frac{\partial}{\partial y'}, \quad \frac{\partial}{\partial z} = \frac{\partial}{\partial z'}$$

(why is this true?) and the inverse of (**6.11**) to transform (**1**), thus obtaining (**2**) in
S', for a start.]

6.5 What is the path of an electrically charged particle which is projected into a uniform
(parallel) magnetic field at an angle other than a right angle to the lines of force?
Support your answer with exact arguments.

6.6 In a frame S there is a uniform electric field $\mathbf{e} = (0, a, 0)$ and a uniform magnetic
field $\mathbf{h} = (0, 0, 5a/3)$. A particle of rest mass m_0 and charge q is released from rest
on the x axis of S. What time elapses before it returns to the x axis? [*Answer:*
$75\pi cm_0/32aq$. *Hint:* Look at the situation in a frame in which the electric field
vanishes.]

6.7 What is the motion of an electrically charged particle which is projected in a
direction parallel to the lines of force into a uniform (parallel) electric field?

6.8 There are good reasons for saying that an electromagnetic field is "radiative"
if it satisfies $\mathbf{e} \cdot \mathbf{h} = 0$ and $\mathbf{e}^2 = \mathbf{h}^2$. Show that the field of a very fast moving charge
($v \approx c$) is essentially radiative in a plane which contains that charge and is orthog-
onal to its motion.

6.9 In a frame S, two identical particles with electric charge q move abreast along lines
parallel to the x axis, a distance r apart and with velocity v. Determine the force, in
S, that each exerts upon the other, and do this in two ways: (i) by use of (**6.21**) and
(**6.1**), and (ii) by transforming the Coulomb force from the rest frame S' to S by use
of the four-vector property of (**5.31**), with $dm'/dt' = 0$ (why?). Note that the force
is smaller than that in the rest frame, while each mass is evidently greater. Here we
see the dynamical reasons for the "relativistic focusing" effect whose existence we
deduced from purely kinematic considerations before. (See the last paragraph of
Section 2.13.) Do these dynamical arguments lead to the expected time dilation of an
"electron cloud clock"? Also note from (i) that as $v \to c$ the electric and magnetic
forces each become infinite, but that their effects cancel.

6.10 Instead of the equal charges moving abreast as in the preceding exercise, consider now two *opposite* charges moving at the same velocity but one *ahead* of the other. By both suggested methods determine the forces acting on these charges, and show that they do *not* act along the line joining the charges (e.g., a nonconducting rod) but, instead, apparently constitute a couple tending to turn that join into orthogonality with the line of motion. [Trouton and Noble, in a famous experiment in 1903, unsuccessfully looked for this couple on charges at rest in the laboratory, which they presumed to be flying through the ether. The fact that force and acceleration are not necessarily parallel was unknown then, and the null result seemed puzzling. However, it contributed to the later acceptance of relativity.]

6.11 According to Maxwell's theory, charge is not only *invariant* (i.e., each charge has the same measure in all frames), but it is also *conserved* (i.e., the total charge involved in an experiment remains the same at all times). Explain why it is nevertheless possible for a current-carrying straight wire to be electrically neutral in one frame while possessing a net charge in another (as we saw in Section 6.4).

6.12 We have asserted that the components of the electromagnetic field, arranged in the pattern (**6.13**) in every inertial frame, constitute a tensor transforming according to the scheme (**4.24**). Test this assertion by so transforming to the usual frame S' a pure electromagnetic field of intensity e in the z direction in S, and checking the result by use of the transformation formulae (**6.11**) and (**6.12**).

6.13 Illustrate, by an example, Van Dam and Wigner's theorem asserting that the total energy and momentum of a system of particles that interact *at a distance* cannot remain constant in all inertial frames. It is for this reason that *field theories* (e.g., Maxwell's) allow the field itself to carry energy and momentum, so that the conservation laws can be strictly satisfied by the *total* system of field-and-particles. [*Hint*: Consider two identical charged particles simultaneously released from rest at different points in a given inertial frame.]

Chapter 7

7.1 Say which of the following are intrinsic relative to a two-dimensional surface:
    (i) The property of a line being straight.
    (ii) The angle at which two curves intersect.
    (iii) The property of two curves being tangent to each other at a given point.
    (iv) The length of a curve between two of its points.
    (v) The shortest distance, in the surface, between two given points.
    (vi) The area contained within a closed curve.
    (vii) The normal curvature of the surface in a given direction (i.e., the curvature of the section obtained by cutting the surface with a plane containing the normal at the point of interest).
    (viii) The "geodesic" curvature of a curve on the surface (i.e., the curvature, at the point of interest, of the projection of the curve onto the tangent plane at that point). [In fact, it can be shown that the geodesics on a surface are precisely those curves whose geodesic curvature vanishes everywhere.]

7.2 A line l is common to a space V and a subspace V' of V (e.g., l may lie on a surface V' in Euclidean 3-space V). Prove that if l is a geodesic in V, it is also a geodesic in V'; give examples to show that the converse is not necessarily true.

7.3   The curvature of a plane curve (or, indeed, any curve) is defined as the rate of turning of the tangent (in radians) with respect to distance along the curve; it can be shown to equal $\lim(2z/r^2)$ as $r \to 0$, where r is the distance along the tangent at the point of interest and z is the perpendicular distance from the tangent to the curve. Now, the equation $z = f(x, y)$ of an arbitrary surface, if sufficiently well-behaved, can be expressed as a Taylor series $z = a + bx + cy + dx^2 + exy + fy^2 + \cdots$ for small x, y. If we choose a tangent plane to the surface as the (x, y) plane and the point of contact as the origin, that equation assumes the simpler form $z = Ax^2 + Bxy + Cy^2$ (neglecting cubes and higher powers). Use this fact to prove that the maximum and minimum of the normal curvature of a surface [cf. Exercise 7.1(vii)] occur in orthogonal directions. Gauss proved the extraordinary theorem that the product of the two (evidently nonintrinsic) extrema of the normal curvature is equal to the (intrinsic) curvature K. Thus if the spine and the ribs of the horse that fits the saddle of Figure 7.3 locally approximate circles of radii a and b, the curvature K at the center of the saddle is $-1/ab$.

7.4   In the plane, the total angle Δ through which the tangent of a closed (and not self-intersecting) curve turns in one circuit is evidently always 2π. On a curved surface, the corresponding Δ (increments of angle being measured in the successive local tangent planes) generally differs from 2π. According to a deep and beautiful theorem (Gauss–Bonnet), $2\pi - \Delta$ equals $\iint K dS$, the integral of the Gaussian curvature over the enclosed area. Consider a sphere of radius a and on it a "geodesic triangle" formed by three great-circular arcs making a right angle at each vertex. Test the theorem for *both* the areas that can be considered enclosed by this triangle. [Note that there is no contribution to Δ along a geodesic, as follows from the bracketed remark in Exercise 7.1(viii).]

7.5   Consider a sphere of radius a and on it a "geodesic circle" of radius r, as in Figure 7.4. To find the total angle Δ through which the tangent of this circle turns in one circuit (increments of angle always being measured in the local tangent plane), construct the cone tangent to the sphere along the given circle and then "unroll" this cone and measure the angle at its vertex. Thus prove $\Delta = 2\pi \cos(r/a)$. Then verify that this accords with the Gauss–Bonnet theorem of the preceding exercise —as applied to *both* possible "insides" of the circle.

7.6   It is often thought that the sphere is the only two-dimensional surface of *constant* curvature that is both finite and unbounded. Nevertheless. a plane surface, too, can be finite and unbounded. One need merely draw a rectangle in the plane, discard the outside, and "identify" opposite points on opposite edges. The area of the resulting surface is evidently finite, yet it has no boundary. Each point is an internal point, since each point can be surrounded by a circle lying wholly in the surface: a circle around a point on an edge appears in two halves, yet is connected because of the identification; around the vertices—all identified—such a circle appears in four parts. Unlike the plane, however, this surface has the topology of a torus, and thus lacks global isotropy. (Consider, for example, the lengths of geodesics drawn in various directions from a given point.) Still, it is *locally* isotropic in its planeness.

A similar though rather more complicated construction can be made on a surface of constant negative curvature. Issuing from one of its points, we draw eight geodesics of equal length r, each making an angle of $45°$ with the next. Then we draw the eight geodesics which join their endpoints. What can you say about the angles at the vertices of the resulting "geodesics octagon" in terms of its total area? Deduce that r can be chosen to make these angles $45°$, and chose r so. Labeling the vertices successively A, B, C, D, E, F, G, H, identify the following directed geodesics: $AB = DC$, $BC = ED$, $EF = HG$, $FG = AH$. Draw a picture and verify that (i) each point on an edge other than a vertex is an internal point, (ii) the eight vertices are all identified and constitute an internal point, (iii) at all these points the curvature is the same as "inside." Can the same trick be worked with a four- or six-sided polygon?

Analogous constructions exist, obviously, for flat three-space and also, much less obviously, for three-spaces of constant negative curvature.

7.7 To illustrate that geodesics in pseudo-Riemannian spaces generally have neither minimal nor maximal length, consider the x axis of Minkowski space M_4. Having linear equation $x = \sigma$, $y = z = t = 0$ (σ being a parameter), it *is* a geodesic. Consider neighboring curves to it, say between $x = 0$ and $x = 2$, consisting of two straight portions: one from $(0, 0, 0, 0)$ to (a, b, c, d), and one from (a, b, c, d) to $(2, 0, 0, 0)$. Show that for suitably chosen a, b, c, d, these neighbors can have greater *or* lesser square interval than -4. But recall that *timelike* geodesics in M_4 *are* maximal.

7.8 Suppose we suspect that spacetime, far from gravitating masses, instead of being Minkowskian, is a four-space of constant nonzero curvature. In principle, how could we test this hypothesis? Comparing the radius and surface of a geodesic two-sphere could be part of the test, but this would *not* probe the spread of timelike or null geodesics, nor of spacelike geodesics with timelike separation. How would you check on all these? [*Hint*: draw a three-dimensional Minkowski diagram to develop your ideas.]

7.9 Gravity evidently affects the rate of a pendulum clock. In a stronger field, does a given pendulum clock go faster or slower? Is this in any way connected with gravitational time dilation?

7.10 Rederive the result (7.13) by appeal to Planck's relation $E = h\nu$ for the energy of a photon: During its "fall" down a gravitational field, a photon should gain energy just like a particle, since photons and particles are convertible into each other, or simply because we regard photons as limiting particles. But since the photon's speed cannot increase, its frequency must. [*Note*: The weak EP enters this argument when we equate the photon's inertial mass $h\nu/c^2$ with its gravitational mass.]

7.11 A satellite is in circular orbit of radius r around the earth (radius R), satisfying Equation (7.29). A standard clock C on the satellite is compared with an identical clock C_0 on earth. Prove that the ratio of the rates of these clocks is given by

$$\frac{\nu}{\nu_0} \approx 1 + \frac{Gm}{Rc^2} - \frac{3Gm}{2rc^2},$$

and note that this exceeds unity only if $r > 3R/2$ (i.e., $r - R > 3184$ km).

7.12 Consider three synchronized clocks A, B, C at rest in a static gravitational field, as in Figure 7.7. Suppose the coordinate times for light to travel along AB, BC, CA, respectively, are 1, 1, and 5 seconds—which would, of course, be impossible in Minkowski space, i.e., in the absence of gravitation. Consider the gravitational field of a single concentrated spherical mass, and in it indicate a "triangle" to which the data might apply, qualitatively. [*Hint*: Remember that standard clocks at low gravitational potential go slow relative to those at zero potential; what does this imply about the *coordinate* speed of light at low potential?]

7.13 In Newtonian theory, every gravitational field relative to AS is completely described by a scalar potential $\varphi = -\iiint (\rho/r)dV$, evaluated instantaneously over all space, the field itself being $-\mathbf{grad}\ \varphi$. If the field is stationary, φ is independent of time, and thus there exist no stationary fields which are not also static. For example, a "gravitational" field like that found by an observer at rest on a uniformly rotating turntable has no analog in Newtonian AS. Nevertheless, it is possible to define a conservative scalar potential on the the turntable, corresponding to the centrifugal force only, since the Coriolis force acts at right angles to a particle's motion and does no work. The centrifugal force being $\omega^2 r$ (ω = angular velocity, r = distance from the center), the potential is $-\frac{1}{2}\omega^2 r^2$. If we substitute this in (7.21), we obtain a frequency shift between a point at radius r and the center which agrees, *to first order*, with our previously found result (3.6). However, the agreement is not exact. This is because we have ignored the fact that the force on a unit rest mass at rest in the field, measured by an observer at rest in the field, is not $\omega^2 r$, but rather $\omega^2 r \gamma^2(\omega r)$. Why? Show that (if $c = 1$) the corresponding potential is $\frac{1}{2}\log(1 - \omega^2 r^2)$, and that this, when substituted in (7.21), agrees exactly with (3.6). The present example, like that of the uniform acceleration field (see Section 8.6), is interesting in that it allows a given noninfinitesimal frequency shift to be interpreted alternatively by purely kinematic or purely gravitational considerations.

Chapter 8

Note: Use the formulae of Appendix I where appropriate.

8.1 For the polar metric of the plane, $dr^2 + r^2 d\theta^2$, compute the Christoffel symbols Γ^i_{jk} (put $r = x^1, \theta = x^2$) and verify that $\theta = $ constant satisfies the geodesic equations (8.15).

8.2 Prove that the metric $y^2 dx^2 + x^2 dy^2$ represents the Euclidean plane. [*Hint:* Verify that *all* components of the curvature tensor vanish; do not forget the symmetries (8.21).]

8.3 Prove that the metric $ydx^2 + xdy^2$ represents a *curved* space. [*Hint:* Find at least *one* component of the curvature tensor different from zero.]

8.4 A coordinate x^μ is said to be timelike, null, or spacelike, if the corresponding coordinate lines $x^\nu = $ constant, $\nu \neq \mu$, are timelike, null, or spacelike, respectively. Reexpress the Minkowski metric $dt^2 - dx^2 - dy^2 - dz^2$ in terms of *two* timelike and two spacelike coordinates. [*Hint:* Refer to a Minkowski diagram to suggest two timelike coordinate axes. *Note:* There will now be "cross terms" in the metric, since two timelike directions cannot be orthogonal, cf. Exercise 4.7.] Prove that, if x^μ is the *only* null or timelike coordinate in a metric, it cannot be constant along a particle's worldline.

8.5 A curve $x^\mu = x^\mu(u)$ is a geodesic if it satisfies Equations (8.15) with u in place of the arc s. The parameter u is then called an "affine" parameter along the geodesic and it can be shown to be a linear function of s unless the geodesic is null. Prove that a geodesic in the metric $ds^2 = dt^2 - d\sigma^2$, where $d\sigma^2$ is a 3-metric independent of t, follows a geodesic (spatial) track in the metric $d\sigma^2$. [*Hint*: Without loss of generality, assume $d\sigma^2$ is diagonal.]

8.6 By reference to (7.24), and the fact that every 3-metric can be diagonalized, it is seen that the most general static metric can be brought into the form $ds^2 = A(dx^1)^2 + B(dx^2)^2 + C(dx^3)^2 + D(dt)^2$, where A, B, C, D are functions of x^1, x^2, x^3 only and $D = \exp 2\varphi$ ($c = 1$). Prove that the four-acceleration $A^\mu = D/ds(dx^\mu/ds)$ of a particle at rest in the spatial coordinates is $(-\varphi_1/A, -\varphi_2/B, -\varphi_3/C, 0)$. By putting $A = B = C = -1$ for LIF coordinates, we see that this equals (**grad** φ, 0). Reference to (4.13) now shows that the three-acceleration **a** of such a particle is *exactly* **grad** φ, as was already established approximately in Section 7.6.

8.7 Establish the symmetry of the Ricci tensor, (8.32).

8.8 Prove that for a concentrated spherical mass m, the *Newtonian* "escape velocity" (i.e., the minimum velocity at which a particle must be projected so as to reach infinity) is c at distance $r = 2Gm/c^2$ from the center. In the units of the metric (8.43) this corresponds precisely to the Schwarzschild radius $r = 2m$. [However, the physical analogy with the Schwarzschild situation is rather tenuous.]

8.9 The metric of any surface of revolution, like that of Figure 8.1, can be written (with a redundant coordinate z) in the form $d\sigma^2 = dz^2 + dr^2 + r^2 d\phi^2$ (justify). By equating this to $dr^2/[1 - (2m/r)] + r^2 d\phi^2$, find z as a function of r for Flamm's paraboloid. Compare with (8.44). Also show that, if m were negative, *no* real surface of revolution would exist with that metric.

8.10 By applying Formula (7.6) to neighboring meridians (plane sections containing the axis) of Flamm's paraboloid (8.44), find its curvature as a function of r. [$-m/r^3$.]

8.11 By applying Formula (7.6) to neighboring particles falling along the same radius vector, find approximately the curvature of Schwarzschild spacetime in an orientation determined by dr and dt. [*Hint*: Use the Newtonian tidal force $df = (2m/r^3)dr$ ($G = 1$).]

8.12 Use (8.68) to find the proper time $\Delta\tau_0$ elapsed per revolution on a particle in circular orbit of radius r around a mass m. Find the ratio of $\Delta\tau_0$ to $\Delta\tau_1$, the corresponding proper time elapsed at a point at rest on the surface of the central mass, if that has a radius R. Compare this with the approximate formula derived in Exercise 7.11.

8.13 Suppose that at the instant a particle in circular orbit around a spherical mass passes through a certain point P, another freely moving particle passes through P in a radially outward direction, at the precise velocity necessary to ensure that it will fall back through P when also the orbiting particle next passes through P. By using the results of the preceding exercise, prove that it is certainly possible for two such particles to take *different* proper times between their two coincidences. (The precise amount of the difference is of little interest in the present context.) Note that here we have a version of the twin "paradox" in which both twins are permanently unaccelerated in their rest frames.

8.14 What is the *coordinate velocity* of light in the Schwarzschild space (8.43) at coordinate r, (i) in the radial direction, (ii) in the transverse direction? [$1 - 2m/r$, $(1 - 2m/r)^{1/2}$.]

8.15 Obtain the integral

$$\Delta t = 2 \int_p^{r_0} \{\alpha^{-2} + \alpha^{-1}p^2(r^2 - p^2)^{-1}\}^{1/2} dr, \qquad \alpha = 1 - \frac{2m}{r},$$

for the coordinate time it takes a light signal to travel along a path in the Schwarzschild space (**8.88**) which starts and ends at radial coordinate r_0, whose closest approach to the origin is at radial coordinate p, and whose equation is approximated by $r \cos \theta = p$, $\phi = 0$, i.e., by a line which is "straight" relative to the coordinates. Neglecting squares of m/p, one can quite easily show that this integral has the value

$$2x_0(1 - m/r_0) + 4m \log\{(r_0 + x_0)/p\}, \qquad x_0 = (r_0^2 - p^2)^{1/2}.$$

Verify that this exceeds $2x_0$, i.e., the value when $m = 0$, and considerably so when p is small. Show how one can *immediately* modify the answer if, instead, the signal travels between two *unequal* values of r, say, r_0 and r_1. [The time delay discussed here is due to the slowing down of standard clocks *and* to the modified space geometry near the central mass—unlike the gravitational Doppler shift, which is due to the former only. This has led to a proposal by I. I. Shapiro [*Phys. Rev. Lett. 13*, 789 (1964)] for a "fourth test" of GR, which consists in comparing the time it takes for a radar signal to be bounced off a planet when the signal passes close to the sun, with the time it takes when the signal is practically unaffected by the solar field. [For results, which now verify the predictions with an accuracy of 5%, see *ibid. 20*, 1265 (1968), *26*, 1132 (1971).]

8.16 Explain, *in principle only* but nevertheless in full detail, how you would solve a collision problem involving "test particles" (i.e., particles whose mass is too small to affect the *field*) in curved spacetime. Consider, for example, a particle momentarily at rest at $r = 9m/4$ in the metric (**8.43**). It is struck head-on by a second particle falling towards the origin. If the collision is elastic, and if the coordinate velocities dr/dt of the particles immediately after impact are equal and opposite, $\pm 4/45$, what was the coordinate velocity of the incoming particle just before impact? [$-40/369$.]

8.17 Define a four-vector ξ_μ by its components in the Schwarzschild metric (**8.72**) thus: $\xi_\mu = (0, 0, 0, \alpha)$. Prove that $\xi_{\mu;\nu} = -\xi_{\nu;\mu}$. (Vectors satisfying this equation are called *Killing vectors*.) If $U^\mu = dx^\mu/ds$, deduce that the quantity $U^\mu \xi_\mu$ is constant along *any* geodesic in the Schwarzschild metric. [*Hint*: Differentiate absolutely with respect to s.] Use this fact to relate the γ factors, as measured with standard clocks and rulers at rest in the metric, at any two points through which a freely falling particle passes. Compare with the energy equation of Exercise 5.13(ii). Generalize this result to the standard static metric (**7.24**), and compare with the Newtonian energy equation $\frac{1}{2}v^2 + \varphi = $ constant.

8.18 Write down the geodesic equations for Schwarzschild space (**8.72**), and derive the forms given in the text, namely (**8.52**), (**8.53**), and (**8.69**).

8.19 Consider the metric $ds^2 = X^2 dT^2 - dX^2 - dY^2 - dZ^2$ (which is, in fact, fully discussed in Section 8.6) purely on its own merits. By treating it as a particular static metric (**7.24**), with $c = 1$, obtain the frequency shift ν_1/ν_2 between any two points in the metric, one having $X = X_1$, the other $X = X_2$. Compare with Exercise 3.5. From the form of the potential φ deduce that the force felt by an observer at rest at coordinate X is $1/X$. Recall, from Section 2.16, that this is indeed the observer's proper acceleration.

8.20 In the situation illustrated in Figure 8.6, show that a particle dropped from rest on the rocket at "level" $X = X_1$ to a lower level $X = X_2$ ($< X_1$) always takes the same *proper* time, namely $(X_1^2 - X_2^2)^{1/2}$, and also the same *coordinate* time T, namely $\tanh^{-1}\{(X_1^2 - X_2^2)^{1/2}/X_1\}$. [*Hint*: The particle worldline will be tangent to the X_1 hyperbola. Why? Consider a LT which transforms the hyperbola into itself and the tangent into the vertex tangent.]

8.21 For the metric of Exercise 8.19 work out the Christoffel symbols and hence the geodesic equations of motion in the X direction. Translate the solutions of these equations by use of **(8.88)** into the Minkowskian coordinates x and t, and show that they are linear in x and t. [This calls for considerable manipulative trickery!]

8.22 Prove, by any method, (i) that light tracks in the space with the metric of Exercise 8.19 are semicircles, with typical equation $X^2 + Y^2 = a^2$, $X > 0$; (ii) that the *coordinate velocity* of light at all points with coordinate X is X in *all* directions; (iii) that there is no upper limit to the *coordinate time* for a light signal to travel in the X direction from X (> 0) to X_1 ($> X$), or vice versa, provided that X is sufficiently close to zero.

8.23 In **(8.88)**–**(8.90)** we showed how Minkowski space M_4 "looks" from a rocket. Analogously, by applying a rotation of constant angular velocity ω to the x and y axes, derive a *stationary* metric of M_4 as "seen" from a uniformly rotating platform. Relate the coefficient of the squared "platform" time differential in this metric to redshift and centrifugal force. [*Hint*: Section 7.5, and Exercise 7.13.] Generalize to arbitrary stationary metrics.

8.24 Suppose a field of four-force F_μ (in flat *or* curved spacetime) were derivable from a scalar potential Φ as follows: $F_\mu = \Phi_{,\mu}$. Prove that the rest mass of particles moving in this field would be variable according to the relation $m_0 = \Phi/c^2 + \text{constant}$. [*Hint*: Exercise 5.9.]

8.25 Prove that every Riemannian 2-space is an Einstein space. [*Hint*: Every two-dimensional metric can be diagonalized.] Prove also, by use of **(8.134)**, that **(8.142)** implies $\Lambda = \text{constant}$ *except* for a 2-space.

8.26 By methods analogous to those following **(8.59)**, and under the assumption that $e \ll 1$, derive the extra perihelion advance **(8.154)** due to a Λ term in Einstein's equations. Equation **(8.153)** can be assumed.

Chapter 9

9.1 In connection with our cosmological scale illustrations (e.g., dimes for galaxies), prove the following theorem of Newtonian theory concerning scale models of systems of bodies moving under their mutual gravitation, in which each body is scaled down to one of *equal density*: The orbits of the bodies in the model, under *their* mutual gravitation, are similar in shape and identical in time to those of the original system.

9.2 Consider the analog of Milne's model in Kruskal space: let *test dust* move along the (geodesic) lines $T = \text{constant}$ in the quadrants II and IV of the Kruskal diagram

Figure 8.7. This "universe" is not isotropic, but it *is* homogeneous: all fundamental worldlines are equivalent. (As we have seen, they can be transformed into each other by Lorentz transformations.) Describe the geometric development of this universe, i.e., the succession of public spaces R = constant, from a beginning on E_1 to an end on E_2. How is cosmic time related to R?

9.3 By considering the volume and mass of a small sphere of constant *coordinate* radius r in the metric **(9.28)** of the steady state theory, prove that the mass dM created in volume V per time dt is given by $dM = 3H\rho V dt$. Assuming, in cgs units, that $H = 1.5 \times 10^{-18}$ and the density $\rho = 2 \times 10^{-31}$, and given that one year is $\sim 3.2 \times 10^7$ and the mass of a hydrogen atom is $\sim 1.7 \times 10^{-24}$, prove that one new hydrogen atom would have to be created in a volume of one km^3 about every sixty years to maintain the assumed constant density.

9.4 We may assume, from symmetry, that free particles can execute radial geodesic motion (θ, ϕ = constant) in the RW metric **(9.27)**. By reference to Appendix I, with $(x^1, x^2, x^3, x^4) = (r, \theta, \phi, t)$ and $c = 1$, obtain the geodesic equation $t'' - (g_{11}\dot{R}/R)r'^2 = 0$, where each prime denotes d/ds. Use the metric itself to reduce this equation to the form $t''/(t'^2 - 1) = -\dot{R}/R$, and hence integrate it to find $t'^2 - 1 = A/R^2(t)$, with A an arbitrary (positive) constant, as the general differential equation for timelike radial geodesics. Observe that the radial coordinate does not explicitly enter this equation, and so it must apply to the alternative form of the metric **(9.26)** also.

9.5 Explain *in principle only*, but nevertheless in full detail, how a problem such as Exercise 2.28 (concerning the distance covered by a rocket moving with large constant proper acceleration) would have to be worked if one assumed an RW background and were interested in voyages between galaxies whose cosmic proper distance increases significantly during the trip. [*Hint*: Exercise 4.6.]

9.6 In an infinitely expanding RW universe **(9.26)**, a particle is projected from the origin with velocity v_0. Prove that this particle ultimately (as $t \to \infty$) comes to rest in the substratum. [*Hint*: Cast the geodesic equation of Exercise 9.4 into the form $\gamma v = A^{1/2}/R(t)$, where $\gamma = (1 - v^2)^{-1/2}$, v being the local velocity relative to the substratum.] In particular, if $k = 0$ and the particle was shot out at $t = t_0$, prove that it comes to rest at coordinate ρ_0 given by

$$\rho_0 = \int_{t_0}^{\infty} \frac{adt}{R(R^2 + a^2)^{1/2}}, \quad a = \frac{R(t_0)v_0}{(1 - v_0^2)^{1/2}}.$$

What happens if the integral diverges, e.g., if $R = t^{1/2}$?

9.7 Use Figure 9.1 to illustrate *graphically* in the case of Milne's universe the result of Exercise 9.6 according to which any freely moving particle eventually comes to rest relative to the substratum. What is the coordinate ρ of the fundamental particle that will be reached only in the infinite future by a freely moving particle emitted at the spatial origin at time $\tau = a$ with velocity u? [$\rho = u(c^2 - u^2)^{-1/2}$.]

9.8 According to Planck's law of "blackbody" radiation, the energy density of photons in the frequency range v_0 to $v_0 + dv_0$ is given by

$$du_0 = 8\pi h v_0^3 c^{-3}(e^{hv_0/kT_0} - 1)^{-1} dv_0, \tag{1}$$

where k is Boltzmann's constant and T_0 the absolute temperature. We have inserted the subscripts zero to indicate that we look at the various quantities at some given cosmic instant t_0. Now suppose that the radiation in question uniformly fills an RW universe with expansion function $R(t)$, and suppose we can neglect its interaction with the cosmic matter. As the universe expands or contracts, will the radiation maintain its "blackbody" character (1), and, if so, with what temperature law? Fill in the details of the following argument: $v/v_0 = R_0/R$, $du/du_0 = R_0^4/R^4$, hence (1) without the subscript, provided $T/T_0 = R_0/R$.

9.9 If an event horizon exists, prove that the farthest galaxy in any direction to which an observer can possibly travel if he starts "now," is the galaxy which lies on his event horizon in that direction "now."

9.10 If a model universe has both an event horizon and a particle horizon, prove that the farthest galaxy, on any line of sight, from which a radar echo can be obtained (theoretically), is that at which these horizons cross each other.

9.11 By reference to the rubber model, and taking into account universes with finite or infinite past or future, give an explicit argument for Property (ix) at the end of Section 9.6, namely that, when a model is run backwards in time, the event horizon and particle horizon exchange roles.

9.12 By reference to the rubber model, describe the most general (isotropic and homogeneous) universe in which it is possible for two given galaxies, seen simultaneously by a given observer in the same direction at one given instant, to exhibit the same nonzero red shift and the same distance by apparent size, and yet for one to be at twice the proper distance as the other.

9.13 Consider a Robertson–Walker universe with $R = ct$ and $k = 1$. In theory, an observer can see each galaxy by light received from two diametrically opposite directions. Prove that the red shifts D_1, D_2 in the light arriving simultaneously from opposite directions but from the same galaxy satisfy $D_1 D_2 = \exp(2\pi)$.

9.14 Fill in the details of the following derivation of a condition for the avoidance of Olbers' paradox in RW models: From **(9.40)** and **(9.43)**, the radiant energy density at the origin at time t_0, due to a source of intrinsic luminosity B, is given by $BR^2/c\rho^2 R_0^4$ (dropping the subscript 1 and writing R for R_1 etc.). Let n be the number of galaxies per unit comoving coordinate volume (strictly constant in conservative models), and $B(t)$ the intrinsic luminosity of an average galaxy as a function of cosmic time. Then the radiant energy density at the origin, due to a shell of cosmic matter at coordinate radius ρ with coordinate thickness $d\rho$, is given by

$$du = \frac{B(t)R^2(t)}{c\rho^2 R_0^4} \times \frac{n4\pi\rho^2 d\rho}{(1 - k\rho^2)^{1/2}}.$$

When we sum over all shells which contribute light to the origin at that instant, we must of course allow for the finite velocity of light, and count each contribution according to its emission time. From **(9.26)**, $-cdt = R(t)d\rho/(1 - k\rho^2)^{1/2}$ for incoming light. Hence the total radiant energy density at the origin is given by

$$u = \frac{4\pi n}{R_0^4} \int_{-\infty, 0}^{t_0} R(t)B(t)dt,$$

where the lower limit of integration must be chosen according as the model exists for all negative cosmic times or only from some definite cosmic instant ($t = 0$)

onward. If B is taken as constant, the condition for the avoidance of Olbers' paradox is thus the convergence of the integral $\int R(t)dt$ for the model's entire past. [*Note*: (i) the spatial curvature k does not affect the result; (ii) in the case of closed universes, each galaxy may contribute several times to the total flux measured at the origin, its light having circumnavigated the universe several times; (iii) no allowance has been made for possible absorption of emitted light on the way to the origin.] Under the present assumptions, which of the Friedmann models (see Section 9.10) have an Olbers problem?

9.15 Adapt the argument of the preceding exercise to prove that in the steady state theory $u = \pi \bar{n} B/H$, where $\bar{n}$ is the constant number of galaxies per unit proper volume, and B is the constant intrinsic luminosity of an average galaxy.

9.16 Prove, from the formulae, that for all RW models with $k = 0$, the proper distance σ (at light emission) and the distance S from apparent size coincide. By use of the rubber models illustrate why this is so, and also why it is *not* so in the cases $k = \pm 1$.

9.17 For Milne's model (**9.13**) [which, in the notation of (**9.26**), has $R(t) = ct$ and $k = -1$], prove that, for a light signal from ρ_1 at t_1 to the origin at t_0, $\Sigma(\rho_1) = \log(t_0/t_1)$, and hence derive the following relation between the three "distances" described in Section 9.7 of an object at cosmic time t_1: $L/S = \exp(2\sigma/ct_1)$. Prove also that S coincides with the "ordinary" distance of the object at *its* proper time t_1 in the Minkowski rest frame of the observer; and that both σ and L for particles near the observer's "boundary" of the universe become arbitrarily large.

9.18 For the steady state model (**9.28**) prove the following relation between S and L: $L/S = (1 - HS/c)^{-2}$. Use this to rederive the distance of the event horizon, $\sigma = S = c/H$. Why does the infinite value of L indicate an event horizon in this model but *not* in the Milne model?

9.19 Fill in the details of the following proof that the only (zero-pressure) Friedmann models which have constant curvature *as four-spaces* are precisely the six empty ones: Consider two particular neighboring timelike geodesics issuing from the origin, namely the worldlines of two galaxies at equal radial coordinate r, separated by a small comoving coordinate distance dl. The cosmic time t, being proper time at each fundamental particle, serves as "distance" along these geodesics if we take $c = 1$. All their spacelike joins at fixed t, of magnitude $\eta = R(t)dl$, are orthogonal to them. Evidently $\ddot{\eta} = \ddot{R}(t)dl = (\ddot{R}/R)\eta$. Comparison with (**7.6**) now shows that we cannot have constant curvature unless $\ddot{R} = -KR$, i.e., $\dot{R}\ddot{R} = -KR\dot{R}$, i.e., $\dot{R}^2 = -KR^2 +$ constant. Comparison with (**9.73**), and the fact that ρ varies as $1/R^3$ [cf. (**9.75**)] finally leads to $\rho = 0$ (and, incidentally, to $K = -\Lambda/3$). Hence only the empty models *can* have constant curvature. Reference to the list following (**9.84**) shows that, in fact, they all *do* have constant curvature. It can be shown quite apart from the restriction of the field equations that these six RW metrics are the only ones with constant curvature. This is intuitively clear from inspection of Figure 9.6 (for what other substrata could be painted onto the two 4-spaces of constant curvature, S_4^- and S_4^+?) and from a similar consideration of the possible substrata in M_4.

9.20 A qualitative analysis similar to that illustrated in Figure 9.8 can be made also for models *with* pressure. The main task is again to plot the curves $\dot{R}^2 =$ constant on a (Λ, R) diagram. To this end, deduce from (9.79) the relation $d\rho/dR = -3(\rho + pc^{-2})/R$ and then, from (9.73), derive the condition

$$\frac{d\Lambda}{dR} = -\frac{6k}{R^3} + \frac{24\pi Gc^{-2}(\rho + pc^{-2})}{R} = 0$$

for the extrema of these curves. If $k = 0$ or -1, the loci of $\dot{R}^2 = 0$ have no extrema and look quite similar to those of the pressure-free case, if we assume $\rho, p \geq 0$. The models themselves are correspondingly similar: oscillating if $\Lambda < 0$, and inflectional if $\Lambda \geq 0$. If $k = 1$, however, the analysis is complicated because $(\rho + pc^{-2})$ is now not known explicitly as a function of R. It can be shown that there may be several extrema for positive R instead of the single one when $p = 0$. (See, for example, R. C. Tolman, *Relativity, Thermodynamics, and Cosmology*, Section 157, Oxford University Press, 1934.)

Subject Index

Texts and Monographs in Physics

Series edited by W. Beiglböck

The Theory of Photons and Electrons
The Relativistic Quantum Field Theory of Charged Particles with
Spin One-Half
Second Expanded Edition
By **J.M. Jauch** and **F. Rohrlich**
1976. xix, 553p. 55 illus. ISBN 0-387-07295-0

> One of the standard references and texts on quantum electrodynamics since its first publication in 1955. It provides a particularly systematic presentation starting with first principles, dealing with the general theory as well as with its applications to the important physical processes; both scattering and bound state problems are solved. The new edition brings this book up to date by means of an addendum in which new contributions are presented and appropriate references to recent literature are given.

Polarized Electrons
By **J. Kessler**
1976. ix, 223p. 104 illus. ISBN 0-387-07678-6

> This monograph is an introduction to the field of spin-polarized free electrons, a relatively new and rapidly developing field of modern physics. After carefully introducing basic principles, the author discusses the polarization effects in electron scattering from unpolarized targets, exchange processes in electron-atom scattering, numerous spin polarization effects in ionization processes, and electron polarization in solids. A final chapter is devoted to the applications and future prospects in this field.

Inverse Problems in Quantum Scattering Theory
By **K. Chadan** and **P.C. Sabatier**
1977. approx. 360p. 32 illus. ISBN 0-387-08092-9

> The authors cover radial problems at fixed angular momentum, radial problems at fixed energy, one-dimensional problems, three-dimensional problems, construction of the scattering amplitude from the cross section, and exact and approximate methods.

The Concepts and Logic of Classical Thermodynamics as a Theory of Heat Engines
Rigorously Constructed upon the Foundation Laid by S. Carnot and F. Reech
By **C. Truesdell** and **S. Bharatha**
1977. approx. 176p. approx. 15 illus. ISBN 0-387-07971-8

> This volume develops completely and rigorously the consequences of the existence of a thermal equation of state, the 18th-century Doctrine of Latent and Specific Heats, and Carnot's General Axiom: the work done in a Carnot cycle is positive and is determined by the operating temperatures of that cycle and the heat it absorbs.

A Springer-Verlag Journal

Communications in Mathematical Physics

Editor-in-Chief J. Glimm, New York, New York

Editorial Board H. Araki, Kyoto, Japan
R. Geroch, Chicago, Illinois
R. Haag, Hamburg, Federal Republic of Germany
W. Hunziker, Zurich, Switzerland
A. Jaffe, Cambridge, Massachusetts
J.L. Lebowitz, New York, New York
E. Lieb, Princeton, New Jersey
J. Moser, New York, New York
R. Stora, Marseille, France

Communications in Mathematical Physics is devoted to such topics as general relativity, equilibrium and nonequilibrium statistical mechanics, foundations of quantum mechanics, Lagrangian quantum field theory, and constructive quantum field theory. Mathematical papers are featured if they are relevant to physics.

Contact Springer-Verlag for subscription information.

Lecture Notes in Physics

Managing Editor **W. Beiglböck**

This series reports on new developments in physical research and teaching— quickly, informally, and at a high level. The type of material considered for publication includes preliminary drafts of original papers and monographs, lectures on a new field, or presenting a new angle on a classical field, collections of seminar papers, and reports of meetings.

Vol. 10   J.M. Stewart, **Non-Equilibrium-Relativistic Kinetic Theory.** 1971. iii, 113p.

Vol. 13   M. Ryan, **Hamiltonian Cosmology.** 1972, vii, 169p.

Vol. 14   **Methods of Local and Global Differential Geometry in General Relativity.** Edited by D. Farnsworth, J. Fink, J. Porter, and A. Thompson. 1972. v, 188p.

Vol. 44   R.A. Breuer, **Gravitational Perturbation Theory and Synchrotron Radiation.** 1975. vi, 196p.

Vol. 46   E.J. Flaherty, **Hermitian and Kählerian Geometry in Relativity.** 1976. viii, 365p.

 Springer-Verlag New York Inc.
175 Fifth Avenue
New York, NY 10010

Printed in the United States
By Bookmasters